CALIFORNIA NATURAL HISTORY GUIDES

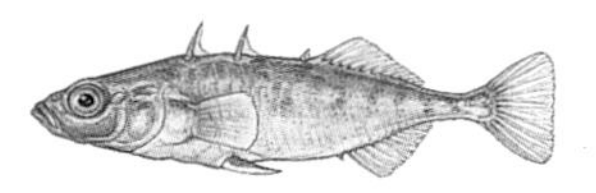

FIELD GUIDE TO FRESHWATER FISHES OF CALIFORNIA

California Natural History Guides

Phyllis M. Faber and Bruce M. Pavlik, General Editors

UNIVERSITY OF CALIFORNIA PRESS

Berkeley Los Angeles London

FIELD GUIDE TO FRESHWATER FISHES OF CALIFORNIA

REVISED EDITION

Samuel M. McGinnis

Illustrated by Doris Alcorn

This book is dedicated to Dr. Robert C. Stebbins, master natural history writer, superb illustrator, outstanding scientist, devoted teacher, and friend.

University of California Press, one of the most distinguished university presses in the United States, enriches lives around the world by advancing scholarship in the humanities, social sciences, and natural sciences. Its activities are supported by the UC Press Foundation and by philanthropic contributions from individuals and institutions. For more information, visit www.ucpress.edu.

California Natural History Guide Series No. 77

University of California Press
Berkeley and Los Angeles, California

University of California Press, Ltd.
London, England

ISBN-13, 978-0-520-23728-5 (cloth, alk. paper),
ISBN-10, 0-520-23728-5 (cloth, alk. paper)

ISBN-13, 978-0-520-23727-8 (pbk., alk. paper),
ISBN-10, 0-520-23727-7 (pbk., alk. paper)

Cataloging-in-Publication Data for this title is on file with the Library of Congress.

Manufactured in China
10 09 08 07 06
10 9 8 7 6 5 4 3 2 1

The paper used in this publication meets the minimum requirements of ANSI/NISO Z39.48–1992 (R 1997) (*Permanence of Paper*). ♾

Cover photograph: California Golden Trout, the state's official freshwater fish. Painting by Doris Alcorn.

The publisher gratefully acknowledges the generous contributions to this book provided by

the Gordon and Betty Moore Foundation in Environmental Studies
and
the General Endowment Fund of the University of California Press Foundation.

CONTENTS

ACKNOWLEDGMENTS

This manuscript received exceptionally painstaking and detailed prepublication reviews by Dr. Robert J. Behnke, Colorado State University; Dr. Peter B. Moyle, University of California at Davis; and Mr. Theodore W. Wooster and Mr. Kenneth Aasen, California Department of Fish and Game. Their input from several areas of ichthyological expertise has greatly enhanced this revision.

The author and illustrator express deepest thanks to acquisitions editor Laura Cerruti, project manager Scott Norton, and project editor Kate Hoffman of the University of California Press California Natural History Guide series. Their keen interest and enthusiasm over this book were highly contagious and made the revision process most enjoyable.

We also thank the Wayne Alcorn family for the photo of the Golden Trout, and Dr. Chris Kitting for his photo of Kokanee Salmon and for sharing his underwater photography technique with us. The author wishes to acknowledge the stimulation and assistance received from students in his annual ichthyology class at California State University at Hayward (1964 through 2002). The results from the thousands of seine hauls we shared provided much of the information in this book.

Finally the author thanks Molly M. McGinnis for her computer skills for the production of this manuscript plus her continuing willingness to apply her culinary skills to the many fish species that I have deposited on her kitchen counter over the past four and one half decades and which are discussed in the Cooking the Freshwater Fishes of California section.

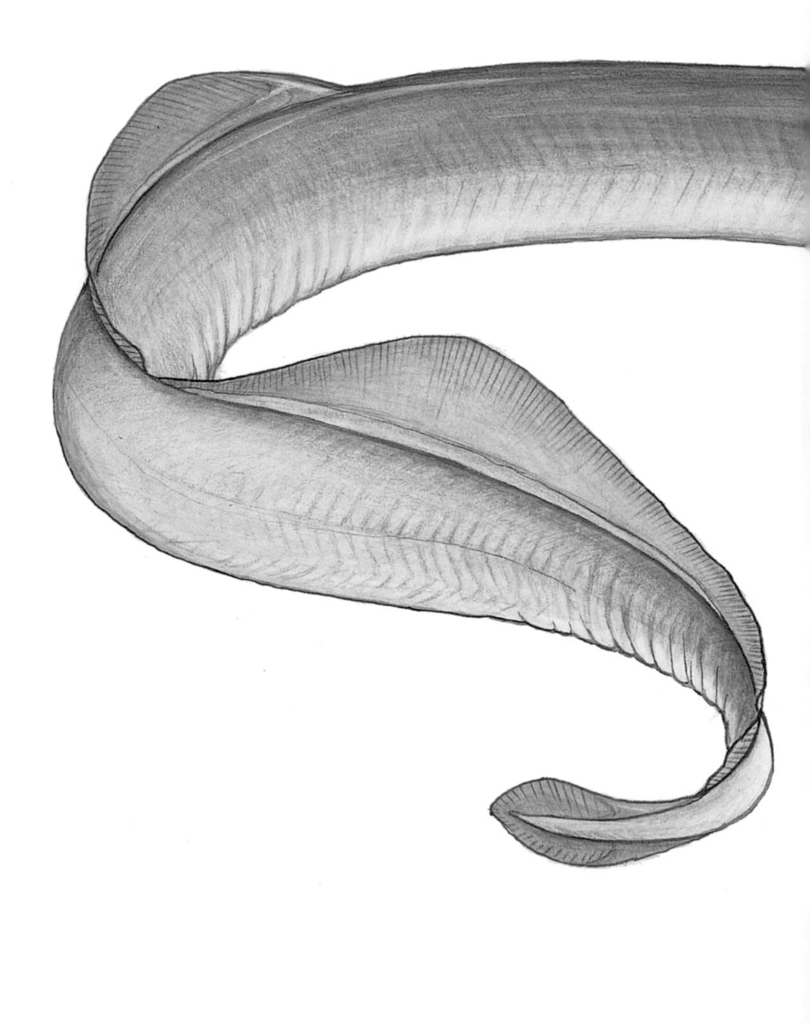

INTRODUCTION

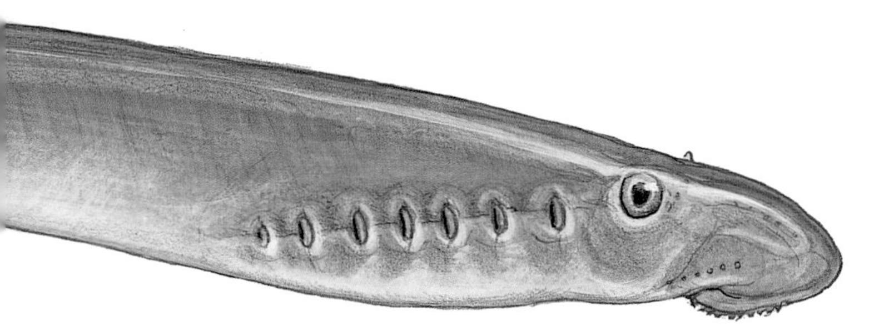

FRESHWATER HABITATS

Once upon a time there was a beautiful land bounded on the north and east by high, rugged mountains and large lake basins, on the south and east by desert hills and flats, and on the west by a deep, clear sea. Up from the sea rose coastal hills, and in the center of this land, between these hills and mountains spread a wide valley traversed by meandering rivers and dotted with floodplain ponds, marshland, and a trio of large shallow lakes. This was California.

Because of the relative isolation of this land from the rest of the continent, many unusual species of plants and animals lived there, and among the vertebrates the most numerous were the fishes. Habitats for the fishes were also more varied than those found in most states. Only a few fishes could adapt to the many cold, fast, mountain streams, fed most of the year by melting snow or summer thundershowers. These were mainly species of trout and salmon (Salmonidae), suckers (Catostomidae), minnows (Cyprinidae), and sculpins (Cottidae).

From their mountain headwaters the streams entered the foothills, where their currents slowed and their waters warmed. Eventually they joined one of two large rivers, now called the Sacramento and the San Joaquin. These rivers flowed toward each other from the north and the south, respectively, and joined through a series of tortuous connections to form a huge delta, which meandered west for over 50 miles before emptying into a large bay with a narrow channel to the sea. Although the temperature of this delta habitat was warmer and its flow slower than the mountain streams that fed it, its nature still imposed many restrictions on fish life. The heavy seasonal flow of water in the upper reaches of this system continuously cut away at the banks and scoured the inshore vegetative areas where most fishes spawned. As the delta flow neared the great bay, it spread out to produce an extensive, shallow, well-vegetated zone and marshy backwater areas. An even more restrictive element was the daily tidal invasion of saltwater, which changed the lower delta from fresh to brackish. Relatively few species evolved the necessary physiological adaptations to cope with these changes in salinity.

The Sierra Nevada and its foothills did not have a monopoly on streams: numerous small waterways wound throughout the inner coast and the Coast Ranges. Except for a few large flowages such as the Russian, Klamath, and Eel Rivers, their flow was highly variable, unlike that of most mountain streams. They became raging torrents during the periods of heavy winter rain, but without the large, melting snow fields and high-volume springs to feed them throughout the long, dry summer, many withered to beds of dry gravel interspersed with small, wet potholes. Fish residents of such streams had to cope with high winter flow on one hand, and warm, oxygen-poor summer pools on the other. Conditions such as overpopulated potholes selected for one or two species of minnow, sticklebacks (Gasterosteidae), and an occasional sucker or sculpin. Streams that flowed to the ocean or a coastal bay hosted young salmon and Steelhead *(Oncorhynchus mykiss irideus)* as well.

In post-Pleistocene California, lakes were confined mostly to the mountains. Thousands of small lakes were sprinkled throughout the Sierra, along with a few large bodies of water such as Lake Tahoe, Eagle Lake, and Honey Lake. Lowland temperate lakes were scarce. The most familiar of these today is Clear Lake, formed many thousands of years ago as a result of a succession of geological events. A blocked basin filled with water from small streams that once ran through it, and several stream-adapted fishes began the long process of genetic adaptation to a lake form of life. Also at that time three enormous shallow lakes lay at the southern end of the great Central Valley. In very wet years, they merged to form one continuous body of freshwater more than three times as long and twice as wide as Lake Tahoe.

Of all California's pristine aquatic habitats, perhaps the most unusual were the widely spaced desert springs and pools in the southeastern portion of the state, remnants of an ancient aquatic complex that existed during a wetter period in the Southwest's history. As this large chain of lakes slowly dried up over the ages, only those few fish species that could tolerate the rigorous conditions of a desert pond remained. The lower Colorado River, reduced to a slow, meandering stream, also became the center of evolution of some unique river fishes.

This, then, is a brief sketch of the freshwater habitats of Cali-

Plate 1 (above). Rae Lakes (10,500 ft) in the Sierra Nevada.

Plate 2 (right). A Sierra Nevada stream in later summer.

Plate 3 (below). Kern River.

fornia as they existed before the arrival of Europeans. It was a land of fast streams, a few deeper rivers, shallow intermittent creeks, desert springs, marshy backwaters, and a sprinkling of small and large lakes (pls. 1–13). The unique fish faunas present in these habitats were far more restricted with respect to species variety than the freshwater fish complexes of the eastern United States. There, a more varied complement of basic fish types, or families, produced greater species diversity because of periodic segmentation and reconnection of large freshwater systems. However, the California fishes were well adapted to their varied habitats, and like well-adapted faunas everywhere, lived in harmony with their environment. During the past century, all this drastically changed.

Plate 4. Lower Stanislaus River.

Plate 5. Clear Lake.

Plate 6. San Pablo Reservoir and Ranch Pond.

Plate 7 (above). Upper Delta area.

Plate 8 (right). Mormon Slough in the San Joaquin River.

Plate 9 (below). Tidal marsh shore in the lower Delta.

Plate 10 (above). California aqueduct.

Plate 11 (left). Coastal Creek mouth.

Plate 12 (below). Mojave Chub sanctuary at Zzyzx Springs.

Plate 13. Soda Lake bed in the Mojave Desert.

The Golden Spike, the Golden Trout, and the Golden State

As human societies everywhere made the monumental transition from hunting to agriculture, we have followed the habit of taking our favorite animals with us wherever we settled. Along with domestic animals, we have taken wild species valued for food, sport, or commercial uses, and nearly all of them have been detrimental to the new environment.

As the American pioneers journeyed westward, they found no real loss in leaving behind eastern game species. On the shortgrass prairie, the Sharp-tailed Grouse replaced the Prairie Chicken, and the Pronghorn Antelope filled in for the scarce deer. The Rockies had the Mule Deer instead of the Whitetail Deer of the East. On their westward trek, the pioneers saw the Sage Grouse, then the Spruce Grouse, and finally the rich California Quail populations of the Coast Ranges. Although eventually a few game bird species were introduced in the West, only the Ring-necked Pheasant and Wild Turkey may be viewed as successes.

Aside from half-wild boars, burros, and horses, the West has escaped any introductions of large mammals. Even the small mammal and bird faunas of the West have remained relatively pure, with only a few introductions, such as the House Mouse, Norway Rat, Fox Squirrel, Muskrat, and Eastern Red Fox, arriving with any assistance from humans. No intentional introductions of eastern reptiles into the West have been recorded, and with a few exceptions (such as the Bullfrog and the African Clawed Frog) the amphibian fauna likewise remains relatively unaltered. However, the native freshwater fishes of California have sustained extensive manipulation, and in some cases total destruction, through the introduction of numerous foreign species, as well as extensive habitat destruction.

An event in Utah in 1869 marked the beginning of the end for the pristine California fish fauna: the completion of the first transcontinental railroad, commemorated by driving a gold spike to secure the last piece of track. Prior to that time, it was extremely difficult to bring eastern species of fishes to the far West. Unlike birds and mammals, fishes do not travel well by wagon or horseback for any great distance or time, and it is doubtful that any made it to California in that fashion. Long ocean voyages on wind-powered vessels likewise were not conducive to live-fish transport. But the blossoming of the railroads in the West in the 1870s made possible the fulfillment of the western fish importer's dreams. Indeed, as early as 1873 the first specially built "aquarium car" took to the rails with the specific purpose of bringing eastern fishes to the West. This allowed many ecologically unsound whims and schemes to become reality, and eastern species poured into California waterways.

Today, freshwater fish introductions from both outside and inside the state continue. One recent example is the purposeful introduction of a species of smelt from Japan, the Wakasagi *(Hypomesus nipponensis),* into foothill reservoirs from which it had easy access to the San Francisco Estuary. Now it is in direct ecological and genetic competition with the threatened native Delta Smelt *(H. transpacificus).* In addition to continual foreign fish introductions, cultigens of the native Rainbow Trout *(Oncorhynchus mykiss)* have been introduced into many California freshwater habitats, in some cases replacing native Golden Trout subspecies *(O. m. aguabonita* and *O. m. whitei)* and Lahontan Cutthroat Trout *(O. clarki henshawi).*

The "Dam-nation" of California

At this point, you may be wondering about the ability of introduced species to survive in California aquatic habitats. Wouldn't the restrictive conditions prove hostile to eastern North American species that are, for the most part, lake-dwelling or slow-water fishes? That would be the case for many introduced species; however, shortly after fish importation began, Californians embarked on a complementary pastime—lake building. Some of the first successful attempts in the late 1800s were modest-sized impoundments in which small valleys were blocked with earthen dams. Lake Chabot in Alameda County is an example that persists today. Anthony Chabot solidified the earthen dam for his reservoir by driving horse teams back and forth on each layer of soil as it was applied. These initial attempts at lake construction were mostly private ventures, and the extent of the impoundment depended on how much land the individual owned. Then the Army Corps of Engineers, electric power companies, and the state of California began to lend a hand, and instead of one small canyon here and there, major valley systems, some containing entire towns, were covered with billions of gallons of water. Today 116 large dams have blocked the direct flow of all major streams in the state to create these massive reservoir stores of water, primarily for agricultural use.

The destruction of native habitats wrought by this process is comparable to that observed at the center of an atomic blast. Extensive riparian woodlands, streams, and adjacent hillside woodlands disappear. The well-adapted native fishes described earlier live in such streams. Most fish species can survive the initial flooding of stream watersheds, but then they face water depths hundreds of times that of the original woodland creeks in summer. In addition, reservoirs are mostly steep sided with hardly any inshore, or littoral, zone, and without shallow water, the major food source of creek fishes—aquatic insect larvae—does not flourish. Even where some gradually sloping shores are present, the highly fluctuating water levels usually caused by summer irrigation and hydroelectric drawdown prevent the establishment of any sort of permanent littoral zone community. Given thousands of years for gradual genetic adaptation to slowly increasing water depths, most stream species would probably evolve into well-adapted lake forms. The natural evolutionary process and timescale are not, however, part of the state's aquatic master plan.

Although the new, deep lakes are foreign and, for the most part, hostile habitats for California's native stream fishes, they are fairly well suited to many of the newly arrived eastern lake species. It is this great expansion of reservoir lakes, along with the channelization of Central Valley rivers and hundreds of miles of irrigation canals, that has made possible the successful establishment of an extensive eastern fish fauna in the West.

You might still argue that although a new reservoir completely alters a portion of a creek basin and its immediate watershed, the remaining creek segments and the native fish species up- and downstream from the new lake will remain the same. Unfortunately, that is not usually true. The most dynamic and disastrous effects have been on California's most economically important freshwater-dependent fishes, the anadromous salmonids. Many of the dams, especially on the San Joaquin River, were constructed with no regard to migratory salmon and Steelhead. Poorly designed fish ladders, the complete blockage of some tributary streams, and insufficient or ill-timed water releases were equally disastrous to most migratory salmonid populations in the Sacramento River drainage. As a result, the premier commercial and sport salmonid in this state, the Chinook, or King, Salmon *(Oncorhynchus tshawytscha)*, has lost most of its spawning and juvenile-rearing habitat.

Native fish populations in both the inlet and outlet creeks of the new reservoirs also rarely escape impact. Water release from most dams usually occurs near their base, and cold water from the reservoir's hypolimnion zone supplies the downstream creek channel. This release is often relatively constant throughout most of California's half-year-long dry season. Now instead of a late-summer creek condition consisting of relatively warm pools connected by shallow riffle flows or cool pools connected by underground flow, a permanent, warm, deep-water pool and flow state exists. A number of the eastern fish species introduced into a new reservoir are well adapted to these new stream conditions, with the result that they often replace the more drought-adapted native species.

This scenario may also occur in the lower reaches of the inlet streams to a new reservoir because water backs up in them through much of the dry season. Perhaps the most drastic effect occurs in the upper segments of inlet streams when an exceptionally heavy winter rainfall causes successive scouring of the creek basin and shoreline. Under natural conditions, resident native fishes are either killed or washed downstream to large "sanctuary pools" from which they or perhaps their offspring eventually emigrate, repopulating the upper-creek areas. When, however, such sanctuary pools are replaced by a deep, new lake, the poorly adapted, displaced natives are soon lost, and recolonization of the upper-creek areas is accomplished, or at least attempted, by introduced species.

Little Dogies and Big Drains

The post–gold rush immigrants to California were for the most part farmers and ranchers, not instant fortune seekers, but the destruction of freshwater habitats wrought by their interests was often more devastating than the ecological messes the placer miners left behind. The Spanish, with their open-range cattle, led the advance of European immigrants into California, disrupting the tribal way of life of the native Californians and eventually leading to their decline. The Spanish open range cattle were not confined to one area, however, until most edible greenery was gone. Soon, the American rancher followed and had to coexist with his counterpart, the American farmer. The latter lay claim to the flat lands of the great Central Valley, where soil tillage was the easiest, while cattle ranchers began to fence the upland annual grasslands of the Coast Ranges and their foothills. These fences, which allowed ranchers to control the movement of their herds, meant that cattle were often confined in one area until late summer, foraging along the many small creeks in these areas, eating green and brown leaves alike until the landscape was denuded and the immediate creek watershed severely eroded. Entire pristine creek ecosystems were lost in this way. In the degraded creeks that remained, even the shallowest summer pools were disappearing, and the few that persisted became wallowing sites for the ever-present cattle. Today many Californians regard such creek remnants as dry gulches or miniature badlands, never realizing that until recently they were beautiful riparian communities with a full complement of animal life, including a well-adapted native fish fauna. A similar scenario occurred in the Sierra where heavy sheep grazing in the late nineteenth century was followed by intensive cattle grazing throughout the twentieth century, again leading to the destruction of creek ecosystems by devegetation and erosion.

By no means did livestock grazing have a monopoly on rural aquatic habitat destruction. Many farmers came to the pristine Central Valley well versed in the art of draining shallow lakes and marshes, and farming the newly exposed rich soils. The great shallow lakes of the southern San Joaquin Valley (Tulare, Kern, and Buena Vista) were prime targets for such activity. Besides being relatively shallow, they underwent a partial drying each summer, thus exposing potentially tillable fertile land to the eager eyes of farmers. As a result, all three were completely drained by the early twentieth century.

To get some idea of the magnitude of this freshwater habitat loss, let us for a moment invoke the time machine fantasy of science fiction writing and take an imaginary drive south from the center of the state to the end of the Central Valley along modern Interstate 5, which has been transported in time to the mid-1800s. We would of course be traveling through saltbush scrub semidesert habitat instead of alfalfa and cotton fields, but just south of the Coalinga area, where the large stock yards are now situated, we would become aware of a large body of water to our left—Tulare Lake. If we happened to stop somewhere along its 33-km (20-mi) length, we would find a small commercial fishery for larger native minnows such as the Sacramento Blackfish *(Orthodon microlepidotus)*. If our time-warp trip happened to take place during spring following a heavy-rainfall winter, all three lakes would be temporarily connected to form one continuous freshwater habitat approximately 125 km (75 mi) long and 40 km (24 mi) wide.

The northern end of the great valley was also the focus of massive freshwater drainage projects. If we continue our imaginary nineteenth-century visit by driving north from the point where Interstate 5 crosses the San Joaquin River, we would expect to pass through the town of Stockton, but instead, we would encounter a small marker announcing that we were entering the town of Tuleberg, named for the massive tule bulrush marsh surrounding the town and stretching as far as the eye could see. In fact, most of our imaginary drive northward in the Central Valley would be on an elevated highway much like the present-day Yolo Causeway portion of Interstate 5 north of Sacramento. Today the need for such structures has been eliminated by the construction of high levees along the upper banks and immediate flood plains of all valley rivers and by draining the extensive marsh land beyond. These were the rich, permanent backwater areas of the original San Francisco Estuary, where many of our native fishes lived and spawned. The habitat loss for this group has been catastrophic, especially for a once-abundant species, the Thicktail Chub *(Gila crassicauda)*, a mainstay in the diet of several Native American tribes. It is now extinct.

Native versus Introduced Species

The enormous impact of species introduction on the California freshwater fauna requires an in-depth review. As defined here, native species are those freshwater anadromous and catadromous fishes that occur or once occurred naturally within the borders of California. Introduced species are those that have been brought to this state from other parts of North America or the world. The terms "foreign" and "exotic" are also often used in other writings to refer to fishes from other countries. A second type of introduction occurs when a native species from one area or drainage within California is moved to another drainage that already contains its own complement of native fishes. Examples of this type of introduction are less common. In this book the term "introduced" always refers to the introduction of species from outside California.

When the golden spike was driven in 1869, California had approximately 67 native fish species that spent all or part of their lives in freshwater, and there were no introduced forms. After the completion of the transcontinental railway system, however, it took less than two years for the California Fish Commission to arrange the purchase of 15,000 American Shad *(Alosa sapidissima)* from a fish culturist in the state of New York, bring them West, and in June 1871 release this popular anadromous eastern seaboard species into its new reproductive habitat, the Sacramento River. William Dill and Almo Cordone presented a detailed account of this first introduction in their 1997 California Department of Fish and Game monograph. The introduction of American Shad marked the beginning of the end of California's pristine freshwater fish complex. Today, a total of 50 freshwater species have been introduced into the state—six more than had been introduced when the first edition of this book was published in 1984—and the number will surely continue to increase.

Of the 67 native species, four are now extinct, and approximately 43 native species and subspecies have been designated as endangered, threatened, or a species of special concern by state and federal agencies.

California is second only to Florida in freshwater fish introductions. With its many semitropical aquatic habitats, Florida is now home to scores of tropical species that have been purposely or accidentally released. When compared to all other states, however, California wins the dubious introductions contest with species to spare. For instance, Minnesota, a state noted for its 10,000 lakes and good fishing, has 135 native fishes and only 14 introduced species. Of the latter, four are anadromous salmonids stocked in Lake Superior, and Brown Trout *(Salmo trutta)* and Rainbow Trout have been added to some streams. Of course the ever-present Common Carp *(Cyprinus carpio)* and Goldfish *(Carassius auratus)* are there, too, along with a few uninvited immigrants such as the Alewife *(Pomolobus pseudo-harengus)* and the Sea Lamprey *(Petromyzon marinus)*, which probably arrived via human-made connections between the Great Lakes.

From the beginning, the majority of California introductions were legal or sanctioned by state or federal agencies or closely allied organizations. Most introductions were approved by the California Fish Commission (later to become the California Fish and Game Commission). But why? To answer that question, you must first be aware of the near-complete lack of any ecological consciousness in this country through the nineteenth century and well into the twentieth century. Indeed, it was not until 1949 that Aldo Leopold laid the foundation for the concepts of conservation ethics in his landmark book, *A Sand County Almanac.* This lack is best illustrated by another event initiated by the connection of the transcontinental railroad in 1869. The railroad spelled doom for the great American Bison herds, because it meant that hides could be shipped to eastern markets in great quantities. The new immigrant American hunters joined many of their Native American counterparts and began the near-total elimination of what many mammalogists now agree was the largest population of a single large mammal species that ever existed. Between 1872 and 1874, 3.7 million were killed in the southern part of their range alone. Only 10 years later in 1882, however, Northern Pacific records show that only 300 hides were shipped east, and by 1883 the last remaining small herd was all but wiped out. The very few that escaped this final slaughter constituted the base of the present-day American Bison preservation effort.

Given this and other concurrent wildlife destruction efforts that would be unthinkable today, it is little wonder that the early decision makers for the importation of nonnative fishes into California most likely never considered the ecological consequences of such acts. The accepted belief of that era was that most California native fishes were "trash" or worthless species, and the stated policy of the original U.S. Fish Commission, the California Fish Commission, and other associated state agencies was to propagate and stock "valuable" nonnative fishes. In fact, in 1889 the U.S. Fish Commission employed the "father of California ichthyology," Dr. David Starr Jordan, to survey western waters specifically to look for opportunities to introduce "valuable" nonnative species.

Commissioners such as Dr. Jordan perceived it to be their duty to bring better commercial fishing and sportfishing to California, a state with relatively few native large piscivorous freshwater fish species. If nonnative species introductions had been well thought out in advance and perhaps tried first in small numbers at isolated sites, the current mishmash of introduced and native species might have been avoided. Instead, the policy of "introduce first and maybe ask questions later" has persisted to the present. California freshwater fish habitats can be likened to a giant community aquarium that began as a balanced ecosystem with a moderate number of well-adapted species but quickly degenerated as the custodians of this imaginary fish tank began to frequent tropical fish stores full of a variety of intriguing species that they simply could not resist. As a result, new, large, introduced piscivorous species ate smaller ones, and small, seemingly harmless new species began to outcompete the original inhabitants by expressing their superior reproductive capacity. Such events did not dampen the enthusiasm of our fish fanciers, however; and their solution to each new problem was always the same: introduce more new fishes. As a result, some of the original inhabitants of the aquarium perished, while others that were never intended to occupy a dominant position in this ever expanding community did just that. This is the state of the California freshwater fish bowl today.

But now for the specifics. The principal native game fishes of California include two species of sturgeon (White Sturgeon *[Aci-*

penser transmontanus] and Green Sturgeon *[A. medirostris]*), two subspecies of trout (Rainbow Trout, including Steelhead and Golden Trout, and two subspecies each of Cutthroat Trout *[Oncorhynchus clarki]* and Golden Trout), two species of salmon (Chinook Salmon and Coho Salmon *[O. kisutch]*), two species of pikeminnow (Colorado Pikeminnow *[Ptychocheilus lucius]* and Sacramento Pikeminnow *[P. grandis]*), and one species of sunfish (Sacramento Perch *[Archoplites interruptus]*). All of these, with the exception of the Golden Trout, the Lahontan Cutthroat Trout, and the Sacramento Perch, are anadromous species, which breed in freshwater but grow up to spend their adult life in the sea. The prevailing opinion of the late nineteenth century, however, was that this modest group of large sport fishes would be far too minimal to satisfy the needs of anglers through the early twentieth century. In the absence of any concept of ecological niche requirements, introduced species were believed to be additions to, not competitors of, the original freshwater fish fauna.

The introduction of nonnative fishes falls under two categories: sanctioned and unsanctioned. As already mentioned, most intentional introductions were sanctioned by the California Fish and Game Commission and the California Department of Fish and Game for the purpose of improving fishing by bringing new game or predatory species into the ecosystem (examples include Striped Bass *[Morone saxatilis]* and Largemouth Bass *[Micropterus salmoides]*). Once these and many other species were established, a second wave of sanctioned introductions took place to supply what was perceived as much-needed new prey species (Threadfin Shad *[Dorosoma petenense]* and Wakasagi) for the previously introduced piscivores. Additional introductions have been made as attempts for biological control of mosquitoes, gnats, and other undesirable insects (Western Mosquitofish *[Gambusia affinis]* and Inland Silverside *[Menidia beryllina]*); for biological control of algae and other aquatic plants (cichlid family [Cichlidae] and Grass Carp *[Ctenopharyngodon idella]*); and to provide additional fishing opportunities and a new source of food for people (bullhead and catfish).

Most ichthyologists now feel that all fish introductions, sanctioned or not, are detrimental, and that a moratorium should be placed on any future sanctioned introductions in California. Proposed introductions are at last being carefully screened by the California Department of Fish and Game, and so far very few have been accepted.

Unsanctioned introductions include (1) "escapees" from cultivation at tropical fish farms in southern California (Sailfin Molly *[Poecilia latipinna]* and Oriental Weatherfish *[Misgurnus anguillicaudatus]*); (2) contamination of legal interstate fish shipments by unwanted species (Bigscale Logperch *[Percina macrolepida]* and Rainwater Killifish *[Lucania parva]*); (3) stowaways in the water ballast tanks of foreign ships (Yellowfin Goby *[Acanthogobius flavimanus]* and Shimofuri Goby *[Tridentiger bifasciatus]*); (4) "dumping" of bait buckets (Golden Shiner *[Notemigonus crysoleucas]*, Red Shiner *[Cyprinella lutrensis]*, and Fathead Minnow *[Pimephales promelas]*); (5) private importation of a favorite "back home" species (Tench *[Tinca tinca]*); and (6) "kind-hearted" releases (Goldfish and numerous tropical species). Some accidental introductions will likely remain as isolated populations, but others, such as the Yellowfin Goby, Bigscale Logperch, and Shimofuri Goby, are rapidly spreading throughout the Delta and Central Valley and into southern California.

Troubled Waters

To put the problem of introductions in perspective, we must remember that this is only one of many factors affecting California freshwater fish fauna today. Already mentioned are the massive habitat destruction caused by dams and drainage projects, and the devegetation and erosion of creek riparian zones caused by intensive stock grazing practices. Other impacts include (1) loss of larval fishes and fry in the pumps that supply the California Aqueduct, Delta-Mendota Canal, and large power plants; (2) excessive water diversions from the Delta and most major river systems in California to supply agricultural and urban needs to the south; (3) siltation of creek gravel spawning beds and fry-rearing pools by soil eroded from upstream watershed areas where careless logging or residential development has occurred; (4) gravel mining within a stream floodplain and often within the stream basin itself; (5) pollution, usually spills or releases of toxic materials into small waterways; (6) confinement of most urban creeks to concrete channels and underground culverts; and (7) a massive California trout and salmon hatchery system that often creates cultigens better adapted to life in concrete runway tanks than in streams and creeks, and diverts funds and efforts from the restoration and management of natural habitats.

The primary problem, of course, lies in the ambient medium in which these fishes live—freshwater. On the "water planet" Earth, 96.5 percent of all water is saltwater, and almost all of the remaining 3.5 percent (freshwater) is tied up in ice, snow, and groundwater reserves. This leaves less than .25 percent in rivers and lakes, a minuscule amount of which occurs in California where in most areas it does not rain for over half the year. Freshwater—this state's most precious resource—is what many interest groups want, if not demand, and all too often the welfare of those vertebrates that live in it, the freshwater fishes of California, is overlooked.

The Evolution of Native Freshwater Fishes

The study of the origin of California's native freshwater fishes requires a knowledge of the state's geological history because new species usually arise when an initial species is separated into two or more populations by some geological event. For instance, a giant landslide, perhaps triggered by an earthquake, may completely block a canyon's large stream, converting the upper stream area into a long, deep lake. Those individual fishes in the isolated population that already possess some characteristics conducive to lake life survive the drastic habitat change and pass those attributes on to their offspring. In each succeeding generation, selection favors lake rather than stream adaptations, and after thousands of generations, major differences in anatomy, physiology, and behavior will exist between the geographically isolated lake population and the original, downstream species. Many of the changes that may occur affect reproduction, such as spawning time, spawning temperature, substrate preferences, and breeding display and color. Such changes serve to isolate the lake population reproductively from the stream population. Should some future geological or human-caused event reconnect the stream to the lake, the two original populations may once again intermingle but will not interbreed or reproduce fertile offspring, thus complying with the most basic species definition.

To take into account the diversity of all native freshwater fishes of California, ichthyologists must examine those areas of the state whose geological history has resulted in partial or total aquatic isolation for a substantial period of time. Such study is a type of systematics, and those who pursue it are more akin to the paleontologist than, say, the fish physiologist or behaviorist because they can only make best guesses as to what happened where and when in the past. Thus if you discuss the evolution of fish species with several fish systematists, you will usually hear several similar yet slightly differing views, each of which is probably close to the truth.

Study of the evolutionary history of fishes in California has led to the designation of evolutionary areas, called fish provinces, from which our present-day species originated. As you might suspect, there are as many designations and names for fish provinces as there are fish systematists; however, it is not the intent of this book to delve deeply into such scholarly areas of ichthyology. We are not ever likely to glimpse highways signs announcing "You Are Leaving the Klamath Fish Province" or "Welcome to the Sacramento Fish Province." For an introductory treatment of this subject, I present six basic designations suggested for the first edition of this book by fellow state university ichthyologist Dr. John Hopkirk. These still work well today. Each province has at least one isolated or endemic species that does not occur naturally anywhere else in the world (fig. 1). For instance, the Golden Trout evolved in high mountain streams of the Sacramento Fish Province and is endemic to that area. Most native California species, however, are not endemic to any one province but occur throughout several. For instance, the Rainbow Trout, which has wide access to the great river systems and coastal waters of the state, is found in several provinces both within and outside California. Given its ability to go to sea and return to coastal streams to spawn, it has been able to distribute via marine routes.

Sacramento Province

The Sacramento Province is the largest freshwater fish province in California. The many tributaries and interconnections of the pristine Sacramento and San Joaquin River systems provided access to aquatic subhabitats of this vast complex in which the present-day native species evolved. Despite its great size, the Sacramento

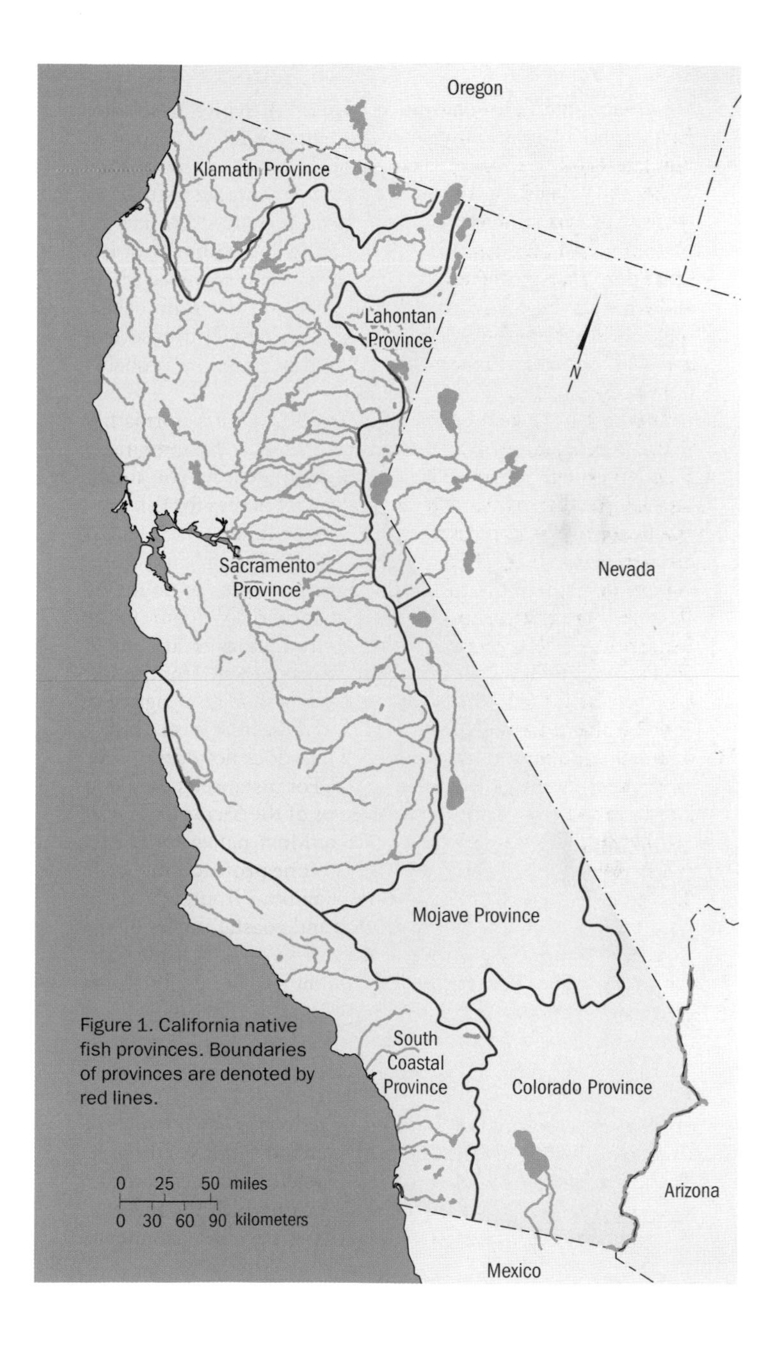

Figure 1. California native fish provinces. Boundaries of provinces are denoted by red lines.

Province was still aquatically isolated by the Klamath Mountains and associated ranges to the north, the Sierra Nevada to the east, and the Tehachapi and San Rafael Mountains to the south. Today this isolation has been breached by a number of human-derived fish-movement pathways. The 17 endemic forms in this province include six species of minnow (Sacramento Blackfish, Hardhead *[Mylopharodon conocephalus]*, Hitch *[Lavinia exilicauda]*, Sacramento Pikeminnow, Sacramento Splittail *[Pogonichthys macrolepidontus]*, and California Roach *[Lavinia symmetricus]*); a sunfish (Sacramento Perch); two suckers (Modoc Sucker *[Catostomus microps]* and Sacramento Sucker *[C. occidentalis]*); a smelt (Delta Smelt); a trout (Golden Trout); a viviparous perch (Tuleperch *[Hysterocarpus traski]*); three species of sculpin (Rough Sculpin *[Cottus asperrimus]*, Riffle Sculpin *[C. gulosus]*, and Pit Sculpin *[C. pitensis]*); and two species of lamprey (Pit-Klamath Brook Lamprey *[Lampetra lethophaga]* and Kern Brook Lamprey *[L. hubbsi]*).

Klamath Province

The Klamath Province is isolated and is composed of two interconnected subprovinces, Upper Klamath and Lower Klamath, that were formerly separated in the vicinity of Klamath Falls. The Upper Klamath Subprovince, which includes Upper Klamath Lake in Oregon, was apparently connected at one time with the Lahontan Province and other provinces of the Great Basin, as indicated by the presence of suckers of the genus *Chasmistes* and chubs of the genus *Gila*. It contains large and relatively shallow lakes and slow-flowing streams, whereas the Lower Klamath Subprovince originally contained only swift-flowing streams and the Klamath River estuary. Because of its connection to the sea, the Lower Klamath Subprovince is dominated by anadromous species. This basic fish complement was altered by the construction of Iron Gate Dam and the resulting large reservoir behind it. This lake was initially populated by native species from the lake-dominated portion of the upper subprovince. Endemics for the entire province are represented by the Lost River Sucker *(Catostomus luxatus)*, the Shortnose Sucker *(Chasmistes brevirostris)*, and the Klamath Largescale Sucker *(Catostomus snyderi)*. These and other indigenous species are primarily river fishes, well adapted to the fast streams and cool lakes of the area.

Lahontan Province

The Lahontan Province drains the eastern side of the Sierra Nevada and includes ancient Lake Lahontan and its peripheral basins. Lake Lahontan was a giant body of water that covered much of central Nevada during the Pleistocene, or Ice Age. Pyramid and Walker Lakes, in western Nevada, are the present-day remnants of Lake Lahontan. The most prominent lake in this province is now Lake Tahoe, which, along with smaller mountain lakes and cold, fast streams, has brought forth one endemic species in California, the Tahoe Sucker *(Catostomus tahoensis).* Some authors now list the Lahontan Province as a subprovince of the expansive Great Basin Province, which ranges eastward to the Rocky Mountains; however, given the influence of the original Lake Lahontan watershed on the evolution of northeastern California native fish species, the Lahontan designation still seems appropriate.

Mojave Province

The Mojave Province, like the Lahontan Province, includes a Pleistocene lake basin, Lake Mojave, which formerly covered the Mojave Desert region of California. Included in this province is the Death Valley region, which also contained Pleistocene lakes. This area is known primarily for its endemic pupfish species but also includes the Owens Sucker *(Catostomus fumeiventris).* The relatively high percentage of endemism in this province is a result of the development of an arid climate and the resulting microgeographic isolation of populations in desert springs. Visualization of the process by which gexographic isolation promotes speciation is far easier in this province than all others. To view, for instance, Salt Creek and its endemic pupfish species of that name (pl. 100) after driving for hours through desert habitats of Death Valley makes it clear that this is a unique and therefore endemic fish species. Indeed, it is difficult to imagine that about 14,000 years ago the valley floor was about 250 m (about 800 ft) below the surface of a great lake.

South Coastal Province

The South Coastal Province is composed of small coastal drainages south of Monterey Bay and contains streams whose flows vary greatly throughout the year. A contiguous group of inner-coast mountain ranges separates this region from the Sacramento, Mojave, and Colorado Provinces. This has permitted the evolution of two endemic species, the Arroyo Chub *(Gila orcutti)* and the Santa Ana Sucker *(Catostomus santaanae).* Two other natives, the Speckled Dace *(Rhinichthys osculus)* and the Unarmored Threespine Stickleback *(Gasterosteus aculeatus williamsoni),* are also present. Because this province includes the Los Angeles basin, where water temperatures are hospitable to the many tropical species that human residents of this area release each year, the freshwater fish diversity of this area continues to grow. Populations of the Oriental Weatherfish, Guppy *(Lebistes reticulatus),* Green Swordtail *(Xiphophorus helleri),* and Porthole Livebearer *(Poeciliopis gracilis)* are already well established.

Colorado Province

Within the confines of California, the Colorado Province contains two species of minnows, the Bonytail *(Gila elegans)* and the Colorado Pikeminnow, the Razorback Sucker (formally called the Humpback Sucker) *(Xyrauchen texanus),* and the Desert Pupfish *(Cyprinodon macularius).* Included in this province is the largest inland body of water in California, the Salton Sea. The present lake was created in 1905 when the Colorado River changed course and flooded through an early irrigation canal into the Imperial Valley. The water began to spill into an ancient saline sink—at one time the delta of the Colorado River—and continued to do so until 1907, when the river regained its original course. Since that time, the water level of the Salton Sea has been maintained and has actually increased as a result of the inflow of excess irrigation water from the Imperial Valley.

If not for its salinity, the Salton Sea would now be one of the most important freshwater fish habitats in the state, and a good portion of this book would be devoted to it and its fauna. Because of leaching and the continuous input of salts via irrigation runoff, which concentrate through evaporation, the Salton Sea has changed from entirely freshwater in 1907 to having a salinity in excess of 45 ppt at the beginning of the twenty-first century (seawater is 33 ppt). Accompanying this chemical change was an equally dynamic progression of fish species, beginning with Colorado Province freshwater endemics, such as the Bonytail and the Razorback Sucker; then introduced brackish-water species, such as the Western Mosquitofish and Sailfin Molly; and finally introduced marine sport fishes, such as the Orangemouth Corvina *(Cynoscion xanthulus)*, Sargo *(Anisotremus davidsonii)*, and Gulf Croaker *(Bairdella icistia)*. During the first decade of this century, these sport fishes will cease to exist here because the salinity is already above that which their eggs and larvae can tolerate. Soon only three species will remain: the Mozambique Tilapia *(Oreochromis mossambicus)*, the Sailfin Molly, and the Longjaw Mudsucker *(Gillichthys mirabalis)*. Small numbers of Desert Pupfishes may also linger in associated lake inlet channels. The three lake dwellers will persist well into this new century, during which time the salinity will continue to increase at the rate of .5 ppt per year.

The Mozambique Tilapia, with an upper salt tolerance limit of about 69 ppt, will be gone by 2050. By 2084, the Sailfin Molly will have reached its upper limit of 87 ppt. But I have collected Longjaw Mudsuckers from salt evaporation ponds in South San Francisco Bay in which the salinity was approaching three times that of seawater, so this marine species may be the sole survivor into the twenty-second century. Ironically, this little goby was brought to the Salton Sea as a bait fish, but it will persist long after its predators have passed on. However, even this diminutive champion of osmotic regulation cannot indefinitely withstand the relentless progression of salinity increases, which will be hastened as irrigation drains put less freshwater into this sea because of diversions of Colorado River water to other states. Thus as early as 2200 the Salton Sea will become "ichthyologically extinct."

PHYSICAL AND BIOLOGICAL ENVIRONMENT

Physical Environment

A sound understanding of the physical environment of an organism is crucial to an overall appreciation of its biology. This is especially true of fishes because their environment is so different from our terrestrial one. Perhaps one of the best examples of this difference is the thermal and oxygen stratification that occurs in lakes and ponds throughout the warmer months of the year (fig. 2). As warm weather sets in, highly productive (eutrophic) lakes stratify into three distinct thermal zones: an upper warm layer, the epilimnion; a zone of rapidly changing temperature, the thermocline; and a lower zone of cool water, the hypolimnion. Such stratification occurs because warm water is lighter than cold water, and nearly all heat in a lake is acquired by the absorption of solar radiation near the surface. During the calm summer months the density difference and absence of vertical or eddy currents permit the warm water to float on top of the cooler, lower water. Fish species are very set in their thermal preferences at any one time of the year and are reluctant to venture from a preferred thermal zone. Thus temperature alone can be a very restrictive force within a given habitat.

The dissolved oxygen concentration is also stratified during the warmer months. The surface waters contain the highest levels because it is here that green plants and surface wave action provide most of the lake's oxygen. Conversely, the bottom waters—though physically capable of holding more oxygen because of the cooler temperature—are usually deficient in this gas as a result of depletion by bacteria. The lack of vertical mixing in summer perpetuates this effect and, in eutrophic lakes, results in a sharp vertical oxygen stratification (fig. 2). Because most fishes are very rigid in their oxygen requirements, this is an additional restrictive force in lake habitats. When temperature and oxygen are

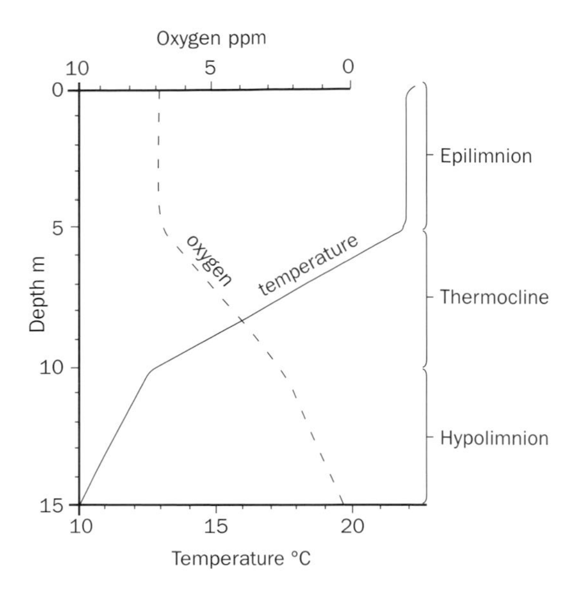

Figure 2. Thermal and oxygen stratification patterns in a temperate lake.

considered together, the problem is further complicated. For instance, consider a Rainbow Trout *(Oncorhynchus mykiss)* introduced into a lake in summer. This is one of the most oxygen-demanding species of freshwater fishes, but it also prefers cool water. As fig. 2 demonstrates, there is no place in the lake habitat where both items occur at optimal levels for this species; the fish must therefore make certain physiological compromises. I and a former graduate student, Peter Michael, explored this compromising situation by releasing experimental hatchery-reared put-and-take Rainbow Trout into Lake Chabot during the summer thermal and oxygen stratification period. We obtained the depth and water temperature for each fish from miniature radio transmitters fed to the fishes before release (pl. 14), and then took water samples at each fish's chosen depths and analyzed them for oxygen content. Our data showed that faced with this temperature versus oxygen dilemma, released hatchery Rainbow Trout equipped with radio transmitters consistently chose oxygen availability over preferred water temperature and remained in

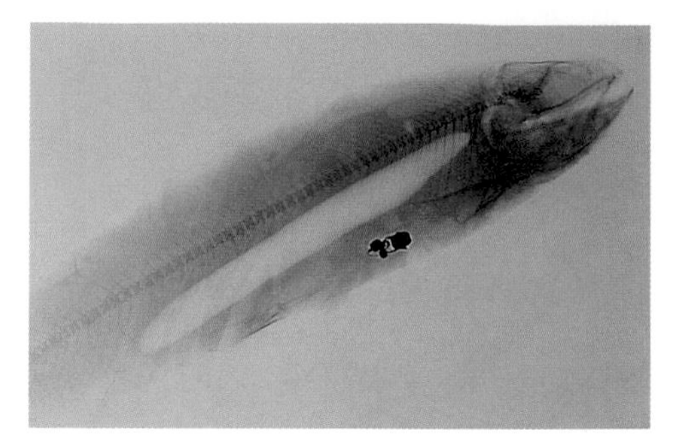

Plate 14. Swim bladder (light area) in Rainbow Trout with ingested radio transmitter below it.

the epilimnion region. Unfortunately, our experiments ended there, and the effects of such compromises on growth and metabolism are still poorly understood.

A working knowledge of these two environmental factors, and of the tolerance of individual species for each, is of prime importance to both anglers and aquarium enthusiasts. Fishing for species that need both moderate oxygen levels and cool temperatures in the warm, inshore waters of a reservoir lake in midsummer will usually net you nothing but wasted time. Likewise, the stocking of Rainbow Trout cultigens in a warm, poorly oxygenated body of water can end in disaster. The section "Anatomy and Physiology of Freshwater Fishes" explains these relationships in more detail.

Thermal and oxygen stratification disappear in late fall as lower solar input and greater reradiation to the atmosphere combine to cool the epilimnion. The change in density due to surface cooling produces eddy currents, which aid in carrying the surface water downward. At 4 degrees C (39 degrees F) the water completely mixes throughout the lake. Ice forms only at the surface because of added cooling from subfreezing air and surface evaporation. Energy supplied by higher wind speeds aids this process. The end result, which is called the "fall turnover," is a complete mixing of a lake's water. Mountain lakes that freeze over for several months also have a spring turnover because little vertical mixing takes place beneath the ice. In lakes that do not freeze, the water in all but the deepest parts is continuously mixing between periods of summer stratification.

Carbon dioxide (CO_2) is another gas that affects fishes both directly and indirectly. The indirect effect is the most pronounced because CO_2 is the major raw material for photosynthesis: low levels of CO_2 mean low algal and aquatic vascular plant growth. Much of the CO_2 in freshwater habitats enters as calcium carbonate ($CaCO_3$). Later the CO_2 is utilized and then released by plants. Calcium for this process comes from the rock and soil layers surrounding a lake or stream, which is washed in during rainy periods or carried in from more distant areas by incoming waters. The relationship of calcium to carbon dioxide, and ultimately to basic productivity, is so important that lakes are generally classified as hard (200 to 300 ppm Ca), medium (100 ppm Ca), or soft (0 to 50 ppm Ca) water lakes. Numerous studies have shown a direct relationship between fish abundance and calcium levels. Most of California's freshwater habitats, with the exception of mountain lakes, are high in calcium.

California freshwater lakes show great variation in nutrient levels. The supply of nitrates, phosphates, silicates, and other nutrients must be enhanced periodically to maintain high productivity, and the watershed of an aquatic habitat is important in determining the lake's nutrient content. Mountain lakes generally have low nutrient levels, whereas Central Valley ponds and irrigation ditches tend to have high levels. New reservoir lakes appear to undergo a nutrient explosion when fertile valleys are initially flooded, resulting in rapid fish growth and excellent fishing during the first few years.

In lakes, an additional source of nutrients is the bacterial decay of dead organisms. The turnover period is extremely important in that it redistributes these nutrients to the photosynthetic zone from the summer accumulation in the hypolimnion. In streams, nutrients from large amounts of decaying woody debris are dispersed by current flow. Current also produces an unequal distribution of nutrients in stream and river systems. Areas of fast flow usually exhibit very low productivity, whereas backwater areas and deep potholes may act as nutrient traps and decomposition sumps and thus approach the high productivity of some pond and lake systems. However, most nutrients in a stream or river system are deposited at mouth or delta areas, making habitats such as estuaries some of the most productive aquatic areas in the world.

The overall productivity of freshwater habitats depends, of course, on the interplay of all these factors. For instance, a lake may have high nutrient and CO_2 levels but a very low water temperature so that potential productivity is not realized. Three terms useful in describing productivity are "eutrophic," for a highly productive system, "oligotrophic," for a minimally productive system, and "mesotrophic" for in between situations.

Biological Environment

Algae and vascular plants form the base of the freshwater food chain or pyramid, and with only a few exceptions their presence is necessary for fish survival and growth. Unicellular algae are particularly important in this regard, and their role in the aquatic ecosystem may be compared to that of grass in a pasture. When the grass is lush, cattle and sheep production is high. The same is true of fish production, for the one-celled algae (phytoplankton) provide food for the small, free-swimming invertebrate animals (zooplankton), which, in turn, are eaten by most fish fry and some adults as well. Thus the green, cloudy waters of algae-rich lakes in summer, though undesirable to many people, are the pastures of plenty for fishes.

Of course, too much of a good thing is not always good, and occasionally the production of various algal types greatly exceeds the ability of zooplankton to consume it. A mass algal die-off can result, and aerobic bacteria begin to break down the remains and use up oxygen in the process. This important recycling activity can become so great that it depletes the water oxygen level, mainly at lower depths. As a result, many fishes, especially bottom-dwelling species, begin to suffocate and die if they cannot find an oxygen-adequate refuge. Because this phenomenon usually occurs in June, July, and August, it has come to be known as "summer kill," and its occurrence is usually dramatic: thousands of fishes float dead at the surface. Clear Lake, in Lake County, is noted for occasional summer kills, especially in shallow back bay areas where water circulation is minimal.

The Lake-Pond Environment

Algal production in temperate-zone lakes is not uniform. It often peaks in spring and fall because these periods combine adequate nutrient mixing (due to turnover) with abundant sunlight. Not all lakes exhibit the same sequence of algae production. For instance, in Clear Lake, Lake County, phytoplankton production in spring is not as great as in fall. In Lake Chabot, Alameda County, the reverse is true.

A primary role of phytoplankton is as a food source for zooplankton. Zooplankton, and phytoplankton to some extent, are eaten by fish fry and by the adults of those species adapted to plankton feeding. The growth of zooplankton populations follows the initial rise in phytoplankton numbers, which in turn is preceded by nutrient upwellings or enrichment by drainage into the lake. Zooplankton species have seasonal cycles of abundance. In Lake Chabot, for example, one of the two major rotifer species appears in early April, reaches a population high in mid-June, and essentially disappears from the scene by mid-August. A second species appears in late July, and by August, when the first species is scarce, it is well on its way to an early-October population peak.

Unlike most phytoplankton species, zooplankton are quite mobile and respond to light in a negative manner. Thus they are scarce at the surface throughout the day but mass there in great numbers from dusk to dawn. A knowledge of zooplankton movements can be helpful in fishing for certain filter-feeding sport fishes. White Bass *(Morone chrysops)*, for example, tend to school at the surface at dawn and dusk while feeding on zooplankton. This activity is so vigorous that the scooping action of their mouths continuously breaks the water surface and allows you to locate the school. A small streamer fly cast skillfully into such a feeding frenzy usually yields exciting fishing.

Figure 3 illustrates the role of each trophic group in the biological cycle of a lake or pond. The algae and vascular plants are the primary producers because they are the only organisms in the cycle that can capture the energy in sunlight and store it in the chemical bonds of organic molecules. The zooplankton are the main primary consumers in the system because most feed exclusively on phytoplankton. In North America, adults of only a few fishes, such as the introduced Grass Carp *(Ctenopharyngodon idella)*, graze on aquatic vascular plants. Instead, the fry of nearly all species are secondary consumers because zooplankton and other small invertebrates make up the majority of their diet. The adults of many species, such as the Threadfin Shad *(Dorosoma petenense)*, remain zooplankton feeders throughout life. A few large species, such as the Kokanee Salmon (landlocked Sockeye

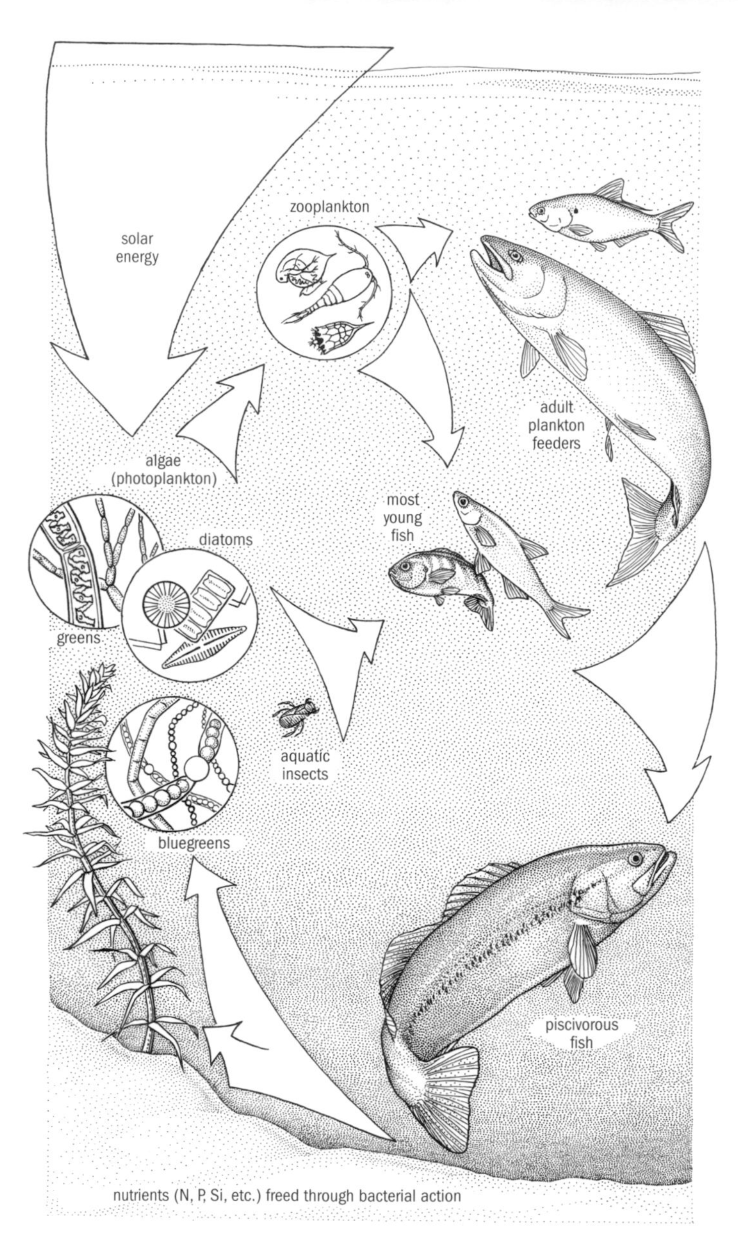

Figure 3. Biological cycle of a lake or pond.

Salmon) *(Oncorhynchus nerka)*, are also lifelong zooplankton feeders and therefore secondary consumers; however, most large freshwater fish species are fish eaters (piscivores) and are classified as tertiary consumers. When members of any trophic level die and are not eaten by scavengers, their bodies undergo bacterial decomposition, usually on the bottom substrate, and the next lake water turnover begins the recycling process.

This concept of energy flow in the biological cycle of aquatic systems is basic and necessary for the proper management of fish populations. For instance, sound sport-fishery management should entail studies of all trophic levels in a habitat to see which, if any, are deficient and thus in need of careful management practices for improvement. Only by such procedures can the desired end product—a sustained annual yield of large sport-fish species—be attained.

The multiple use aspect of most California reservoirs is detrimental to this cycle and thus to good natural fish production. High algal populations of eutrophic lakes may impart distinct tastes and odors to water, and phytoplankton of all types may also clog the screens that filter drinking water from these sources. Thus many impoundments that serve as active or standby sources of water for human consumption are periodically "bluestoned" throughout the productive warm months. Bluestone is copper sulfate ($CuSO_4$), which, when dissolved in water, releases highly toxic copper ions. Intensive plankton sampling in Lake Chabot, a standby water supply for San Leandro, California, was conducted before and after bluestoning. The study revealed that standing crops of both phyto- and zooplankton were cut approximately in half by the treatment. In addition, the fry of certain fish species, such as the Black Crappie *(Pomoxis nigromaculatus)*, apparently were killed in moderate numbers. From the fisheries management viewpoint, this procedure is analogous to a cattle rancher periodically spraying his pastures with a broad-spectrum herbicide.

The steep sides and fluctuating water levels of nearly all California reservoirs cause additional problems for fish production. The inshore zone, which is where vascular plants root and grow and where most species spawn, is minimal. When a small stand of inshore aquatic vegetation does grow, it is usually very ephemeral because of the continuous lake water drawdown through the summer plant growth season. This is the major shortcoming of reservoirs, because rich stands of aquatic vascular plants encircling a lake provide vital cover and food for fry and adults of many fish species. In addition, the inshore, or littoral, zone is the habitat for many aquatic insect larvae, and reduced numbers of this food resource also reduce the fishery potential of a lake.

Entomologists refer to aquatic, immature stages of insects as "larvae," "nymphs," or "naiads," depending upon the order of insects under discussion. For simplicity in this book I refer to all these forms collectively as aquatic insect larvae.

The River-Stream Environment

The biological cycle in rivers and streams (fig. 4) differs from that of lakes and ponds in several respects, primarily because of the continuous one-way movement of the water. Nutrients are distributed horizontally from runoff or from still water decomposition areas upstream. Free-floating algae are usually scarce in fast-flowing water; the major algal populations consist of filamentous species, which cling to the bottom substrate by means of holdfasts, and encrusting diatom and desmid species, which form mats over rocks.

Absent also are most free-swimming zooplankton species. The primary consumer niche is occupied instead by the aquatic larvae of insects such as mayflies, caddis flies, stone flies, damselflies, and dragonflies. Such forms anchor to the bottom, by their claws or suckers or, in the case of some caddis flies, by an adhesive case. Such larval populations can be very large, and in some mountain streams nearly the entire bottom may be covered with caddis fly cases. In addition to these larvae, permanently aquatic insects and other invertebrates such as amphipods, isopods, clams, snails, and crayfishes provide a rich bottom-food resource for fishes, especially in areas of low water velocity and abundant vegetation.

Another fish food source in the river-stream system is commonly referred to as "drift." Drift is composed of terrestrial and flying insects that have fallen into the water and aquatic larvae that have released their hold on the bottom. In many streams, drift provides the vast majority of fish food, especially for trout.

This category of larval and adult insect food is the primary inspiration for a companion hobby to fishing: fly tying. Expert fly tiers rank with master jewelers in their ability to produce finely crafted miniature objects of great beauty. Their materials range

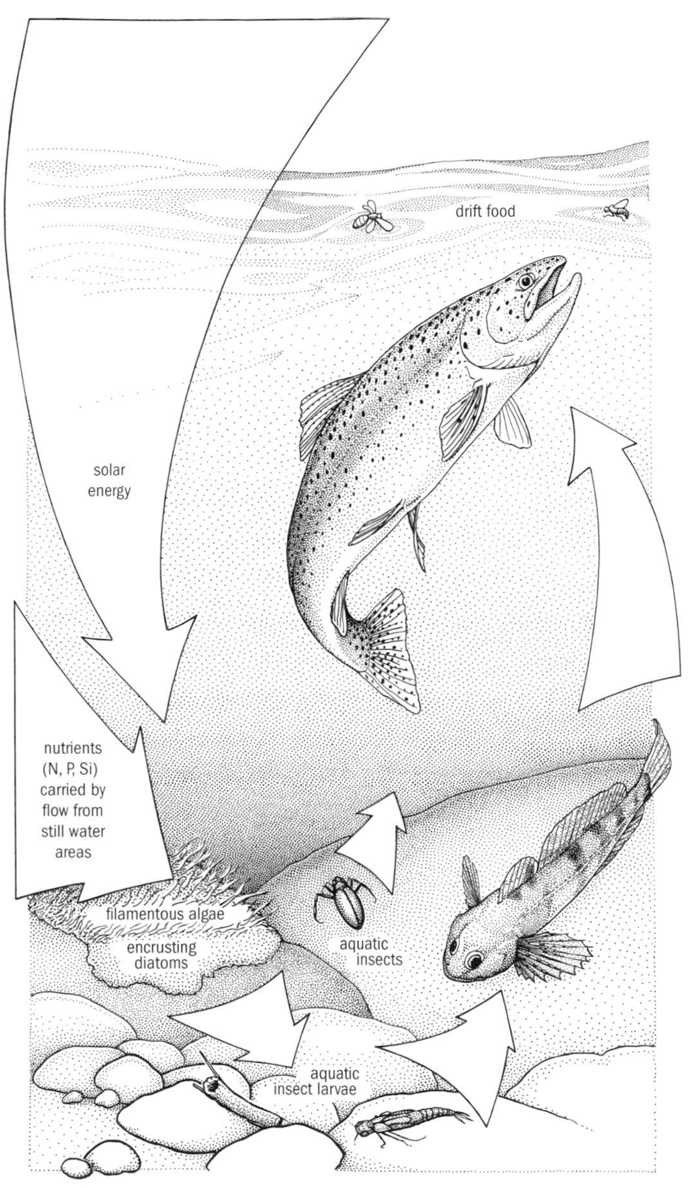

Figure 4. Biological cycle of a river or stream.

from the standard neck feathers or hackles of chickens to hollow hairs of certain hoofed mammals. The successful fly-tying angler must also be an amateur aquatic entomologist, possessing a good working knowledge of the taxonomy and natural history of the insects at his or her preferred fishing area. Upon arriving at a stream, the angler first observes "the hatch"—the newly transformed adult insects that compose a portion of the drift on that particular day. If no hatch is in progress, the angler must make a best guess regarding which fly to use. Once the all-important first fish has been caught and humanely dispatched, its stomach contents can be examined to reveal the fish's preferred insect food that day. The fly-tying angler then chooses the appropriate feathery, hook-loaded substitute and sees how successfully he or she is able to mimic the preferred food item. Anglers who fish catch-and-release streams can use small stomach pumps made especially for fish to quickly retrieve recently consumed items before the fish is released.

Current strength is also a limiting factor for fish species inhabiting fast-flowing water. Permanent residents are either strong-swimming species, such as trout and suckers (Catostomidae), or bottom dwellers, such as sculpins (Cottidae), that sequester between or beneath rocks. Other stream species, such as some minnows (Cyprinidae), must rely on deep pools with slow currents or shallow backwater areas for most of their existence.

One final contrast between lake and stream cycles lies in the disposition of nutrients released through bacterial decomposition. Instead of recycling vertically, as in lakes, the vast majority of nutrients in rivers and streams are carried downstream to the slow or still waters of bays and estuaries. As the current slows and water deepens downstream, the entire biological cycle becomes more lake-like, with pelagic phyto- and zooplankton replacing the rich, sedentary bottom flora and fauna.

ANATOMY AND PHYSIOLOGY OF FRESHWATER FISHES

External Anatomy

Figure 5 illustrates some of the major external anatomical features of fish. Many of these features are important considerations in identifying families and species. Note that there are five major fin types. The dorsal fin may be composed of soft rays only or a combination of spines and rays. In groups that have such a combined dorsal, for example, the sunfish family (Centrarchidae) pictured here, the two segments are fused, but in some other families they are separate. The tail, or caudal fin, may be forked, straight, or rounded at its posterior margin. The anal fin of most California freshwater fishes is composed only of soft rays, but in some families it contains one or more anterior spines. The paired pelvic and pectoral fins rarely contain spiny elements; two exceptions in California are the catfishes and bullheads, which have one anterior pectoral spine, sometimes barbed, and the Threespine Stickleback *(Gasterosteus aculeatus)*, in which the pelvic fins are composed essentially of isolated spines. In addition to these basic fin types, some fishes (such as trout, catfishes, bullheads, and smelt) have a small, fleshy adipose fin posterior to the rayed dorsal.

The form and function of fin types vary greatly among the world's fishes, but in California freshwater fishes, basic fin anatomy and function are quite uniform. The caudal fin provides the power for swimming by means of the lateral sculling action of the caudal peduncle, whereas the paired and median fins function as stabilizing and maneuvering organs. Spines, when present, offer a certain degree of protection against predators. One unusual feature is found in the anal fin of males of the family Poecilidae. The rays have elongated and fused to form an intromittent organ, or gonopodium, which is used to place sperm packets in the vent of the female.

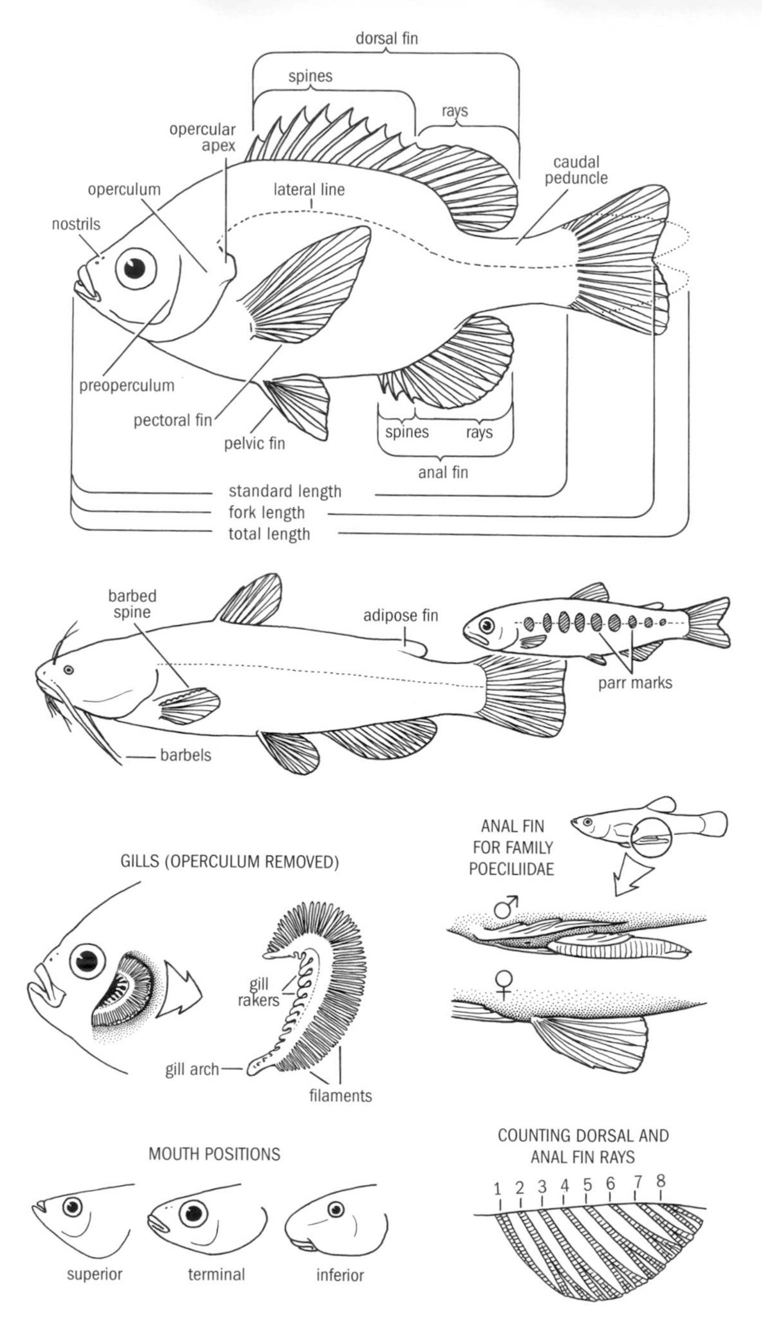

Figure 5. External anatomy of freshwater fishes.

The nostrils of California freshwater fishes do not connect with the mouth cavity but instead are paired holes leading to and from a nasal sac that contains olfactory nerve endings. Water passes continuously through this "in and out" chamber as a fish swims or holds stationary against a creek current. It is therefore always sampling the organic molecular content of the water instead of just taking occasional "sniffs," as we do. The sense of smell is one of the most important sensory inputs in fishes, and it is generally several hundred times more acute than that of humans. Only a few molecules of an organic substance need reach the nasal chamber lining to be detected. This keen olfactory sense not only aids greatly in feeding but also is the basis for homing in salmon and other species that apparently are imprinted with the subtle odors of their hatching or juvenile rearing site.

I was fortunate to be a part of Dr. Arthur D. Hasler's research team at the University of Wisconsin–Madison, which discovered this olfactory homing ability during field research with White Bass *(Morone chrysops)* in Lake Mendota. The lake's White Bass population has two primary spawning areas located only 1.6 km (1 mile) apart along the north shore of this 16-km-long (10-mile-long) lake. We first established that it was sun compass orientation that allowed adult fishes, captured on their respective spawning grounds and then released at midlake, to immediately beginning swimming northward toward "home." However, this type of navigation seemed too broad in scope to permit the necessary fine discrimination between two such closely situated sites. The nasal chambers of a second round of White Bass captured at each spawning site were plugged with Vaseline impregnated cotton, and the fishes were again released at midlake (pl. 15). These also quickly made their way back to the north shore of the lake but once there were unable to correctly identify their home spawning grounds. Because smell was the only sensory input that had been temporarily negated, it had to be olfactory memory that made possible this type of homing behavior. Dr. Hasler took these initial findings to the Pacific Northwest where he and colleagues discovered a similar phenomenon in anadromous salmonids. The ability of the many original populations of these fishes along the Pacific coast to accurately return to their own juvenile home creek or stream was the key factor that allowed for the utilization of the many tributary habitats before the blockage of most by dams.

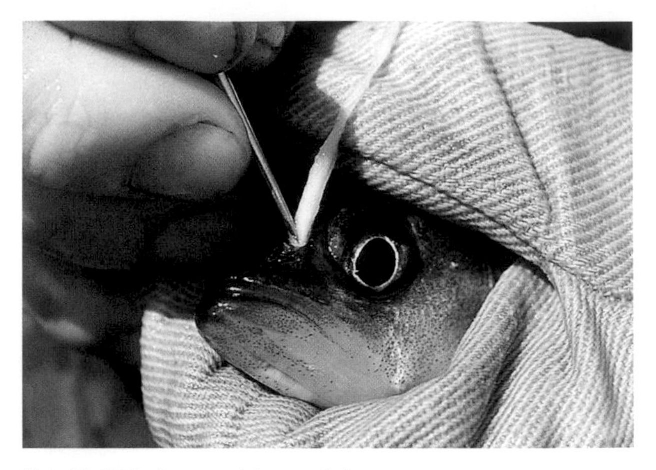

Plate 15. White Bass receiving nasal plug.

Olfaction also functions in communication between individual fishes and among all members of a school. Many minnow species (Cyprinidae) give off an alarm substance when either injured or frightened. When other members of the school smell this substance, they respond by dispersing in all directions, thus reducing the chance of success for an attacking predator.

Vision is also an important sensory input in most species of fishes. The freshwater fish eye differs from ours in that the iris has no muscles; thus the pupil always has the same diameter or, in photographic terms, the same *f*-stop setting. The fish eye lens is also different from ours. Instead of being ellipsoidal and flexible, it is spherical and rigid but free from any spherical aberration or edge focus problems. This sort of lens apparently gives fishes a sharp view of their entire surroundings and is especially good for detecting movement. We can best appreciate what a fish's view of the world is like when we look at photos taken with a "fish-eye lens," which shows objects throughout a 180-degree field.

In most fishes, visual focus is achieved by movement of the lens closer to or farther from the retina. In some sight feeders, such as salmon, the lens is slightly ellipsoidal, permitting sharp focus on an object in the anterior field of view on one portion of the retina while maintaining overall lateral distance vision on another retinal section. Many freshwater fishes, especially shallow-water species, also have good color discrimination. Some minnows, for example, can distinguish at least two dozen different color hues. These findings concerning fish color vision give scientific credibility to the multitude of colors and combinations thereof that fishing lure manufactures have incorporated into their products. Deep-water species do not seem to have good color vision, presumably because most wavelengths of light do not reach their habitat. Instead, these fishes possess very keen dim light vision comparable to that of owls and nocturnal mammals.

A third sensory structure present in most freshwater fishes is the lateral line, which must be ranked in importance with both smell and vision. The lateral line is a barely visible line of pores that runs from the operculum through the caudal peduncle on the midlateral portion of the body. Extensions of this system also lead to the head. It functions as an ultrasensitive current and pressure sensing system, capable of detecting pressure waves generated by other fishes swimming nearby. Apparently it provides a fish with a three-dimensional pressure or water displacement pattern of its habitat. Some piscivorous species such as the Northern Pike *(Esox lucius)* are able to locate and capture their prey in total darkness by using only this sense organ, and blind cave fish exhibit the ultimate development of this system. Thus it is possible that the pressure wave patterns generated by certain fishing lures as they move through the water are as important in attracting fishes as their size, color, or general shape.

The sense of taste is also well developed in fishes, but its importance is hard to separate from that of smell because water passes almost simultaneously through the nostrils and mouth. This sensory separation is more pronounced in the ictalurids, which possess several pairs of external barbels whose surface is covered with taste-sensing organs. By trailing the long barbels over the bottom substrate or waving them in the water column, these fishes are able to taste the presence of food before it actually enters the mouth. With additional taste receptors on the skin and fins, many catfish species are well equipped to recognize any potential food item that they may contact. Suckers (Catostomidae) also have taste receptors on their fleshy lips, as do those minnow species that feed in a grubbing manner. In all such species, taste and smell are far more important than vision in finding food.

Fishes have no external ear opening, middle ear structure, or cochlea, and thus hearing as we know it is greatly limited in most fish groups. Several freshwater families, specifically the cyprinids, ictalurids, and suckers, possess a chain of small bones, the Weberian ossicles, that connect an internal, gas-filled buoyancy organ, the swim bladder, with the inner ear. The taut, thin walls of the swim bladder appear to function rather like our eardrum in the initial reception of sound waves, which are then transmitted through the Weberian ossicles to primitive inner structures. This arrangement permits the reception of middle and higher sound frequencies up to 13,000 Hz. Low frequencies, below 100 Hz, are probably sensed through the lateral line of most fishes.

The position of the mouth is usually a good indicator of the feeding habits of fishes, and their family affiliation. A superior or dorsally opening mouth is found in those species that obtain most of their food by surface skimming, whereas an inferior or ventrally opening mouth denotes bottom foraging. The terminal mouth is by far the most common feeding structure.

One additional external structure of California freshwater fishes is the operculum, a large, bony plate with a thin, flexible posterior edge that completely covers the gill chamber. Fishes breathe water by closing the opercular flap and opening the mouth while lowering the floor of the throat. When water fills the mouth cavity, fish closes its mouth, opens the opercular flap, and raises the floor of the throat. The oxygen-bearing water then passes over the gill surfaces and out the opercular slit. Besides the breathing function of the operculum, the apex of the opercular flap is often the site of display colors, as in the sunfish group.

The gills comprise thin layers of epithelial tissue and capillaries with a bony support, the gill arch. Each gill arch bears many gill filaments, the number of which varies directly with the general degree of activity and metabolism of a species. Each gill filament, in turn, supports hundreds of lamellae, thin-walled passageways that are positioned at right angles to its long axis. The blood flow direction through these lamellae is forward, the opposite direction of the flow of water past them. This is a classic example of a countercurrent exchange system. Experiments with live fishes have shown that the oxygen exchange between water and fish blood is up to seven times greater when the flow of the two fluids is opposite as opposed to parallel. The total number of lamellae in the entire gill chamber may reach several hundred thousand in some species, and the great surface area produced by these structures contributes significantly to the efficiency of the gill as an aquatic respiratory organ.

The gill arch also supports a row of gill rakers, small bony projections that fill in the spaces between gill arches and prevent food items that enter the mouth from escaping through the opercular slit. The size and spacing of these rakers is often a key to the mode of feeding of a species. Long, thin, closely spaced gill rakers are present in filter-feeding fishes; they prevent zooplankton from escaping with the water flow at each respiration. In contrast, predatory fishes tend to have shorter, stouter, more widely spaced rakers that act as jail window bars, keeping prey fishes from injuring the gill filaments and escaping through the opercular slit during the swallowing process.

Many fishes possess teeth, but unlike our heterodont-type dentition, with its four distinct tooth forms, most fish teeth are thin and conical and have a very sharp point. This classic homodont-type dentition is not restricted just to the jaw edges as in the mammalian mouth but instead is distributed in various regions within the mouth cavity. This distribution varies between families and enables each to better capture, hold, and in many cases, even chew their food. Most piscivores have several posteriorly slanting rows of large, sharp teeth on the inside edge of the upper and low jaws that permit the initial capture of prey. Three additional groups, the volmerine teeth on the midline of the roof of the mouth, the palatine teeth that flank them on each side, and the basibranchial teeth on the floor of the mouth, hold slippery prey tightly until the swallowing action is initiated.

All fishes have pharyngeal teeth, which are located on the fifth gill arch. These come in a variety of shapes to match a species' preferred food type. Round, stout pharyngeal teeth like those of the Common Carp *(Cyprinus carpio)* are used to break up the shells and exoskeletons of invertebrate prey (pl. 16). The herbivorous Grass Carp *(Ctenopharyngodon idella)* has long cutting and rasping pharyngeal teeth that allow it to crop aquatic vascular plants. The piscivorous pikeminnows (Colorado Pikeminnow *[Ptychocheilus lucius]* and Sacramento Pikeminnow *[P. grandis]*) have sharp cutting pharyngeal teeth with which they can actually slice up a prey fish. This allows these large but otherwise toothless minnows to occupy a tertiary consumer feeding niche. In

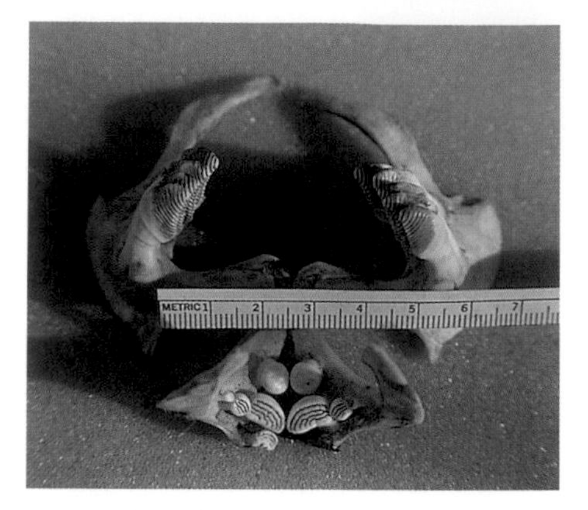

Plate 16. Pharyngeal teeth: Grass Carp (above ruler), Common Carp (below).

contrast to these examples, trout, temperate bass (Moronidae), and pike (Esocidae) swallow their prey whole and thus have only tiny vestigial pharyngeal teeth.

The final and perhaps most obvious portion of a fish's external anatomy is the body covering itself. Most California freshwater fishes possess either smooth-edged, disc-shaped cycloid scales or oval ctenoid scales with toothed or comblike posterior edges. Some groups, such as trout, have very fine or buried scales, and in the ictalurids, scales are completely absent. Slime or mucus, secreted by specialized cells in the epidermal, or upper layer of the skin, is found in all species and appears to be of great importance in protecting the skin against bacterial and fungal infections. When fishes are captured for display in aquariums, this coating must be disturbed as little as possible. A second important function of the mucus coating is the hydrodynamic advantage it imparts to fishes by filling in and smoothing over all rough spots around and between scales to produce maximal streamlining of the body surface. In addition, a minute film of surface mucus is continuously sloughed off during swimming, thus further reducing friction between water and the body.

The skin itself is laden with many chromatophores, color cells that can disperse or contract their pigments through neural stimuli, hormonal stimuli, or both. For this reason, the color of a fish may vary widely at different times and is not often a prime key characteristic in taxonomic descriptions. For example, a male Bluegill *(Lepomis macrochirus)* captured during the spawning season may exhibit beautiful coloration when first caught. After several minutes in a light-colored container, however, it may become a uniform silvery gray with only the dark opercular apex retaining its original color. Light and dark patterns, on the other hand, tend to be less affected during the color change process. Thus the dark, lateral patterns, or parr marks, of young trout and salmon are good characteristics for species identification.

Internal Anatomy

The internal anatomy of fishes is instructive to both the biology student and the amateur naturalist. Although fish anatomy differs markedly from that of humans and other mammals, it displays the basic vertebrate body plan from which all higher forms have evolved. An excellent opportunity to study this anatomy is presented at fish-cleaning time. In a "fishing family," internal fish anatomy often provides a young person's first exposure to the workings of the vertebrate body, and when presented in a scientific manner it may well be an important stimulus for a career in biology or medicine. Because the fish body must be opened in order to view its contents, this section is presented in dissection guide format.

If a selection of fishes is available, either from the angler's creel or the "whole fish" fish market, a large specimen is the best choice for the first dissection attempt. Also, if both piscivorous species (such as the Largemouth Bass *[Micropterus salmoides]*) and omnivores or vegetarians (such as the Common Carp or other cyprinids) are available, choose one of the piscivores, which are usually less confusing because of the much shorter intestine. Both intestinal designs should, of course, be thoroughly explored by the biology student.

To open the fish, use a sharp scissors and cut through the ventral body wall, starting at the vent and proceeding anteriorly to the area below the gills. Keep the tip of the scissors pointed out and away from the fish so that no portion of the internal anatomy is cut during the body wall incision. The second cut should start below the gill area and proceed dorsally just behind the bony arch posterior to the operculum. You will encounter some light bone in this area, but cut right through it, perhaps using an older, dull scissors. At a point just posterior to the apex of the operculum

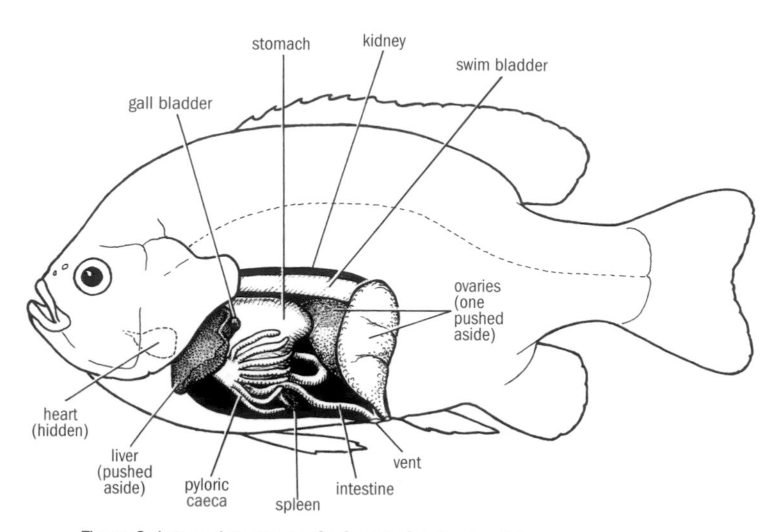

Figure 6. Internal anatomy of a female freshwater fish.

you will feel the tip of the scissors contact the roof of the body cavity. By feeling the boundary between the soft body cavity and the hard dorsal flesh with your fingers, you can cut posteriorly just below this juncture to the vent. Finally, remove the large triangle of body wall to expose most of the internal organs.

With the exception of gravid or egg-laden females, the most prominent organ complex is the stomach and intestine (fig. 6). The stomach is oblong and usually U shaped and is located right behind the gill area. If it is distended with recently ingested food, slit the stomach wall on one side and remove the contents for inspection. Stomach content analysis is one of the most important techniques employed by the fish ecologist, as it can tell much about the habits of a species. At the junction of the stomach and intestine are long, fingerlike projections called pyloric caeca that increase the digestive enzyme producing area of the gut. A specimen in good physical condition should have abundant fat deposits on the thin mesenteries (sheets of tissue) that support the gut.

If the specimen is a gravid female, the ovary with its thousands of eggs must be removed before other organs can be observed. If the specimen is a nongravid female, the ovary appears as a small sac just anterior to the intestine at the point where it empties to the outside through the vent. In a nonreproductive male, the testes appear as two small fingerlike structures in approximately the same area as the ovary in females. In a sexually mature male they are greatly enlarged and, like the mature ovary, may take up much of the body cavity. In a breeding male, a white liquid known as milt will probably pass out of the vent at the time of the ventral incision. This is a secretion of the walls of the sperm ducts, which contains and nourishes the sperm cells.

The liver is the color of beef liver and is located anterior and ventral to the stomach (fig. 6). On the under surface of liver is a small, oval sac, the gall bladder, which stores bile secretions from the liver until needed for digestion of food in the intestine. The spleen is a smaller, dark red organ located near the first bend of the intestine after it leaves the stomach. It is the primary site for red blood cell production in fishes.

The swim bladder, an organ used for buoyancy, is located in the upper portion of the body cavity. In some families, such as Ictaluridae and Cyprinidae, it is a bilobed structure resembling two small, silvery balloons. It is connected to the esophagus by a small duct or tube, but the entire bladder may carefully be taken out intact. This type of swim bladder is called a physostomic bladder and is found primarily in shallow-water species that may easily go to the surface to take in more air through the esophageal duct for additional buoyancy.

In contrast, the physoclystic swim bladder appears as a silvery white roof on the body cavity and is not connected to the esophagus. Instead, a complex bed of capillaries, known as the rete mirabile, or gas gland, employs a countercurrent system to secrete gases from the blood (O_2, N_2, and CO_2) into the bladder. In a reverse manner, excess gases are reabsorbed by another vascular bed, the oval, in the posterior portion of the bladder. This system enables deep-water or pelagic fishes to adjust their buoyancy at different depths without coming to the surface. The X-ray in pl. 14 reveals just how much of the body cavity is taken up by the swim bladder.

Remove the swim bladder and observe a liver-colored strip of tissue lying along both sides of the vertebral column. These strips are the kidneys. In freshwater fishes they function primarily as water pumps and counteract the continuous inflow of water into the blood at the gill surface. A small urinary bladder can be seen dorsal to the gut and ovary or testes at the point where they pass to the outside.

The heart is located far anterior in fishes and lies beneath the gill chamber. It is a two-chambered heart, with the large ventricle the most prominent structure. A large venous reservoir, the sinus venosus, collects blood from the large veins and passes it on to the single auricle. From the ventricle the blood goes to the ventral aorta and then through afferent arteries to the gill capillaries. At this point you may wish to remove the operculum so that the gills may be more easily examined. A detailed study of gill circulation is difficult because of the small size and great number of gill capillaries. Efferent arteries, which take the blood from the gills, join to form the dorsal aorta, which may be seen at the top of the gill chamber. From here the oxygenated blood goes to all tissues of the body under relatively low pressure, and deoxygenated blood eventually returns to the heart via the venous system.

If you happen to dissect a freshly killed specimen, you may find that the heart is still beating. If the fish has been killed according to the procedure outlined in the "How to Clean a Fish" section, this does not mean that the fish is still "mentally" alive and feeling pain. Instead, you are simply observing the intrinsic contractile property of heart muscle that in lower vertebrates continues long after brain death.

Physiology

Although it is not the purpose of this book to explore in depth the complex subject of fish physiology, a few comments on its more important aspects may be helpful in understanding the ecology of these animals. For instance, the type of hemoglobin (oxygen-carrying molecule) that a fish species possesses in its red blood cells determines to a great extent those habitats in which it may or may not survive. The hemoglobin of the Sacramento Blackfish *(Orthodon microlepidotus),* a native California minnow, has a great affinity for oxygen; in other words, it can become saturated with oxygen even when this element is at low levels, as in backwater areas of a warm lake or marsh. In contrast, the hemoglobin found in salmonid species, such as the Rainbow Trout *(Oncorhynchus mykiss),* has a much lower affinity for oxygen and becomes fully saturated only in oxygen-rich habitats such as cool mountain streams. If Rainbow Trout are taken from their preferred habitats and placed in that of the Sacramento Blackfish, they promptly suffocate. Between these two extremes of the fish hemoglobin spectrum lie all other California freshwater fish species hemoglobins.

The temperature requirements of fishes vary as greatly as does oxygen-level tolerance and depend upon the thermal range of the numerous enzyme complexes that direct most life processes. For example, some of California's tropical introductions, such as the cichlids (Cichlidae), die after extended exposure to water at or below 11 to 12 degrees C (52 to 54 degrees F), apparently because of the cessation of most enzyme activity. Several trout species thrive and even exhibit growth well below this range; however, the enzymes of most salmonids begin to break down in the 25- to 30-degree C (77- to 86-degree F) area, a range that is optimal for cichlid growth and reproduction.

Temperate zone freshwater fishes have the ability to adjust or acclimate to changing water temperatures and thus can swim and feed in 2- to 4-degree C water beneath the frozen surface of a mountain lake just as they do in the much warmer epilimnion temperatures of summer. With the exception of some salamander species, activity at these near-freezing body temperatures is found only among fishes, a fact that makes ice fishing possible. Such acclimation appears to take place through a gradual adjustment of major enzyme systems as habitats warm and cool throughout the year. This principle has practical application in freshwater fish aquariums. When wild fishes are brought directly into room temperature tanks, they often die of thermal shock because no time is allowed for a thermal adjustment to take place. If, however, the tank temperature is slowly adjusted over a period of a week or so, 100 percent survival may be obtained.

The energy metabolism of freshwater fishes differs markedly from ours in two major ways. Our muscles and those of birds and our fellow mammals are red in color because they contain the muscle respiratory pigment myoglobin. This molecule receives oxygen from the blood respiratory pigment, hemoglobin, and stores it until needed by contracting muscle cells. Myoglobin accrues to those species possessing it and the ability to engage in sustained, high-energy activity. Most freshwater fishes lack myoglobin and thus have white muscle or "white meat." Without any special oxygen storage ability, such fishes are totally dependent on the direct blood supply of oxygen from the gills. This would still be sufficient to sustain moderate activity if the transport of blood oxygen was fast and efficient, but it is not.

A basic rule of circulatory physiology is that blood loses most of its pressure when passing through a capillary bed such as that found in a lung or gill. Blood leaves the lungs at very low pressure but returns to our four chambered heart where it is then pumped under high pressure to all parts of the body. In fishes, which have the basic two-chambered heart, blood is pumped only once to the gill capillaries, where it gains oxygen but loses most of its pressure. From there it moves slowly and under low pressure to all parts of the body.

Because of the efficient hydrodynamic design of the fish body and the neutral buoyancy state that most fish species maintain, this blood oxygen supply system is adequate unless the need for sustained, high muscular activity arises. This most often occurs when a piscivore chases a prey fish or a territorial male attempts to drive an intruder away. Such actions usually entail only short bursts of activity, as the prey fish is either soon caught or escapes in dense vegetative cover, or a dominant male repels an intruder with one short dash. However, when the predator or intruder is the human angler and the hook is firmly set in the fish's jaw, quick escape is not usually possible, and the shortcoming of the slow oxygen supply system found in white muscle is displayed.

During the first minute or so of "the fight," a fish's resistance to being suddenly tethered to a line is usually quite vigorous and hence one of the great attractions of fishing. All too soon, however, oxygen supply falls far short of demand, and with essentially no stored muscle oxygen to call upon, the entire metabolism switches from aerobic to anaerobic. In the latter process, small amounts of energy are still supplied to the muscle cells, but a toxic byproduct, lactic acid, begins to accumulate within the muscle cells and soon inhibits their ability to contract. Athletes in track events such as the 400m run sometimes experience this same situation as they essentially "run out of gas" (in this case, oxygen) near the end of the race, switch over briefly to anaerobic metabolism, and finally succumb to complete muscle fatigue at or before the finish line. For a fish, this means being pulled in like an inanimate object, unable to mount any further resistance. As for the angler, he or she has skillfully and correctly "played" the fish. In the increasingly popular area of catch-and-release fishing, it is now time for the victor to honor the worthy opponent by carefully removing the hook and then holding the fish in the water so that either natural or created water flow moves into the mouth and over the gills. Usually, after less than a minute of such aquatic artificial respiration, the paramedic feels a surge of power returning to the fish body, indicating that its time for a safe release.

One further physiological phenomenon that is crucial to a number of California fishes is their periodic adjustment to saltwater. To appreciate such adjustments, you must become familiar with the basic relationships that exist between the water in which fishes live and their blood and other body fluids. In the case of freshwater fishes, for example, more salts and protein molecules and fewer water molecules per unit volume are in their body fluids than are in the surrounding lake or stream. These fishes are usually referred to as being hypertonic to their environment, with "hyper" indicating the higher concentrations of materials in the body fluids. The gill is the major site for the passage of water and other molecules into and out of the fish body. Because water always moves from an area where it is in greater abundance (the lake) to an area where it is less abundant (the fish body), there is a continuous influx of water into the blood and other body fluids of freshwater fishes.

Left unchecked, such a situation would soon result in a gross imbalance between water and other molecules in the cells, and the fish would actually drown at the cellular level. To counter this situation, freshwater fishes continuously remove excess water from their blood by means of the kidneys, which act primarily as a physiologic water pump. Oddly enough, this also means that freshwater fishes never drink, because this would only add to the water intake problem.

When a fish enters saltwater, the relative water and salt concentrations of the body fluids and the sea are just the reverse of the freshwater situation. Now, more water molecules and fewer salt molecules per unit volume are in the fish body fluids than are in the sea, and the fish is said to be hypotonic to the sea. Water and other substances still follow the basic rule of moving from an area of greater abundance to one of lesser abundance. As in freshwater, the gill is the major site of water movement, but this time the water is moving out of the fish body, and the fish begins to dehydrate.

When migrating from freshwater to seawater, anadromous

fishes such as salmon and Striped Bass *(Morone saxatilis)* must shut down the mechanisms in the kidney responsible for excreting water. To replenish their body water, they drink seawater and then secrete the resulting excess sodium and chloride ions in the blood back into the sea through the action of special salt secreting cells in the tissue that supports the gill capillaries. In other words, they operate their own blood desalination plant. Each time a migrating fish moves between the freshwater and saltwater environments, one system must be turned off and the other turned on, a physiological feat mastered by relatively few species.Those who can tolerate wide ranges of salinity—called euryhaline species—benefit greatly. The anadromous salmonids, for instance, are able to spawn and spend their early juvenile life in a creek habitat that compared to the ocean, has few serious egg or fry predators. Other euryhaline species such as young Starry Flounder *(Platichthys stellatus)* and Shiner Perch *(Cymatogaster aggregata)* make occasional sojourns into brackish water or even freshwater, where competition for food is presumably less than in the overcrowded inshore areas of the sea. Finally, those miniature masters of osmoregulation such as the Salt Creek Pupfish *(Cyprinodon salinus)* occupy segments of creek pool habitats that may range from brackish (14 ppt) to over twice the salt content of seawater (70 ppt) during the year.

MEASUREMENTS AND DEFINING CHARACTERISTICS

Each species illustrated in this book is accompanied by a brief written description. These descriptions are not intended to be complete taxonomic reviews. Instead, important characteristics of each family have been listed for quick and easy species recognition. In large families with many similar species, detailed characteristics such as dorsal fin rays and anal fin rays are given. In a small family, however, species often differ by one or more pronounced morphological characteristics, and here detailed fin ray counts are usually omitted.

To help keep the verbal accounts relatively short so that they fit the size limits of the page, several abbreviations are used and are defined as follows:

TL = Total length. This measurement is used whenever possible because it is the most widely employed, especially among anglers. Total length is the distance from the anterior tip of a fish to the most posterior tip of the caudal fin when both lobes are pressed together (fig. 5).

SL = Standard length. This measurement is given when a total length is not available in the literature. Standard length is a more reliable scientific measurement and is the distance from the anterior tip of the fish to the posterior tip of the caudal peduncle. Unlike total length, it is not affected by caudal fin breakage or fraying.

FL = Fork length

LL = Lateral line

max. wt. = Maximum weight

ad. = Adult

DF = Dorsal fin

AF = Anal fin

PecF = Pectoral fins

PelF = Pelvic fins

Counts of both spines and rays in the dorsal fin, anal fin, and pelvic fins are often used, especially with closely related species. Spines are relatively easy to count, as they are straight, simple structures. Rays, however, tend to split or branch at the fin margins, so that confusion may result if counts are made from the outer fin edge. Instead, always count fin rays along the fin base. Even if a ray splits soon after it leaves the fin base, such as ray 8 in fig. 5, it should be counted as just one ray.

One final characteristic that is useful in describing both a species and its relative scale size is the number of scales along the lateral line. When using this count for comparative scale size between species, the average adult size must also be considered. Thus if two species both have an average adult TL of 25 cm, but one has 35 to 40 LL scales while the other has 75 to 80 LL scales, the latter has far smaller scales than the former.

In the angling notes section for each species, measurements are given in inches, not centimeters, and weights in pounds, not kilograms, because the vast majority of anglers do not and very likely will never apply the metric system to their favorite pastime. Elsewhere, measurements are provided in both metric and American units.

PROTECTIVE STATUS

In 1966 the U.S. Congress enacted the Endangered Species Preservation Act, the first legislation that afforded some protection to species exhibiting serious population declines and appearing to be on a path to extinction. A second bill, the Endangered Species Conservation Act, was passed in 1969, but neither were forceful enough to prevent some states from ignoring many of their provisions. As a result, a more forceful and encompassing bill, the Endangered Species Act (P.L. 93-205), was passed and signed into law on December 28, 1973, and remains today the federal mandate for endangered species protection and recovery.

Among all states, California led the way in protecting vanishing species with the passage of its 1970 Endangered Species Act, which prohibits the importation, taking, possession, and sale of endangered and threatened species. Both the federal and state acts define "endangered" and "threatened" species as follows.

FE, SE An endangered species in one that is in danger of extinction throughout all or a significant part of its range. A federal endangered fish is noted by FE, and state endangered fish by SE in this book.

FT, ST A threatened species is one that is likely to become an endangered species within the foreseeable future throughout all or a significant portion of its range. A federal threatened fish is noted by FT, a state threatened fish by ST.

The state of California has two additional protective categories.

FPS The fully protected species designation is the most protective of all because unlike federal or state endangered listings, fully protected status precludes any take (killing, collecting, etc.) of a so listed species under any circumstance. A California fully protected species is noted by FPS.

CSSC The California species of special concern designation applies to species whose numbers are beginning to decline or whose habitat is continuously being reduced and thus require special attention and observation. No state law affords legal protection to such species; however, when land use changes near or within their habitat threatens to impact such species, the California Department of Fish and Game usually requests that protective measures be implemented before any such project can proceed. A California species of special concern is noted by CSSC.

One final note is that most endangered and threatened species are really subspecies, and the state and federal laws governing their protection should more correctly be called the "Endangered Subspecies Acts." This has led to some confusion as to why we should be concerned about the loss of one small segment of a population (a subspecies) when the species, represented by other subspecies, is doing fine elsewhere. The answer lies in the basic concept of evolution, whereby natural selection will eventually determine which characteristics within a species' total gene pool will persist through time to eventually emerge as a new species. By having a number of subspecies, each adapted to a different habitat, a species has a far better chance of persisting in the face of the rapid environmental change that our own species continues to create in nearly all of California's freshwater habitats.

ILLUSTRATED KEYS

If you have ever waded through a long, complex, verbal key to an animal or plant group, you may agree that such exercises often tend to suppress interest rather than stimulate it. With this thought in mind, the following pictorial keys and explanatory comments are presented.

Key to Families

First determine to which family the fish belongs by using the family pictorial key (fig. 7) and accompanying explanatory comments. The family key is based on three characteristics: (1) distinct body form, also called the body plan; (2) presence or absence of a forked tail; and (3) number, type, and spacing of the dorsal fins. Thus the recollection of relatively few characteristics permits quick recognition of a family.

Species Illustrations

After determining the family, go to the key that provides species illustrations for that family. Each species of California freshwater fishes is illustrated or described. Start at the beginning of the key's species descriptions for the appropriate family and view the drawings and accompanying text until you find the specimen at hand. It is important to start at the beginning of the species descriptions for that family, because the fishes have been illustrated and described in a definite order to facilitate identification.

The fishes illustrated are adults unless otherwise specified. When comparing a fish to the drawings, remember that body proportions and color can vary within a species according to factors such as age, breeding status, and background color and light intensity if viewed in an aquarium. Most fishes can also change their color or hue in response to the brightness of the day, water temperature, and degree of disturbance it may be experiencing. This is why these keys rarely use color as a defining characteristic. In composing the color illustrations on the following pages, Doris Alcorn has chosen the color scheme and hue that is most often seen in a species immediately after capture or after adaptation to a naturally appointed aquarium. Within each family, general size differences have been implied by drawing some species larger than others, but the small format of this book prevents the illustration of exact differences. A size range for average adult specimens is given in the species description.

Finally, if this is your first attempt at using a fish key, you may need to refer to the illustration of external anatomy (fig. 5) or the glossary if you are uncertain about some terms used in the key.

Aquatic Map of California

Many native California freshwater fishes are still confined to the stream and river drainages where they originated. Thus geographic range is helpful when identifying fishes in the field or preserved specimens with good locality data. In fact, in many cases, range can quickly eliminate closely related species from the determination. The two-page map of major California streams and rivers at the end of the family key (fig. 8) should be referred to when making such geographically based identifications.

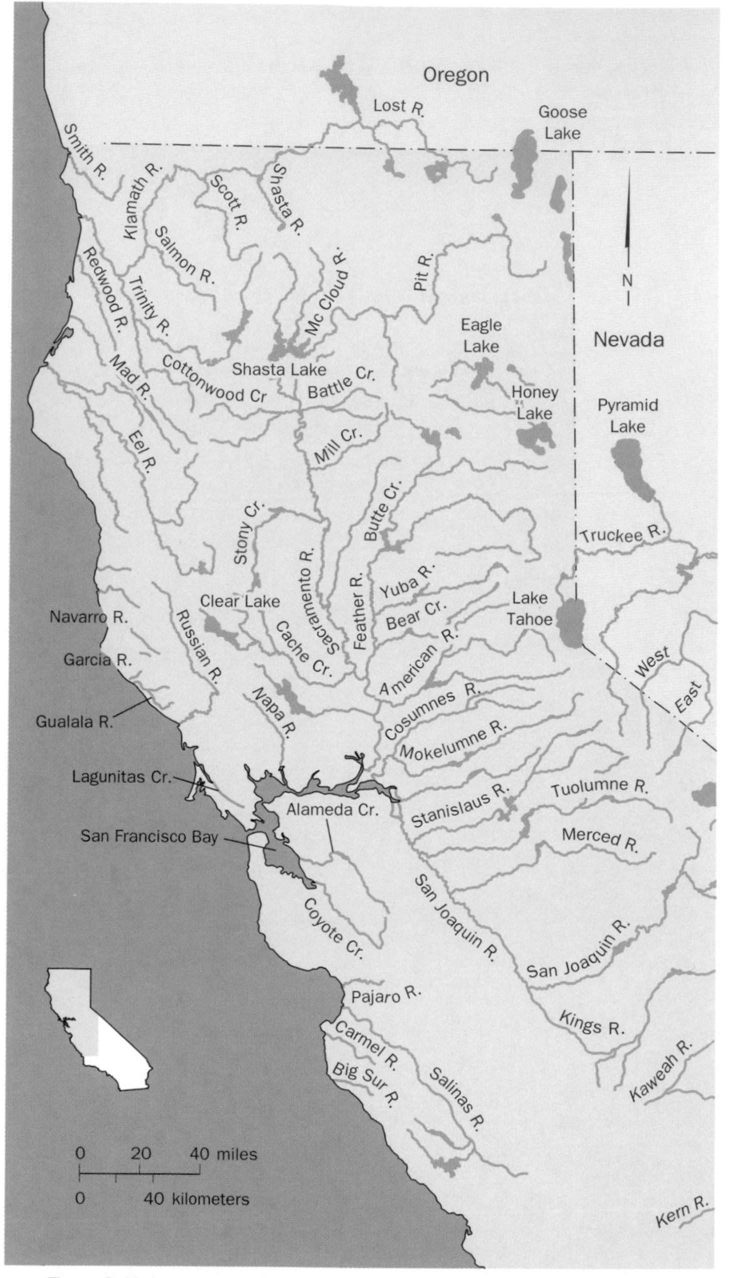

Figure 8. Major streams, rivers, and lakes of California.

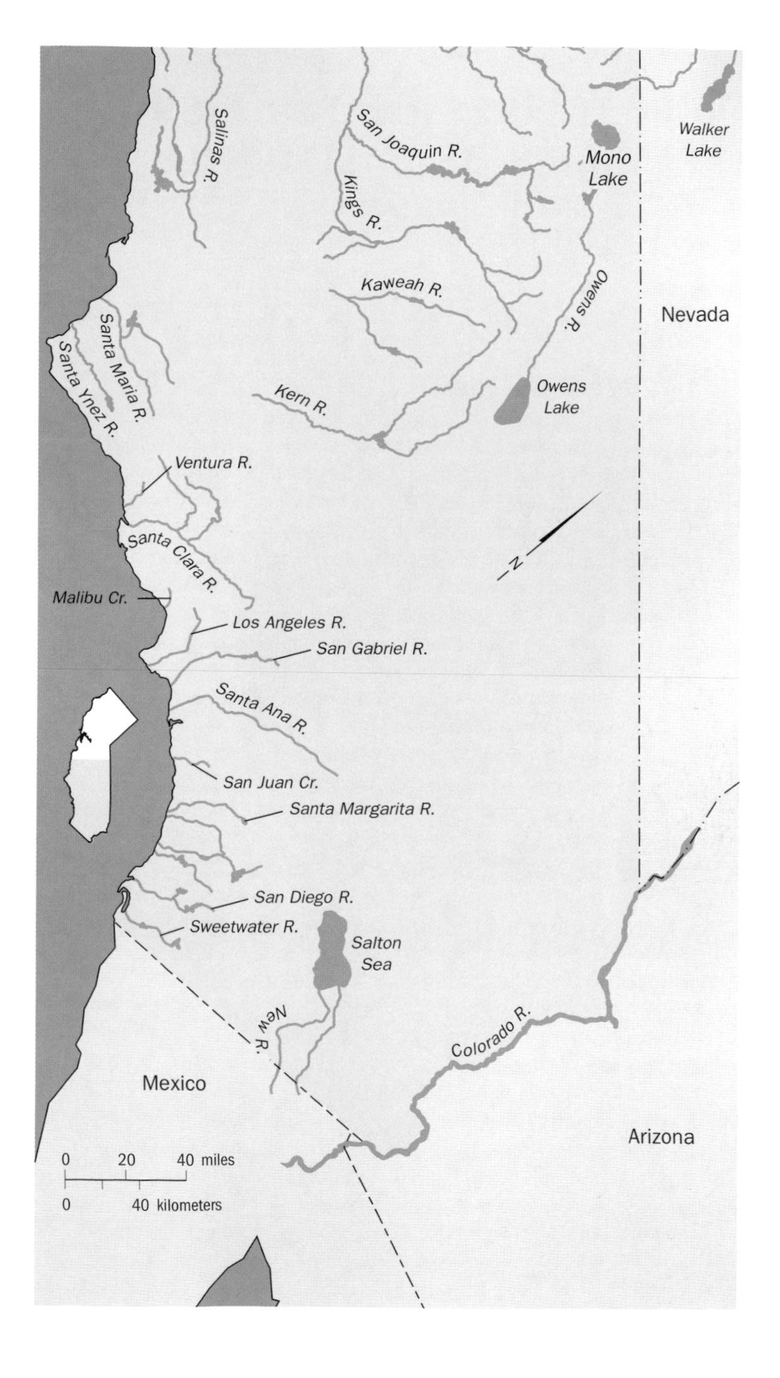

Salinas R.
San Joaquin R.
Kings R.
Mono Lake
Walker Lake
Kaweah R.
Owens R.
Nevada
Santa Maria R.
Santa Ynez R.
Kern R.
Owens Lake
Ventura R.
N
Santa Clara R.
Malibu Cr.
Los Angeles R.
San Gabriel R.
Santa Ana R.
San Juan Cr.
Santa Margarita R.
San Diego R.
Sweetwater R.
Salton Sea
New R.
Colorado R.
Mexico
Arizona
0 20 40 miles
0 40 kilometers

KEY TO FRESHWATER FISH FAMILIES OF CALIFORNIA

Obvious Body Plans

The five body forms in the first group depicted in fig. 7 are so evident that, once observed, they should never be confused with any other freshwater fish families in California. The Starry Flounder (*Platichthys stellatus,* family Pleuronectidae) is the only member of the marine flounder or flatfish group that occurs in freshwater in California and is therefore unique among all freshwater species. The lampreys (family Petromyzontidae) also need little comment because no other fishes in California freshwater habitats have their snakelike appearance. The sturgeons (family Acipenseridae) have a unique head structure, a heterocercal tail in which the vertebral column extends into the upper lobe, and widely spaced rows of large bony plates. The catfishes and bullheads (family Ictaluridae) are scaleless, have an adipose fin, and possess several long barbels or "whiskers" around the mouth area. Finally, the sticklebacks (family Gasterosteidae) have three single, widely spaced dorsal spines in addition to pelvic spines. One word of counsel: the sticklebacks, like many species of fishes, tend to depress their spines and fins when captured and held out of water. Gently squeezing or tapping the fish usually causes fin erection. Fins of dead fishes may have to be erected mechanically with a small probe to observe their true nature.

Rounded or Square Tail (Caudal Fin)

A. SPINY AND SOFT-RAYED DORSAL FIN, CLOSELY SPACED If the head is large in proportion to the body and the eyes bulge out dorsally, it is either a goby or a sculpin. The gobies (family Gobiidae) have fused pelvic fins that form round discs; the sculpins (family Cottidae) do not. If the head is small, without protruding eyes, it is a Bigscale Logperch *(Percina macrolepida).*

B. SPINY AND SOFT-RAYED DORSAL FIN, SEGMENTS FUSED The cichlids (family Cichlidae) also have a double or spilt lateral line.

C. SINGLE DORSAL FIN Three families have rounded tails and single, soft-rayed dorsal fins, and the adult size of most species is quite small (2 to 5 cm [0.8 to 2.0 in.]). The best characteristic by which to distinguish the livebearer family (Poeciliidae) is the presence of a spiked anal fin or "gonopodium" in the male. No such structure exists in the killifish family (Fundulidae), but females of the two groups may appear similar. In female killifishes the third anal fin ray is branched; in livebearers it is not. Like the killifish, the male of the pupfish family (Cyprinodontidae) lacks a gonopodium; however, unlike the elongated killifish body, the pupfish body is quite deep and the caudal peduncle relatively short.

Forked Tail (Caudal Fin)

A. TWO WIDELY SPACED DORSAL FINS Members of both the trout and salmon family (Salmonidae) and the smelt family (Osmeridae) have a fleshy, nonrayed adipose fin as the second dorsal fin. However, trout and salmon have a slightly forked tail, whereas the smelts have a deeply forked tail. If both dorsals are soft-rayed fins, the fish is in the silverside family (Atherinidae). If the first dorsal has spines, it is in the mullet family (Mugilidae).
B. TWO CLOSELY SPACED DORSAL FINS If the fish has thin, dark, horizontal lines on the body, it is a Striped Bass *(Morone saxatilis)* or White Bass *(M. chrysops)*, both in the temperate bass family (Moronidae). If it has dark, vertical bars on the body, it is a Yellow Perch *(Perca flavescens)*, a member of the perch family (Percidae).
C. DORSAL SEGMENTS FUSED OR TOUCHING If a fish has fused or touching dorsal segments, it is very likely a member of the sunfish, crappie, and "black" basses family (Centrarchidae). If it has a row of enlarged scales on the base of the dorsal fin, it is a member of the viviparous perch family (Embiotocidae).
D. SINGLE DORSAL FIN Of the four families in this last group, the pike family (Esocidae) has both the dorsal and anal fins positioned immediately anterior to the caudal peduncle, and the mouth is very large. The remaining three families can appear very similar at first glance. First, feel the ventral aspect of the body to see if it is made up of a knife-sharp row of scales. If so, it is in the shad family (Clupeidae). Next, examine the mouth. Is it strongly inferior, opening directly downward? If so, it is a sucker (family Catostomidae). If a fish with a forked tail and single dorsal fin has neither a sharp keel nor a strongly inferior mouth, it is a true minnow (family Cyprinidae).

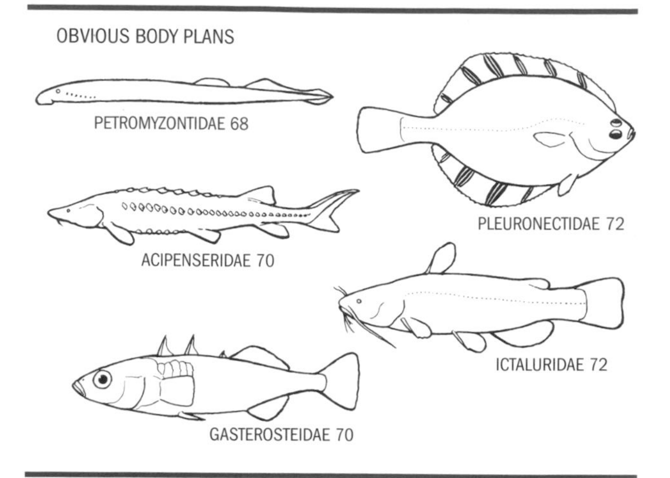

ROUNDED OR SQUARE TAIL

Two dorsal fins, closely spaced (A); single dorsal fin, spiny-soft ray segments fused (B); or single soft dorsal fin (C).

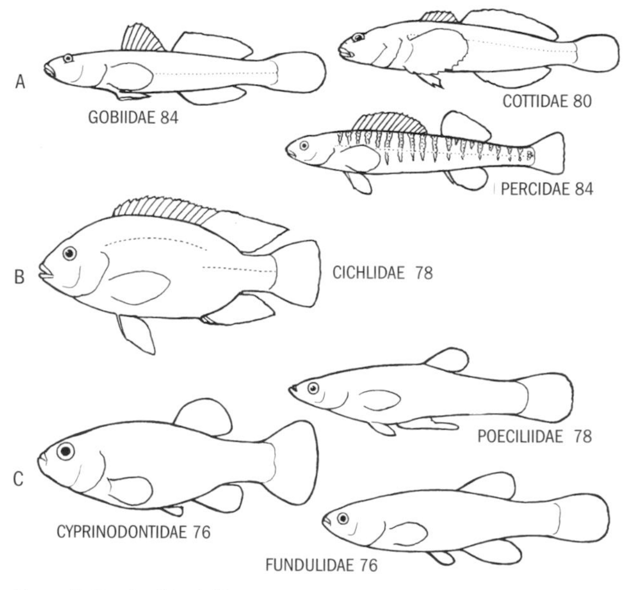

Figure 7. Family pictorial key.

FORKED TAIL

Two dorsal fins, widely spaced (A) or closely spaced (B); single dorsal fin, spiny-soft ray segments fused (C); or single dorsal fin (D).

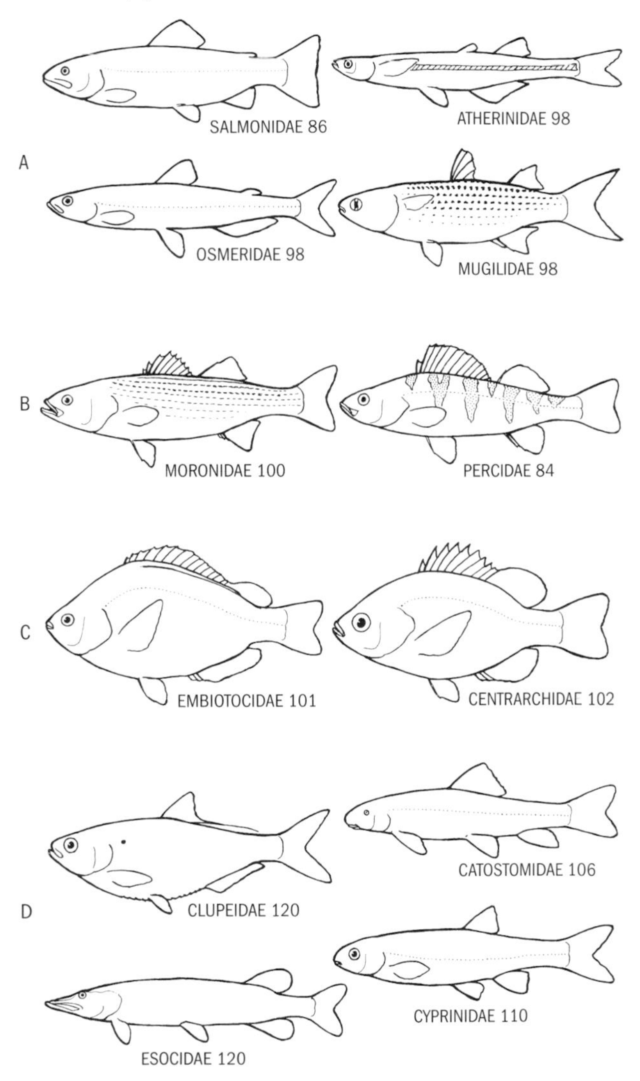

KEY TO CALIFORNIA FRESHWATER FISH SPECIES

Lampreys (Petromyzontidae)

Pacific Lamprey *(Lampetra tridentata)* PAGE 127
STATUS: Native. **DESCRIPTION:** Four pairs of lateral tooth plates (cusps 2-3-3-2); supraoral plate with three sharp cusps; ad. TL over 40 cm (16 in.), max. TL 76 cm (30 in.). **DISTRIBUTION:** Most coastal streams and rivers of California. Dwarf populations (TL 19 to 27 cm [7.5 to 10.5 in.]) occur in the upper Klamath River drainage and Goose Lake.

Pit-Klamath Brook Lamprey *(Lampetra lethophaga)* PAGE 129
STATUS: Native. **DESCRIPTION:** Four pairs of lateral tooth plates (cusps 2-2-2-2); supraoral plate with three dull cusps; ad. TL under 21 cm (8 in.). **DISTRIBUTION:** Pit River, upper Klamath River drainage, and Goose Lake.

Klamath River Lamprey *(Lampetra similis)* PAGE 130
STATUS: Native, CSSC. **DESCRIPTION:** Four pairs of lateral tooth plates (cusps 2-3-3-2); supraoral plate with three sharp cusps; ad. TL under 27 cm (10.5 in.). **DISTRIBUTION:** Lower Klamath and Trinity Rivers.

Kern Brook Lamprey *(Lampetra hubbsi)* PAGE 130
STATUS: Native, CSSC. **DESCRIPTION:** Four pairs of lateral tooth plates (cusps 1-1-1-1); supraoral plate with two sharp cusps; ad. TL under 15 cm (6 in.). **DISTRIBUTION:** Friant-Kern Canal (Kern County), and lower Merced, Kaweah, Kings, and San Joaquin Rivers.

River Lamprey *(Lampetra ayresi)* PAGE 129
STATUS: Native, CSSC. **DESCRIPTION:** Three pairs of lateral tooth plates (cusps 2-3-2); supraoral plate with two sharp cusps; center cusp on transverse lingual lamina large; ad. TL 17 cm (7 in.), never more than 32 cm (12.5 in.). **DISTRIBUTION:** Coastal streams and rivers from San Francisco Bay northward.

Western Brook Lamprey *(Lampetra richardsoni)* PAGE 130
STATUS: Native. **DESCRIPTION:** Three pairs of lateral tooth plates (dull cusps 2-2 or 3-1); supraoral plate with two dull cusps and a large space in the middle; seven to 10 poorly developed cusps on infraoral plate; ad. TL up to 18 cm (7 in.). **DISTRIBUTION:** Portions of the Sacramento basin, Kelsey Creek west of Clear Lake, and coastal streams and rivers from San Francisco Bay northward.

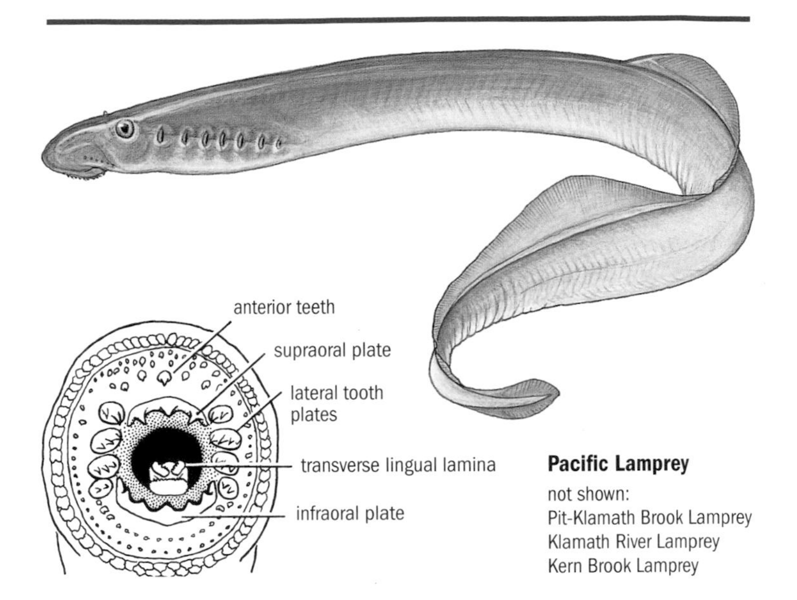

Pacific Lamprey

not shown:
Pit-Klamath Brook Lamprey
Klamath River Lamprey
Kern Brook Lamprey

4 pairs lateral tooth plates

3 pairs lateral tooth plates

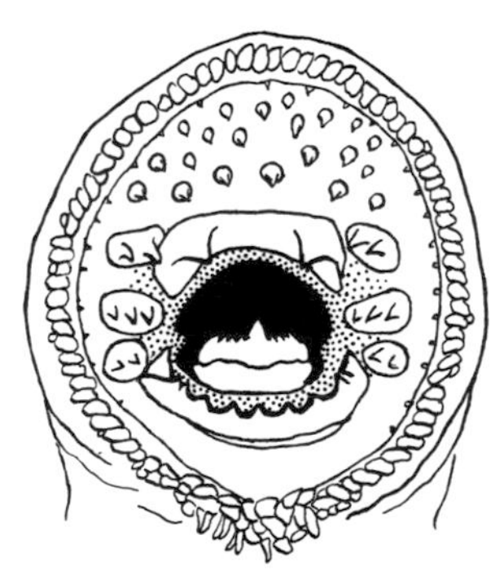

River Lamprey

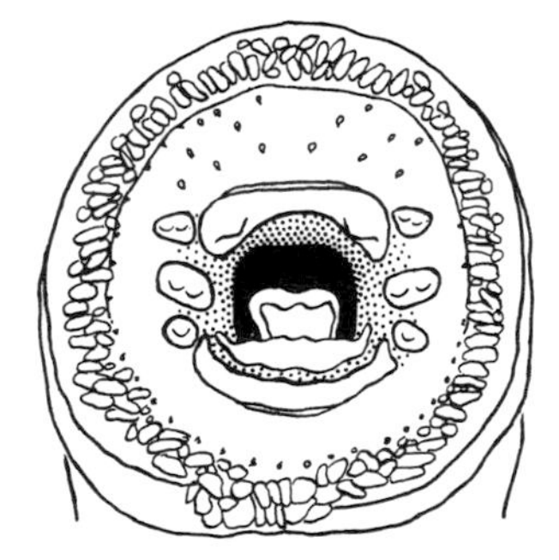

Western Brook Lamprey

Sturgeons (Acipenseridae)

White Sturgeon *(Acipenser transmontanus)* PAGE 134
STATUS: Native. **DESCRIPTION:** Plates in lateral row 38 to 48; double row of four to eight plates each between PelF and AF; barbels usually closer to snout tip than to mouth; color gray brown fading to white ventrally; ad. TL 1 to 1.5 m (3.3 to 5 ft), max. TL about 4 m (13 ft), max. wt. 590 kg (1,300 lb). **DISTRIBUTION:** Large rivers from San Francisco Bay northward.

Green Sturgeon *(Acipenser medirostris)* PAGE 139
STATUS: Native, CSSC. **DESCRIPTION:** Plates in lateral row 23 to 30; single or double row of one to four plates between PelF and AF; barbels usually closer to mouth than to snout tip; color olive green with vague green stripe on lower sides; max. TL 2.7 m (8.8 ft); ad. TL 1.3 to 1.5 m (4.2 to 4.9 ft); max. wt. 175 kg (385 lb); ad. wt. 100 kg (220 lb). **DISTRIBUTION:** Large rivers from San Francisco Bay northward; most common sturgeon in Klamath and Trinity Rivers.

Sticklebacks (Gasterosteidae)

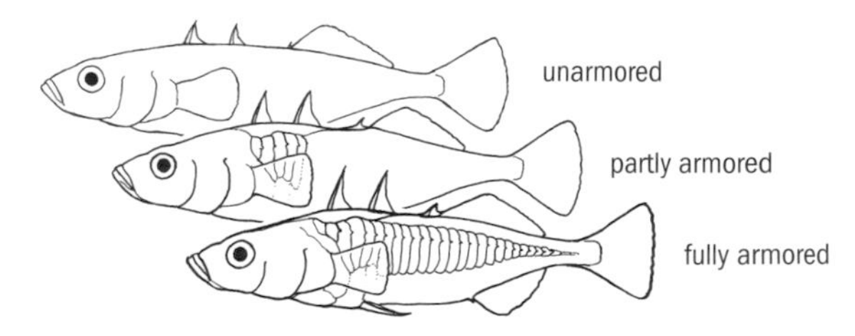

Threespine Stickleback *(Gasterosteus aculeatus)* PAGE 289
STATUS: Native. **DESCRIPTION:** Three sharp spines on back before DF; spinelike PelF; leading edge of AF posterior to leading edge of DF; narrow caudal peduncle; three subspecies defined by number of lateral bony plates: *G. a. aculeatus,* 28 to 35; *G. a. microcephalus,* 1 to 8; *G. a. williamsoni* (FPS, SE, FE), 0; color on back usually greenish, with white to gold on belly, but breeding male has red belly and blue sides; ad. TL 40 to 50 mm (1.5 to 2 in.), max. TL 80 mm (3.2 in.). **DISTRIBUTION:** All coastal streams and rivers, brackish- and freshwater Delta marshes, the Mojave and Owens Rivers, and many low-elevation sites in the Central Valley.

Brook Stickleback *(Culaea inconstans)* PAGE 293
STATUS: Introduced. **DESCRIPTION:** Five spines on back before DF; spinelike PelF and a prominent spine in leading edge of AF; no body plating; leading edge of AF directly below leading edge of DF; narrow caudal peduncle; head noticeably smaller than that of Threespine Stickleback *(Gasterosteus aculeatus);* ad. TL 40 to 70 mm (1.5 to 2.8 in.). **DISTRIBUTION:** Scott River drainage (Siskiyou County).

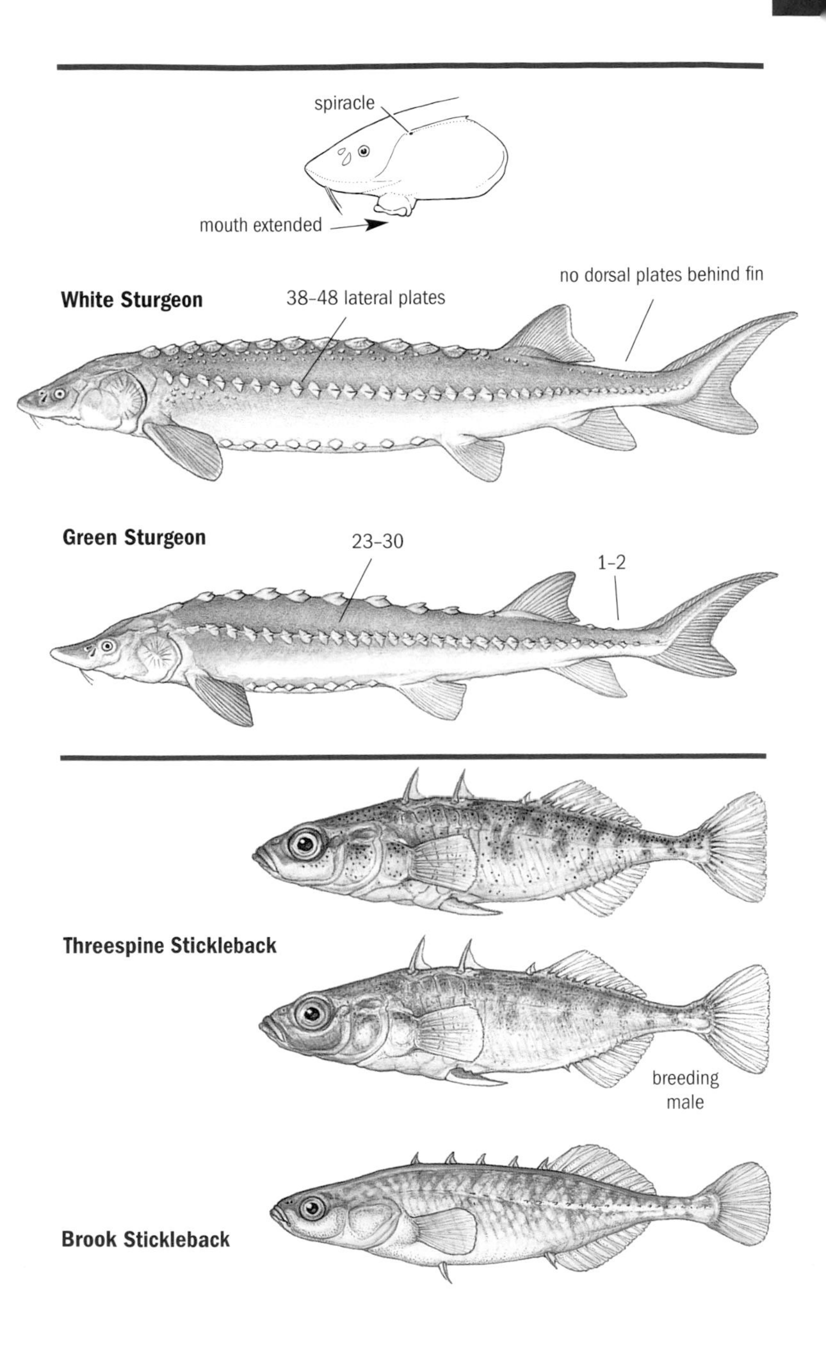
spiracle
mouth extended
White Sturgeon
38–48 lateral plates
no dorsal plates behind fin
Green Sturgeon
23–30
1–2
Threespine Stickleback
breeding
male
Brook Stickleback

Black Bullhead *(Ameiurus melas)* PAGE 263

STATUS: Introduced. **DESCRIPTION:** Entire length of chin barbels darkly pigmented; caudal fin square with slight notch; dark membrane on AF; pectoral spines smooth or slightly rough; ad. color from yellow brown to black, light on belly; ad. TL 20 to 25 cm (8 to 10 in.), max. TL 61 cm (24.4 in.), max. wt. about 3.6 kg (7.9 lb). **DISTRIBUTION:** Most major rivers of California, and some low- and middle-elevation reservoir lakes.

Yellow Bullhead *(Ameiurus natalis)* PAGE 264

STATUS: Introduced. **DESCRIPTION:** Chin barbels white; caudal fin slightly rounded; PecF with sawtooth posterior edge; ad. color yellow brown to black, white belly; ad. TL 25 to 30 cm (10 to 12 in.), max. TL 48 cm (19 in.), max. wt. about 1 kg (2.2 lb). **DISTRIBUTION:** Common only in the Colorado River but occasionally occurring in warm, clear, low-elevation habitats such as southern California reservoirs.

Brown Bullhead *(Ameiurus nebulosus)* PAGE 258

STATUS: Introduced. **DESCRIPTION:** Chin barbels dark except for light portion adjacent to chin; caudal fin straight; dorsal and pectoral spines toothed on posterior edge; ad. color mottled yellow brown, belly white or yellow; ad. TL 25 to 30 cm (10 to 12 in.), max. TL 53 cm (21 in.), max. wt. 2.2 kg (4.8 lb). **DISTRIBUTION:** Warmwater habitats throughout California.

Flathead Catfish *(Pylodictis olivaris)* PAGE 270

STATUS: Introduced. **DESCRIPTION:** Chin barbels light; caudal fin square to slightly indented; pectoral spines rough on both sides; ad. color yellow brown with dark mottling; head flat on top; ad. TL about 80 cm (31.5 in.), max. TL 1.4 m (4.6 ft), max. wt. 56 kg (123.6 lb). **DISTRIBUTION:** Turbid river and reservoir waters of the Imperial Valley and lower Colorado River.

TAIL NOT FORKED

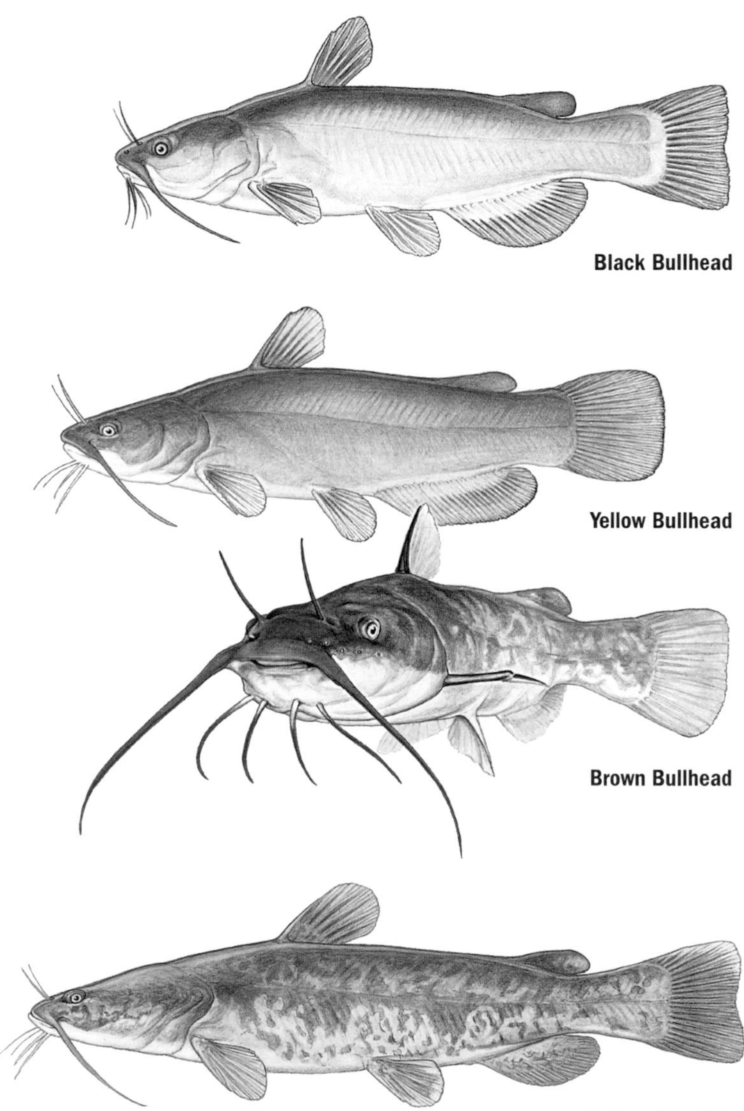

Flathead Catfish

Livebearers (Poeciliidae)

Western Mosquitofish *(Gambusia affinis)* PAGE 282

STATUS: Introduced. **DESCRIPTION:** "Guppylike," small DF with origin posterior to AF origin; no heavy spotting on fins or body; color drab green gray, no melanistic (black) phase; ad. TL 2 to 2.5 cm (0.8 to 1 in.), max. TL 6 cm (2.4 in.) female, 3.5 cm (1.4 in.) male. **DISTRIBUTION:** Most low- and middle-elevation freshwater habitats in California.

Porthole Livebearer *(Poeciliopis gracilis)* PAGE 285

STATUS: Introduced. **DESCRIPTION:** Similar to Western Mosquitofish *(Gambusia affinis)* but with a thicker body; several bold dark spots on a longitudinal row along each side of body; male has a very long gonopodium, reaching to near the end of the caudal peduncle; ad. TL 4 to 6 cm (1.6 to 2.4 in.), max TL 7.5 cm (3 in.). **DISTRIBUTION:** Drainages ditches along north shore of Salton Sea.

Sailfin Molly *(Poecilia latipinna)* PAGE 284

STATUS: Introduced. **DESCRIPTION:** Large DF; spotting on DF, caudal fin, and body forms dark rows; back brown, belly light, DF light blue with yellow edge; melanistic and checkered forms; ad. TL 7 to 9 cm (2.8 to 3.5 in.), max. TL 15 cm (6 in.). **DISTRIBUTION:** Irrigation canals near Salton Sea.

Shortfin Molly *(Poecilia mexicana)* PAGE 285

STATUS: Introduced. **DESCRIPTION:** Similar to Sailfin Molly *(P. latipinna)* but with male DF about one-half and female DF about one-third the length of male Sailfin Molly DF; ad. TL 5 to 7 cm (2 to 2.8 in.), max. TL 10 cm (4 in.). **DISTRIBUTION:** Irrigation canals associated with Salton Sea.

Cichlids (Cichlidae)

Pure forms of the cichlids are illustrated here. Visual identification of these species should be considered tentative because of the tendency for them to hybridize, and also the uncertainty as to whether additional cichlids occur in California.

Redbelly Tilapia *(Tilapia zillii)* PAGE 340

STATUS: Introduced. **DESCRIPTION:** Distinct spot on anterior rayed DF; mouth is horizontal, not oblique; nonbreeding color dark olive with lighter sides; breeding color green back, black head, red and black throat; ad. TL 20 to 25 cm (8 to 10 in.), max. TL 35 cm (13.8 in.). **DISTRIBUTION:** Irrigation canals of Imperial, Coachella, and Palo Verde Valleys, and coastal streams of the Los Angeles area.

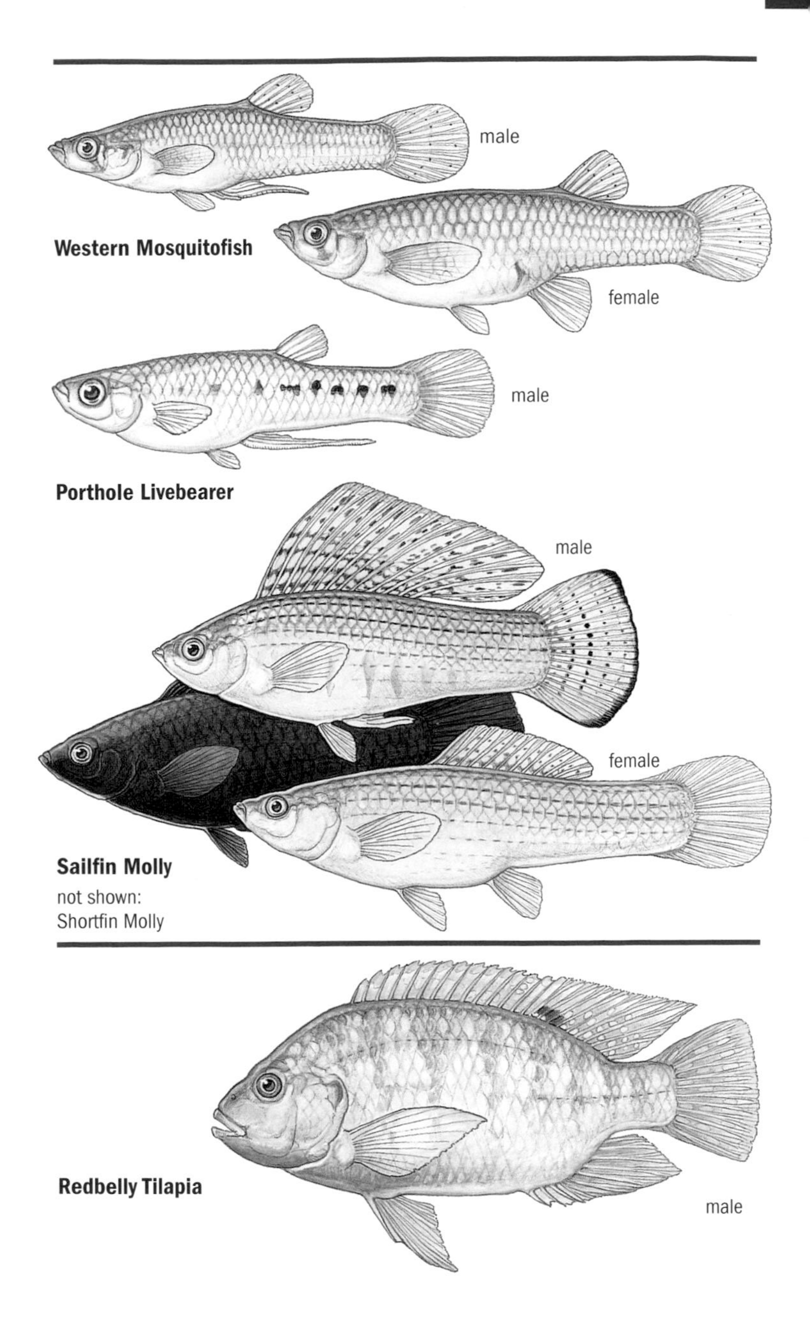
male
Western Mosquitofish
female
male
Porthole Livebearer
male
female
Sailfin Molly
not shown:
Shortfin Molly
Redbelly Tilapia
male

Mozambique Tilapia *(Oreochromis mossambicus)* PAGE 337
STATUS: Introduced. **DESCRIPTION:** No spot on DF; oblique mouth extends to or past outer margin of eye; caudal fin somewhat rounded; color variable; male breeding color blue black body, pale throat, enlarged mouth especially upper lip, red on fins; ad. TL 20 to 25 cm (8 to 10 in.), max. TL 39 cm (15 in.). **DISTRIBUTION:** Irrigation canals of Imperial and Palo Verde Valleys and lower Colorado River.

Blue Tilapia *(Oreochromis aureus)* PAGE 341
STATUS: Introduced. **DESCRIPTION:** Oblique mouth rarely extends to but never past the outer margin of the eye; caudal fin relatively straight at corners as opposed to rounded; females and nonbreeding males have banding on sides; breeding male has dark blue chin and breast and light blue body but does not have enlarged lips; ad. TL 22 to 31 cm (8.7 to 12.2 in.), max. 37 cm (14.6 in.). **DISTRIBUTION:** Lower Colorado River.

Nile Tilapia *(Oreochromis niloticus)* PAGE 342
STATUS: Introduced. **DESCRIPTION:** Oblique mouth rarely extends to or past outer margin of eye; caudal fin with distinct vertical stripes; banding on sides of female and nonbreeding male; breeding male head often flushed with red but without enlarged mouth or lips; can reach much larger size than other known California tilapias; TL 30 to 40 cm (11.8 to 15.8 in.), max. TL 65 cm (25.5 in.); max. wt. 7 kg (15.4 lb). **DISTRIBUTION:** Believed to be in lower Colorado River, possibly hybridizing with Blue Tilapia *(O. aureus)*.

Sculpins (Cottidae)

It is difficult to differentiate the species in this family, even for experts. Many species are very small, and key taxonomic characteristics can vary greatly within a species. The following attempts to isolate those characteristics that best distinguish these fishes. Often, geographic distribution is the best guide.

Rough Sculpin *(Cottus asperrimus)* PAGE 363
STATUS: Native, ST. **DESCRIPTION:** Two preopercular spines, upper well developed, lower blunt; three pelvic supportive elements (1 spine, 2 rays); LL incomplete; 1st DF = 5 to 7, 2d DF = 17 to 19, PelF = 3, AF = 13 to 17; ad. SL 3.5 to 6.6 cm (1.4 to 2.6 in.), max. TL 9.6 cm (3.6 in.). **DISTRIBUTION:** Pit River near Pit Falls, Hat Creek, and Fall River drainage.

Pacific Staghorn Sculpin *(Leptocottus armatus)* PAGE 364
STATUS: Native. **DESCRIPTION:** Preopercular spine large and antlerlike with three to four small spines; angle of jaw extends past eye; head flat; 1st DF = 7, 2d DF = 17, PelF = 4, AF = 17; ad. TL 8 to 12 cm (3 to 4.7 in.), max. TL (California) about 22 cm (8.6 in.). **DISTRIBUTION:** Lower reaches of coastal streams and rivers.

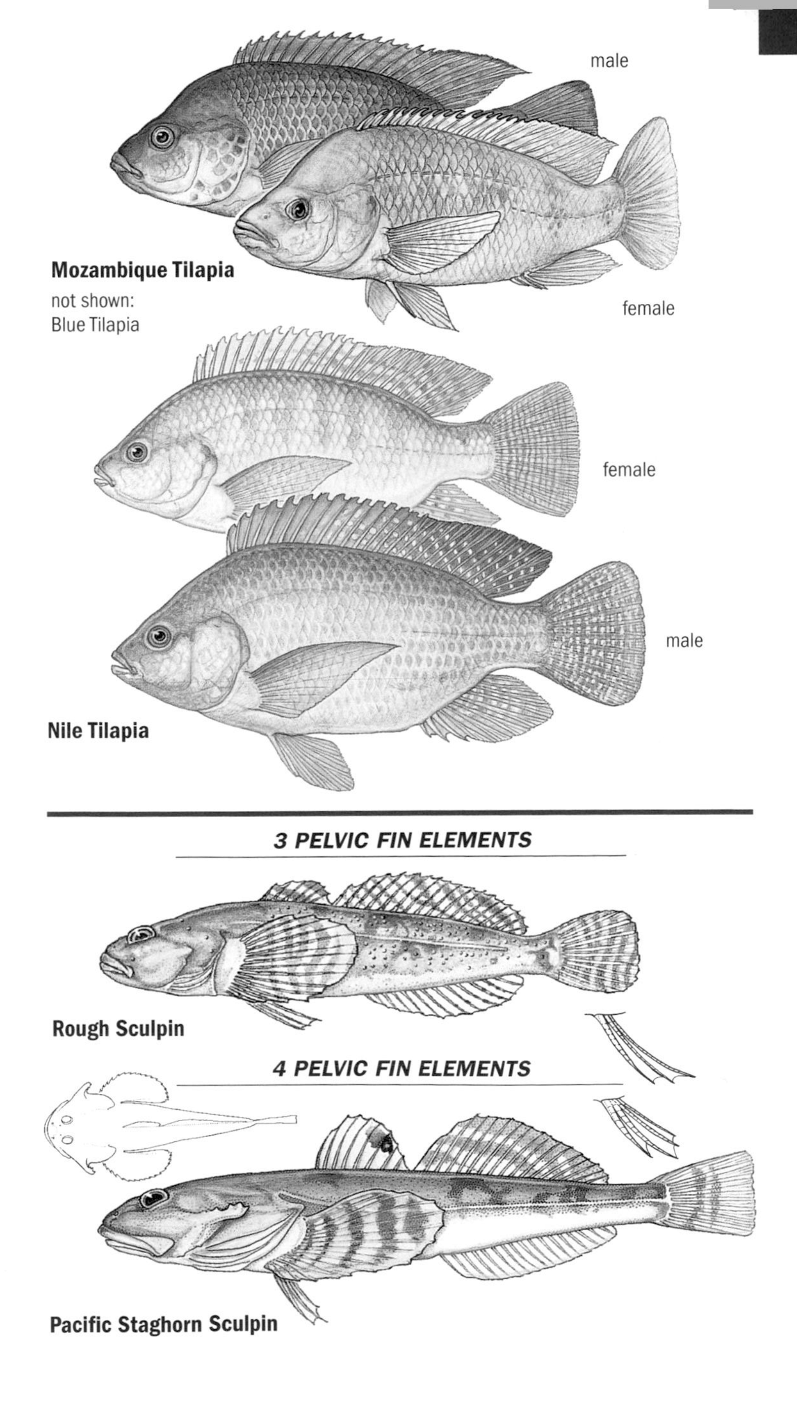
male
female
Mozambique Tilapia
not shown:
Blue Tilapia
female
male
Nile Tilapia
3 PELVIC FIN ELEMENTS
Rough Sculpin
4 PELVIC FIN ELEMENTS
Pacific Staghorn Sculpin

Prickly Sculpin *(Cottus asper)* PAGE 359

STATUS: Native. **DESCRIPTION:** Long AF (16 to 19 rays); prickles on skin behind PecF or over entire body; DF usually joined to one-third height of first soft ray; 1st DF = 8 to 10, 2d DF = 19 to 23, PelF = 4, AF = 17 to 19; ad. SL 6 to 9 cm (2.4 to 3.5 in.), max. TL 30 cm (12 in.). **DISTRIBUTION:** Streams and rivers of coastal California and the Central Valley, Clear Lake, and numerous reservoirs.

Coastrange Sculpin *(Cottus aleuticus)* PAGE 362

STATUS: Native. **DESCRIPTION:** Long PelF reach vent; posterior nostril long and tubular; 1st DF = 8 to 10, 2d DF = 17 to 20, PelF = 4, AF = 12 to 15; ad. TL 5 to 10 cm (2 to 4 in.); max. TL 17 cm (6.5 in.). **DISTRIBUTION:** Coastal streams from Morro Bay northward.

Riffle Sculpin *(Cottus gulosus)* PAGE 361

STATUS: Native. **DESCRIPTION:** Large mouth, angle may reach rear of eye; two to three opercular spines; palatine teeth on roof of mouth; black blotch on posterior 1st DF; 1st DF = 7 to 8, 2d DF = 16 to 19, PelF = 4, AF = 12 to 17; ad. SL 4 to 6 cm (1.6 to 2.4 in.); max. TL 11 cm (4.4 in.). **DISTRIBUTION:** Sacramento–San Joaquin drainage except Pit River; coastal streams from Morro Bay northward to Noyo River (Mendocino County).

Marbled Sculpin *(Cottus klamathensis)* PAGE 363

STATUS: Native. One subspecies, the Bigeye Marbled Sculpin *(C. k. macrops)*, is a CSSC. **DESCRIPTION:** Similar to Riffle Sculpin *(C. gulosus)* but angle of jaw usually does not reach rear of eye; one preopercular spine; 1st DF = 6 to 8, 2d DF = 18 to 19, PelF = 4, AF = 13 to 15; ad. TL 7 to 9 cm (2.7 to 3.5 in.); max SL 11 cm (4.4 in.). **DISTRIBUTION:** Klamath and Lost River drainages and some tributaries of Pit River.

Pit Sculpin *(Cottus pitensis)* PAGE 361

STATUS: Native. **DESCRIPTION:** Similar to Riffle Sculpin *(C. gulosus)* but lacks palatine teeth on roof of mouth; 1st DF = 7 to 10, 2d DF = 17 to 19, PelF = 4, AF = 12 to 15; ad. TL 6 to 8 cm (2.4 to 3 in.); max. TL 13 cm (5 in.). **DISTRIBUTION:** Pit River drainage from Goose Lake to northern Shasta County and upper Sacramento River.

Reticulate Sculpin *(Cottus perplexus)* PAGE 362

STATUS: Native, CSSC. **DESCRIPTION:** Similar to Riffle Sculpin *(C. gulosus)* and Marbled Sculpin *(C. klamathensis)*; small mouth, angle of jaw reaches front of eye; two to three preopercular spines; 1st DF = 7 to 8, 2d DF = 18 to 20, PelF = 4, AF = 13 to 16; ad. SL 4 to 6 cm (1.6 to 2.4 in.); max. TL 10 cm (4 in.). **DISTRIBUTION:** California creek tributaries of the Rogue River (e.g., Elliot Creek).

Paiute Sculpin *(Cottus beldingi)* PAGE 364

STATUS: Native. **DESCRIPTION:** Only sculpin east of the Sierra south of Pit River drainage; 1st DF = 6 to 8, 2d DF = 13 to 16, PelF = 4, AF = 11 to 14; ad. TL 6 to 8 cm (2.4 to 3 in.), max. TL 13 cm (5.2 in.). **DISTRIBUTION:** Streams on eastern side of the Sierra south of Pit River.

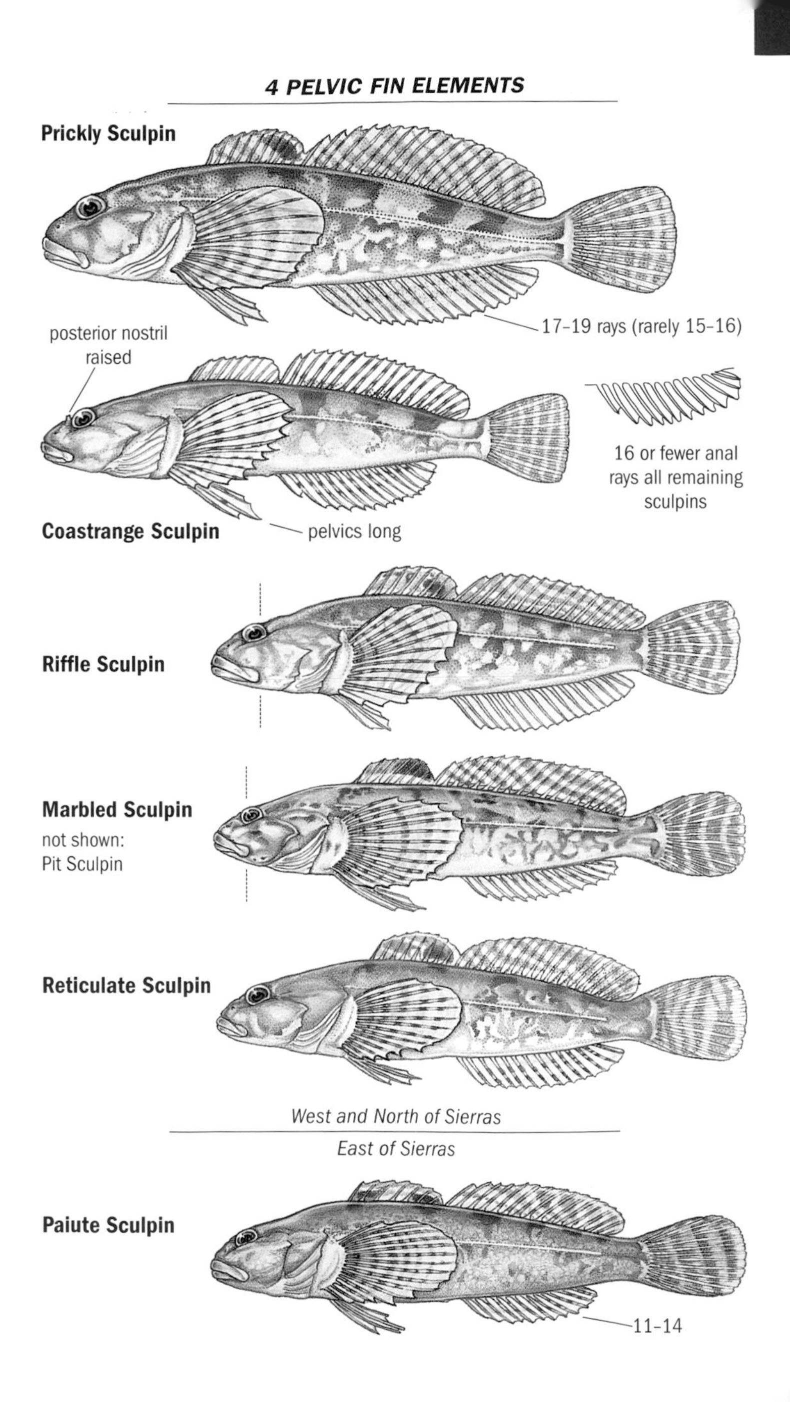
4 PELVIC FIN ELEMENTS
Prickly Sculpin
17-19 rays (rarely 15-16)
posterior nostril raised
16 or fewer anal rays all remaining sculpins
Coastrange Sculpin
pelvics long
Riffle Sculpin
Marbled Sculpin
not shown:
Pit Sculpin
Reticulate Sculpin
West and North of Sierras
East of Sierras
Paiute Sculpin
11-14

Gobies (Gobiidae)

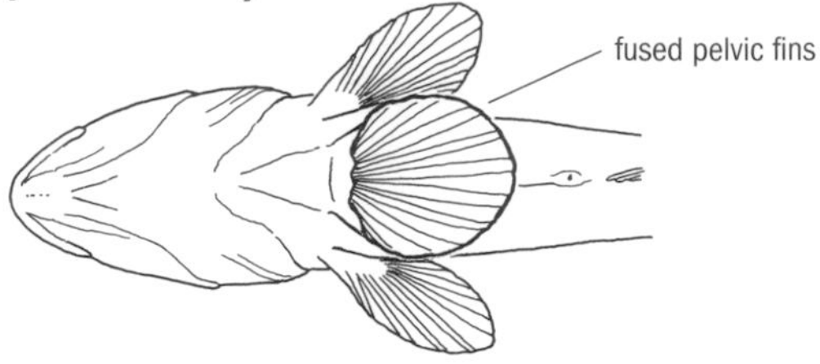

Yellowfin Goby *(Acanthogobius flavimanus)* PAGE 352

STATUS: Introduced. **DESCRIPTION:** Jaw does not reach anterior margin of eye; caudal fin pointed in middle; brown dorsally, pinkish white ventrally; dark blotches form broken horizontal line on side; large goby, ad. TL 10 to 15 cm (4 to 6 in.), max. TL 27 cm (10.6 in.). **DISTRIBUTION:** Throughout middle and lower Delta, Contra Loma Reservoir (Contra Costa County), and San Luis Reservoir (Merced County); continuously expanding range in central coastal California.

Shimofuri Goby *(Tridentiger bifasciatus)* PAGE 355

STATUS: Introduced. **DESCRIPTION:** Jaw reaches to middle but not posterior of eye; caudal fin rounded; light (white in male) crescent at base of PecF; lateral dark stripe from eye to tip of caudal peduncle (masked by dark body color in breeding male); uppermost ray of PecF is attached to the fin web and not elevated as in marine Chameleon Goby; max. TL 10.5 cm (4.2 in.). **DISTRIBUTION:** San Francisco Estuary, California Aqueduct, and Silverwood and Castaic Reservoirs (Los Angeles County).

Tidewater Goby *(Eucyclogobius newberryi)* PAGE 357

STATUS: Native, FE (south of Orange County). **DESCRIPTION:** Jaw reaches posterior margin of eye; upper anterior portion of 1st DF clearer, pale, body dark olive dorsally with mottling on sides and back; small goby, ad. SL 3 to 4.5 cm (1.2 to 1.7 in.), max. SL 5 cm (2 in.). **DISTRIBUTION:** Lower parts of coastal streams; relatively uncommon from San Francisco Bay northward.

Perches (Percidae)

Bigscale Logperch *(Percina macrolepida)* PAGE 334

STATUS: Introduced. **DESCRIPTION:** Small, long, narrow body; 1st and 2d DF well separated; color yellow tan with dark vertical bars and dark spot at base of tail; ad. SL 8 to 10 cm (3 to 4 in.), max. SL 12 cm (5 in.). **DISTRIBUTION:** Throughout the Sacramento–San Joaquin River system and associated low-elevation reservoirs.

Yellow Perch *(Perca flavescens)* PAGE 332

STATUS: Introduced. **DESCRIPTION:** Medium-sized, pelagic fish; 1st and 2d DF barely separated; color green dorsally, yellow white ventrally, with wide, dark vertical bars on the body; ad. TL 20 to 25 cm (8 to 9.8 in.), max. TL 50 cm (19.7 in.), max. wt. 1.9 kg (4.2 lb). **DISTRIBUTION:** Reservoirs of the Klamath River, particularly Iron Gate and Copco.

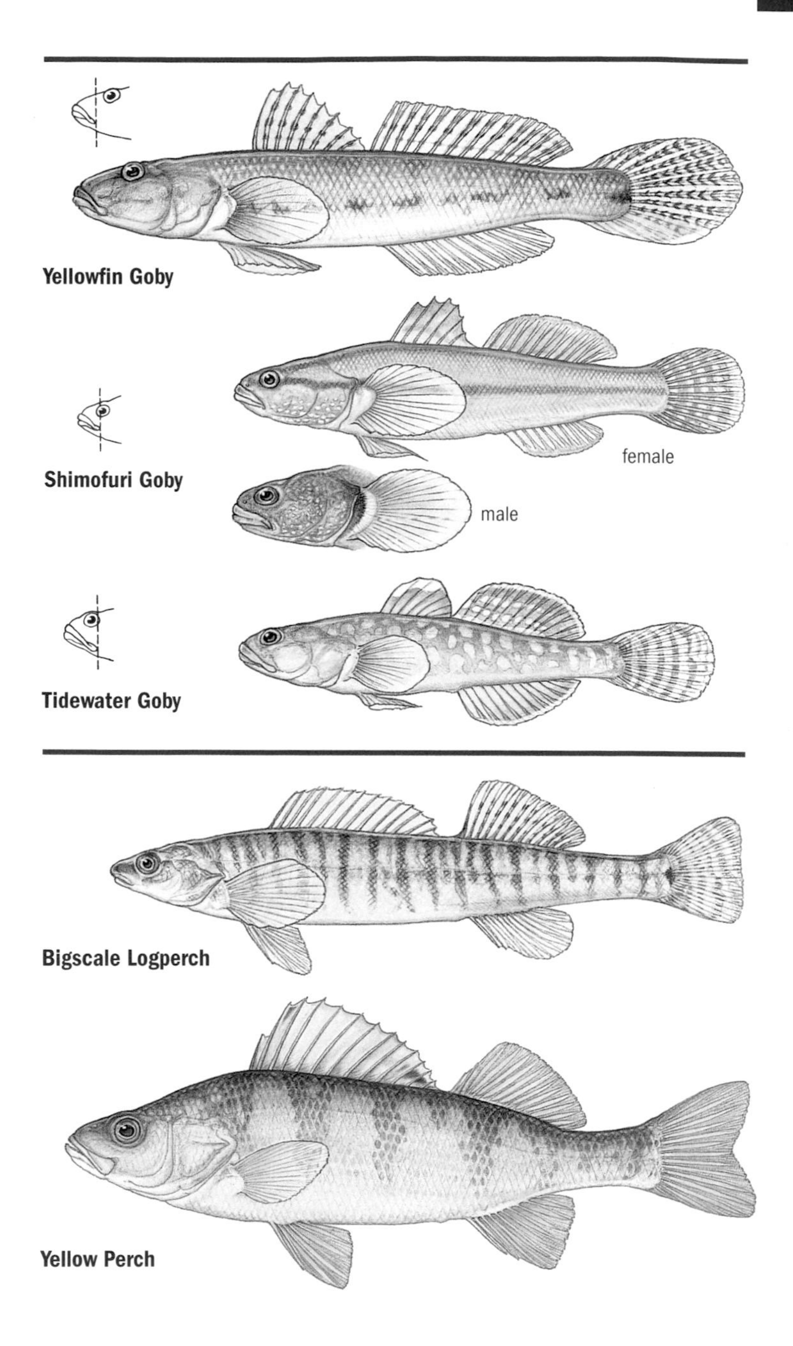
Yellowfin Goby
female
Shimofuri Goby
male
Tidewater Goby
Bigscale Logperch
Yellow Perch

Salmon (Salmonidae)

The color and shape of adult salmon in freshwater change as they approach spawning; therefore, their appearances may be intermediate between the ocean form (illustrated below) and the spawning form (right).

Kokanee Salmon or Sockeye Salmon

(Oncorhynchus nerka) PAGE 160

STATUS: Sockeye native to Pacific coastal streams; Kokanee introduced into California lakes. **DESCRIPTION:** No spots on back, DF, or caudal fin; gill rakers long, slender, and rough; spawning male and female have red bodies and green heads, male has moderate back hump and hooked upper jaw; ad. TL (lake populations) 25 to 40 cm (10 to 15.7 in.), max. wt. (California record) 2.2 kg (4.8 lb) (Kokanee is a dwarfed form); anadromous populations TL up to 80 cm (31.5 in.), max. wt. (outside California) 7 kg (15.4 lb). **DISTRIBUTION:** Introduced populations in numerous reservoirs and lakes; rarely found naturally in California coastal streams.

Chum Salmon *(Oncorhynchus keta)* PAGE 164

STATUS: Native. **DESCRIPTION:** No spots on back, DF, or caudal fin; gill rakers short and smooth; color olive back, maroon sides with wavy vertical green streaks and blotches; moderately hooked upper jaw with prominent front teeth in spawning male; ad. TL 60 to 70 cm (23.6 to 27.6 in.), max. TL 102 cm (40 in.), max. wt. 15 kg (33 lb). **DISTRIBUTION:** Occasionally found in coastal streams and rivers north of San Francisco Bay; small annual run in Sacramento River.

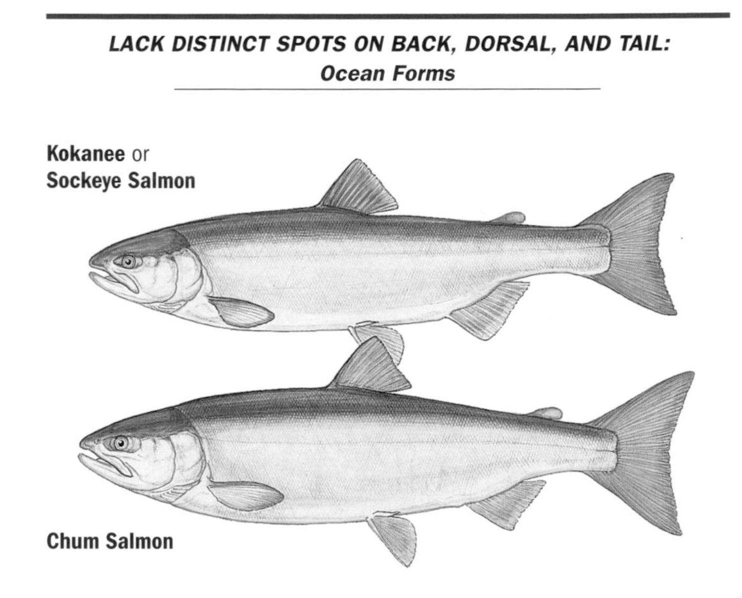

Freshwater Spawning Forms

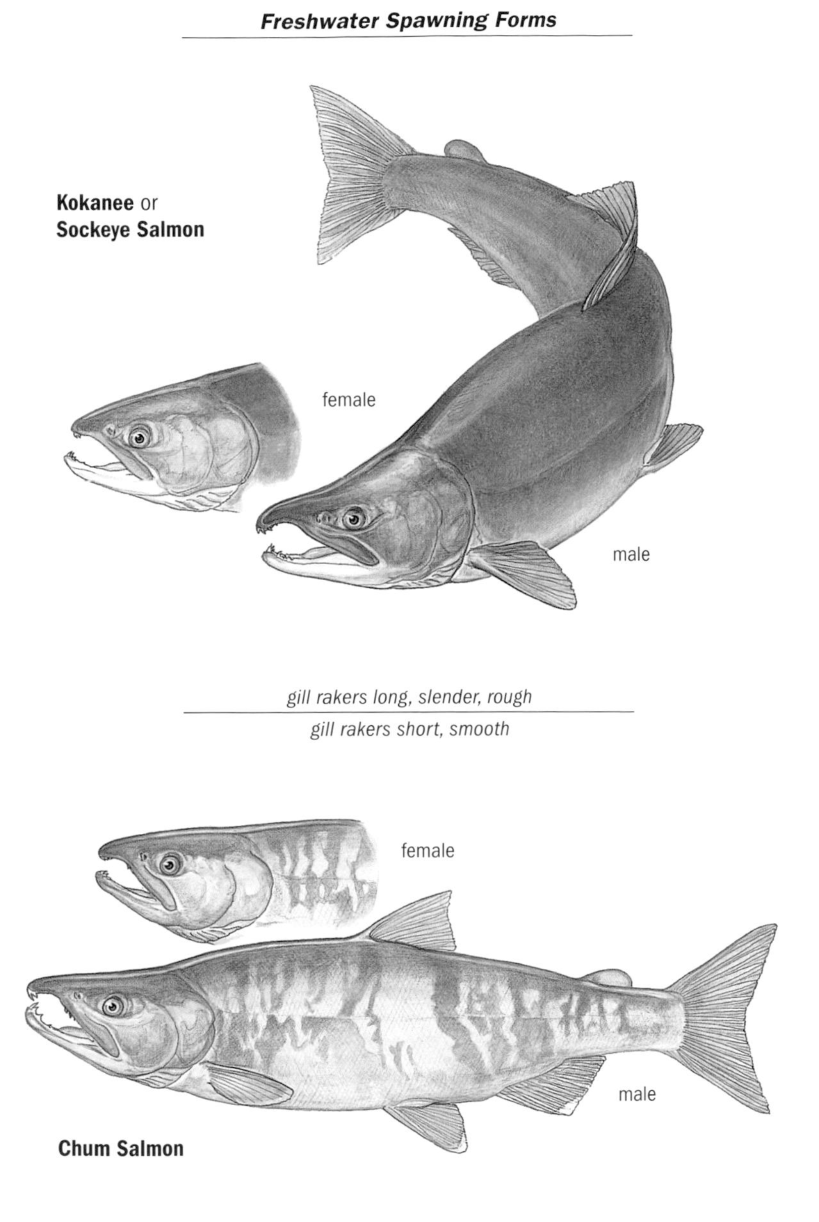

Chinook Salmon or King Salmon
(Oncorhynchus tshawytscha) PAGE 153
STATUS: Native, ST, FT spring run; SE, FE Sacramento River winter run; FT Redwood Creek (Humboldt County) to Russian River (Sonoma County) fall run. **DESCRIPTION:** Round spots on back, DF, and both lobes of caudal fin; gums on lower jaw black; caudal fin rays feel smooth, little change in jaw or body in spawning male; largest of the Pacific salmon, ad. TL about 70 to 80 cm (27.5 to 31.5 in.), max. wt. (California record) 40 kg (88 lb). **DISTRIBUTION:** Large coastal streams and rivers north of San Francisco Bay.

Coho Salmon or Silver Salmon *(Oncorhynchus kisutch)* PAGE 149
STATUS: Native, SE, FT central California coastal streams; FT northern California coastal streams. **DESCRIPTION:** Small round spots on back, DF, and upper lobe of caudal fin; gums of lower jaw gray except usually white at base of teeth, caudal fin rays feel rough when stroked, pronounced hook-shaped upper jaw and red lateral stripe in spawning male; ad. TL 45 to 60 cm (17.7 to 23.6 in.), max. wt. (California record) 10 kg (22 lb). **DISTRIBUTION:** Most coastal streams and rivers from Monterey Bay northward and selected reservoir lakes (introduced into latter).

Pink Salmon *(Oncorhynchus gorbuscha)* PAGE 164
STATUS: Native. **DESCRIPTION:** Large oval spots on both lobes of caudal fin and adipose fin; large purple hump on back and moderately hooked upper jaw in spawning male; ad. TL 40 to 60 cm (15.7 to 23.5 in.), max wt. 5.9 kg (13.1 lb). **DISTRIBUTION:** Only established spawning run in lower Russian River; occasionally taken in coastal streams and rivers north of San Francisco Bay.

SPOTS ON BACK, DORSAL, AND TAIL:
Ocean Forms

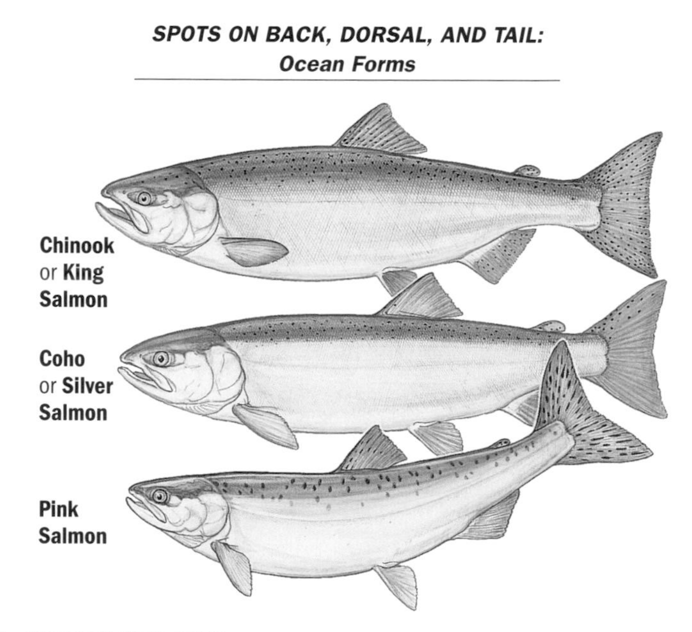

Freshwater Spawning Forms

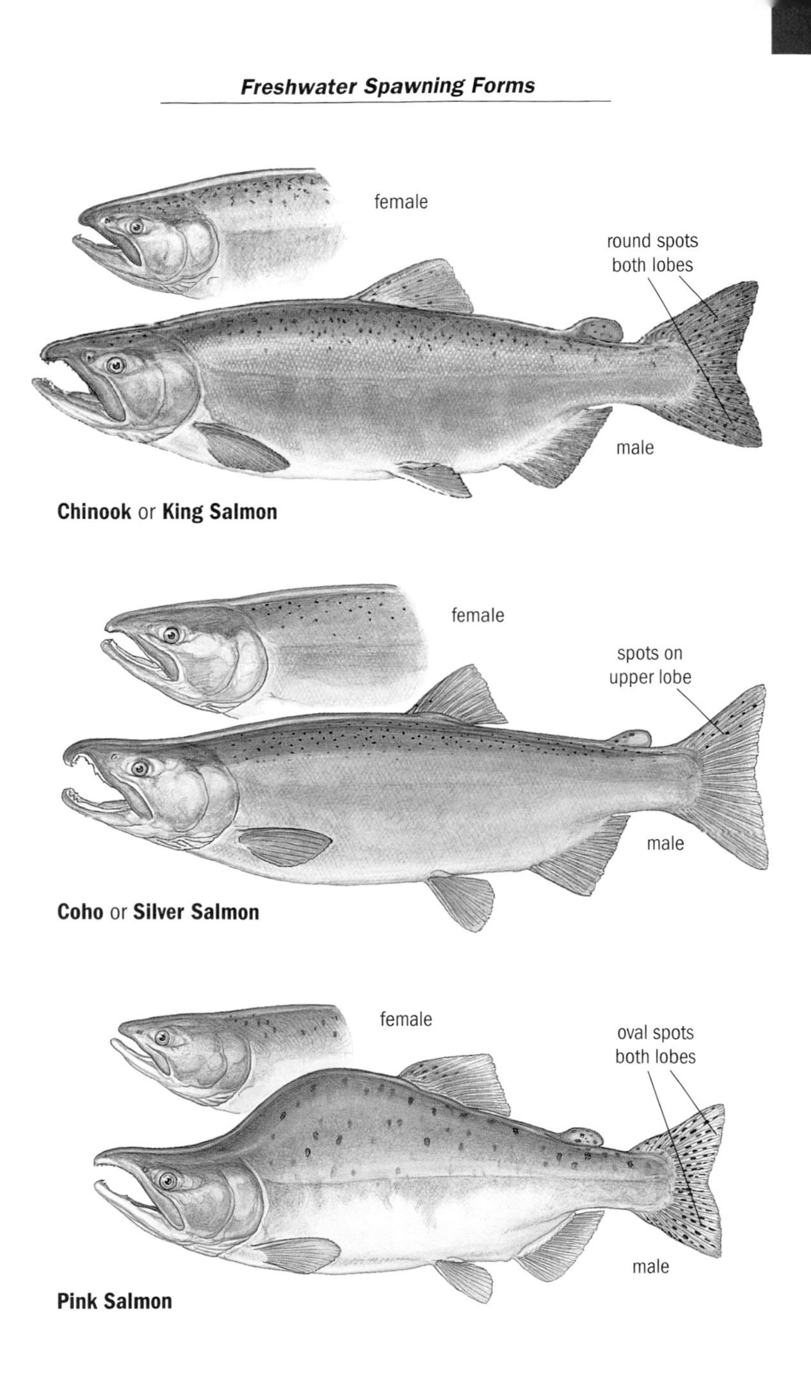

Chinook or **King Salmon**

Coho or **Silver Salmon**

Pink Salmon

Salmon and Trout (Salmonidae): Anadromous Fry

This key is designed for identification of anadromous fry up to 10 cm (4 in.). Nonanadromous fry (whitefishes, trout, and char) are not illustrated, but (like Steelhead *[Oncorhynchus mykiss irideus]*) have eight to 12 major AF rays. This characteristic distinguishes them from salmon fry, which usually have 13 to 19 major AF rays.

Pink Salmon *(Oncorhynchus gorbuscha)* PAGE 164

STATUS: Native. **DESCRIPTION:** Rare in California; no spots or parr marks; blue green back, silver sides; migrates to sea at 5 cm (2 in.) TL.

Chum Salmon *(Oncorhynchus keta)* PAGE 164

STATUS: Native. **DESCRIPTION:** Lacks spots on DF; parr marks faint, not much greater in height than vertical eye diameter, narrower than spaces between, absent below LL; back bright mottled green, sides silver with green iridescence; migrates to sea at 6.5 cm (2.6 in.) TL.

Kokanee Salmon or Sockeye Salmon
(Oncorhynchus nerka) PAGE 160

STATUS: Native. **DESCRIPTION:** Rare in coastal streams; lacks spots on DF; parr marks distinct, not much greater in height than vertical eye diameter, narrower than spaces between, some cut in half by LL; back uniform green, sides silver; mostly introduced in lakes.

Coho Salmon or Silver Salmon *(Oncorhynchus kisutch)* PAGE 149

STATUS: Native. **DESCRIPTION:** Lacks spots on DF; adipose fin uniformly speckled gray; tall parr marks, higher than vertical diameter of eye, narrower than spaces between; white leading edge of AF; to sea at 12 cm (4.7 in.) TL.

Chinook Salmon or King Salmon
(Oncorhynchus tshawytscha) PAGE 153

STATUS: Native. **DESCRIPTION:** May have dark spot at base of front edge of DF; parr marks tall, higher than vertical diameter of eye; adipose fin clear with dark upper edge; forward edge of AF white; to sea at 4 to 8 cm (1.6 to 3 in.) TL.

Steelhead or Coastal Rainbow Trout
(Oncorhynchus mykiss irideus) PAGE 168

STATUS: Native. **DESCRIPTION:** Numerous dark spots on body and DF; parr marks nearly round; adipose fin with black, broken border; to sea at 12 to 25 cm (4.7 to 9.8 in.) TL.

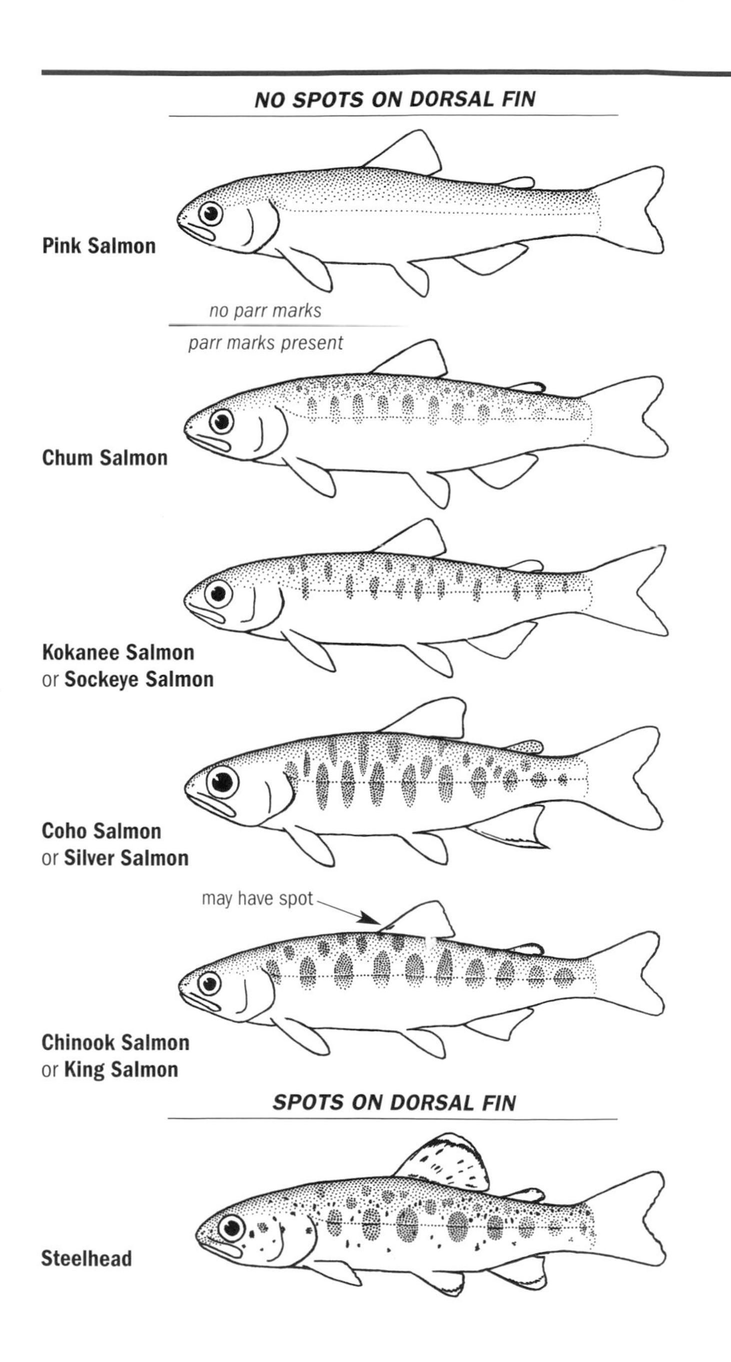
NO SPOTS ON DORSAL FIN
Pink Salmon
no parr marks
parr marks present
Chum Salmon
Kokanee Salmon
or Sockeye Salmon
Coho Salmon
or Silver Salmon
may have spot
Chinook Salmon
or King Salmon
SPOTS ON DORSAL FIN
Steelhead

Whitefish and Trout (Salmonidae)

Mountain Whitefish *(Prosopium williamsoni)* PAGE 194
STATUS: Native. **DESCRIPTION:** Jaw does not extend past center of eye; moderately inferior mouth; large scales; sides silvery, dorsal may be dusky brown; max. TL about 55 cm (22 in.), max. wt. 2.9 kg (6.4 lb). **DISTRIBUTION:** Lake Tahoe and other lakes and streams on the eastern slope of the Sierra.

Bull Trout *(Salvelinus confluentus)* PAGE 187
STATUS: Native, SE, FT. **DESCRIPTION:** No spotting on fins; olive green to gray dorsally, white or orange ventrally; white to yellowish dorsal spots, sometimes pink or red lateral spots; anterior edges of PecF, PelF, and AF are white; max. TL about 1 m (3.3 ft), max. wt. 14.5 kg (31.9 lb), California record 5.1 kg (11.2 lb). **DISTRIBUTION:** Extinct in California.

Lake Trout or Mackinaw Trout *(Salvelinus namaycush)* PAGE 193
STATUS: Introduced. **DESCRIPTION:** Only California trout with large, irregular, light spots over entire body, including DF and caudal fin; tail deeply forked; max. TL 1.3 m (4.4 ft), max. wt. 28.6 kg (63 lb), California record 17 kg (37.4 lb). **DISTRIBUTION:** Lake Tahoe, Donner Lake, Fallen Leaf Lake, and Stony Ridge Lake.

Brook Trout *(Salvelinus fontinalis)* PAGE 188
STATUS: Introduced. **DESCRIPTION:** Introduced. White, wavy lines on a dark olive green back; heavy pale spotting on DF; red spots with blue halos on side; anterior edges of AF, PelF, and PecF white; max. TL 60 cm (23.6 in.), max. wt. (California record) 4.4 kg (9.7 lb). **DISTRIBUTION:** Most mountain streams and lakes in California.

Brown Trout *(Salmo trutta)* PAGE 191
STATUS: Introduced. **DESCRIPTION:** Introduced. Only California trout with both black and red spots on body; black spots on DF, adipose fin, and dorsal body; red spots surrounded by light halos on lower half of body; max. TL about 1 m (3.3 ft), max. wt. 18 kg (39.6 lb), California record 11.9 kg (26 lb). **DISTRIBUTION:** Most trout-type habitats in California.

JAW DOES NOT EXTEND PAST CENTER OF EYE

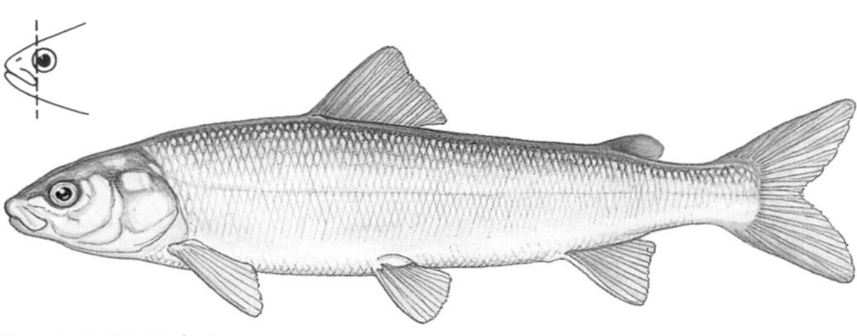

Mountain Whitefish

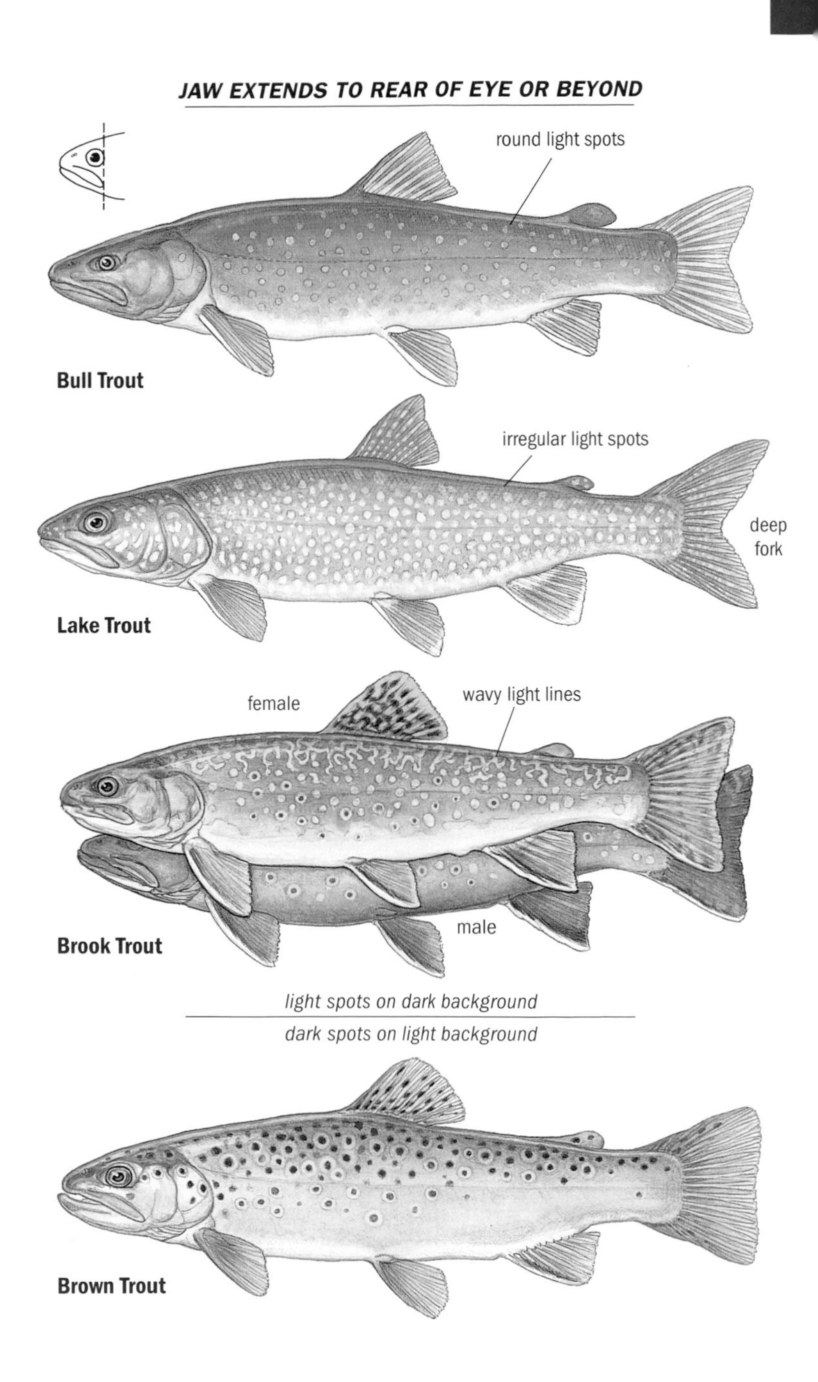
JAW EXTENDS TO REAR OF EYE OR BEYOND
round light spots
Bull Trout
irregular light spots
deep
fork
Lake Trout
female
wavy light lines
male
Brook Trout
light spots on dark background
dark spots on light background
Brown Trout

Coastal Cutthroat Trout *(Oncorhynchus clarki clarki)* PAGE 184

STATUS: Native, CSSC. **DESCRIPTION:** Red slashes on underside of lower jaw; color similar to Rainbow Trout *(O. mykiss)* but with heavier black spotting on lower body; pale lower body color; max. TL 41 to 56 cm (16 to 22 in.), max. wt. 2.7 kg (6 lb). **DISTRIBUTION:** Coastal streams from the Eel River (Humboldt County) northward.

Cutthroat trout usually have

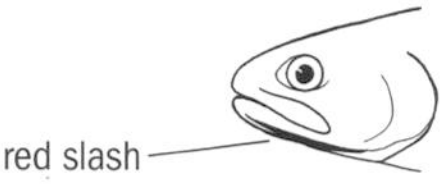

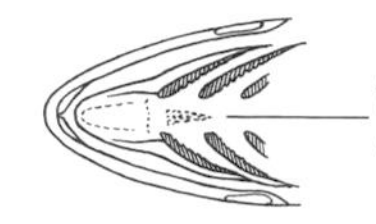

Lahontan Cutthroat Trout
(Oncorhynchus clarki henshawi) PAGE 186

STATUS: Native, FT. **DESCRIPTION:** Red slashes on underside of lower jaw; less spotting than on Coastal Cutthroat Trout *(O. c. clarki)*, spots a little larger; lower body reddish to dark olive; max. TL 1 m (3.3 ft), max. wt. 18.6 kg (41 lb). **DISTRIBUTION:** Streams and lakes on the eastern slope of the Sierra.

Steelhead *(Oncorhynchus mykiss irideus)* PAGE 168

STATUS: Native, FT south-central, central, and northern California coastal areas, and Central Valley; FE southern California coastal area. **DESCRIPTION:** Color ranges from very silver, with only faint dark spotting, to darker dorsal coloration with faint lateral red band and heavier spotting; caudal fin rays feel smooth when stroked, no red slashes on lower jaw; max. TL 109 cm (3.5 ft), max. wt. 19.1 kg (42 lb). **DISTRIBUTION:** Most coastal streams in California.

Coastal Rainbow Trout *(Oncorhynchus mykiss irideus)* PAGE 168

STATUS: Native. **DESCRIPTION:** Conspicuous light pink to red band on lateral body; heavy irregularly shaped black spotting dorsally and on DF and adipose and caudal fins; color variable but generally darker overall than in Steelhead *(O. m. irideus)*; world records for sport fishes list only one weight and length for this species, ad. TL 30 to 41 cm (12 to 16 in.), wt. 0.68 kg (1.5 lb). **DISTRIBUTION:** Most major freshwater habitats in California.

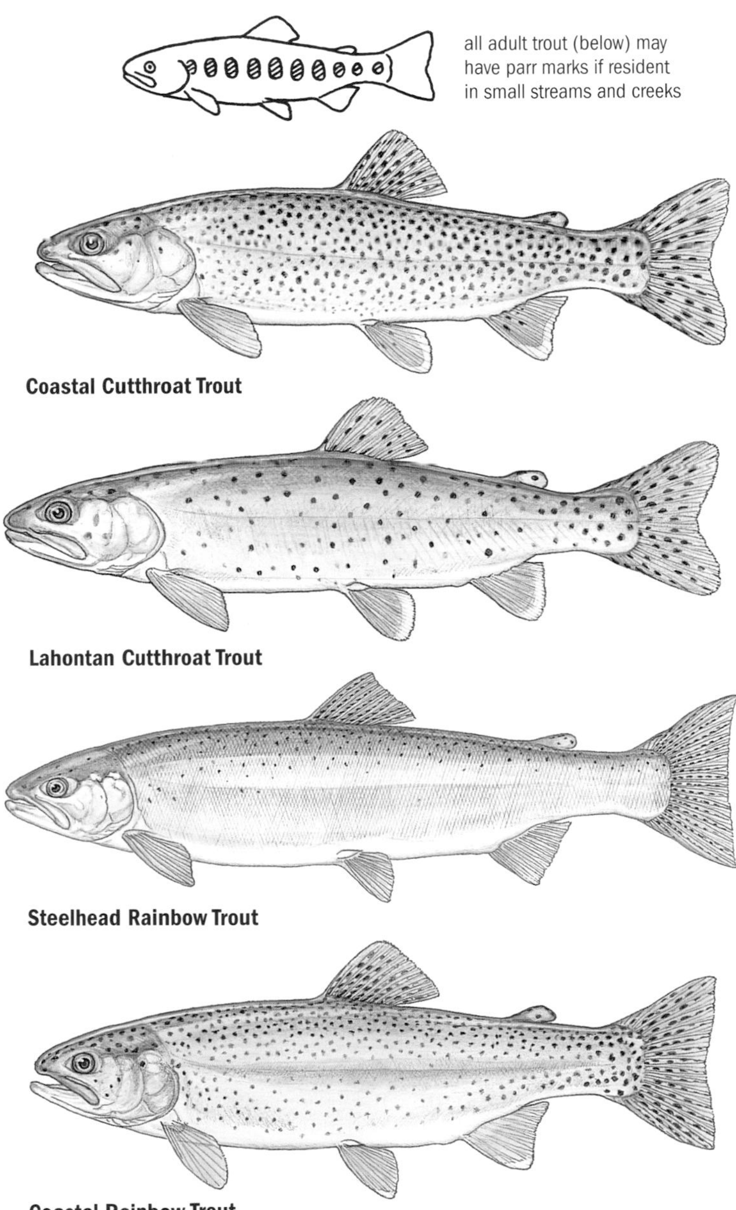
all adult trout (below) may have parr marks if resident in small streams and creeks
Coastal Cutthroat Trout
Lahontan Cutthroat Trout
Steelhead Rainbow Trout
Coastal Rainbow Trout

Eagle Lake Rainbow Trout
(Oncorhynchus mykiss aquilarum) PAGE 182
STATUS: Native, CSSC Eagle Lake. **DESCRIPTION:** Similar in appearance to the Coastal Rainbow Trout *(O. m. irideus)* but with a pinkish rather than red lateral color band, darker dorsal shading, and heavier body and fin spotting; ad. TL 61 to 76 cm (24 to 30 in.), wt. 2.3 to 4.5 kg (5 to 10 lb). **DISTRIBUTION:** Eagle Lake (Lassen County) and other stocked reservoir lakes.

Sacramento Redband Trout
(Oncorhynchus mykiss stonei) PAGE 182
STATUS: Native, CSSC upper McCloud River. **DESCRIPTION:** This name covers a small complex of trout occurring in the headwaters of the McCloud, Pit, and Klamath Rivers, and the Columbia River north of California. Dr. Robert J. Behnke, a world expert on trout, has studied this complex and feels that each drainage may have had its own prototype, and that these have hybridized with one another and with native and introduced Rainbow Trout *(O. mykiss)*. The Sacramento Redband Trout has yellow sides and a brick-red lateral band; ad. retains parr marks in some populations. **DISTRIBUTION:** McCloud, Pit, and Klamath Rivers and Goose Lake.

Kern River Rainbow Trout
(Oncorhynchus mykiss gilberti) PAGE 181
STATUS: Native, CSSC. **DESCRIPTION:** Similar to the Little Kern Golden Trout *(O. m. whitei)* except orange less intense on the lateral band, belly, and anal and pelvic fins; body spots smaller and more irregular in shape, and spots extend well below the lateral line; max TL 71 cm (28.4 in.), max. wt. 3.6 kg (8 lb). **DISTRIBUTION:** Kern River and Lake Isabella, plus Cold Creek and Ninemile Creek tributaries; pure form probably no longer exists because of decades of hatchery rainbow trout stocking.

Little Kern Golden Trout *(Oncorhynchus mykiss whitei)* PAGE 181
STATUS: Native, FT. **DESCRIPTION:** A near twin of the California Golden Trout *(O. m. aguabonita)* but genetically different enough to merit subspecies status; color less intense than in the California Golden, and black spots more numerous over entire body above the lateral line; ad. TL about 15 to 30 cm (6 to 12 in.), wt. 340 g (12 oz). **DISTRIBUTION:** Pure forms most likely in small streams of upper Little Kern River.

California Golden Trout
(Oncorhynchus mykiss aguabonita) PAGE 179
STATUS: Native. **DESCRIPTION:** Parr marks in ad., about 10 along LL; bright golden color on sides below parr marks, red on belly and operculum, red orange lateral band, olive green dorsally with large round spots mainly on caudal peduncle, some spotting on back; max. TL 71 cm (31.5 in.), max. wt. 5 kg (11 lb). **DISTRIBUTION:** Lakes and streams in southern portion of the Sierra, mostly in Fresno and Tulare Counties.

Paiute Cutthroat Trout *(Oncorhynchus clarki seleniris)* PAGE 186
STATUS: Native, FT. **DESCRIPTION:** Red slashes on lower jaw; pale yellow body with orange band much like California Golden Trout *(O. m. aguabonita)*, but very few black spots on body and fins; pale parr marks in ad.; max. TL 46 cm (18 in.). **DISTRIBUTION:** Lakes and streams of the Carson River drainage on eastern Sierra slope.

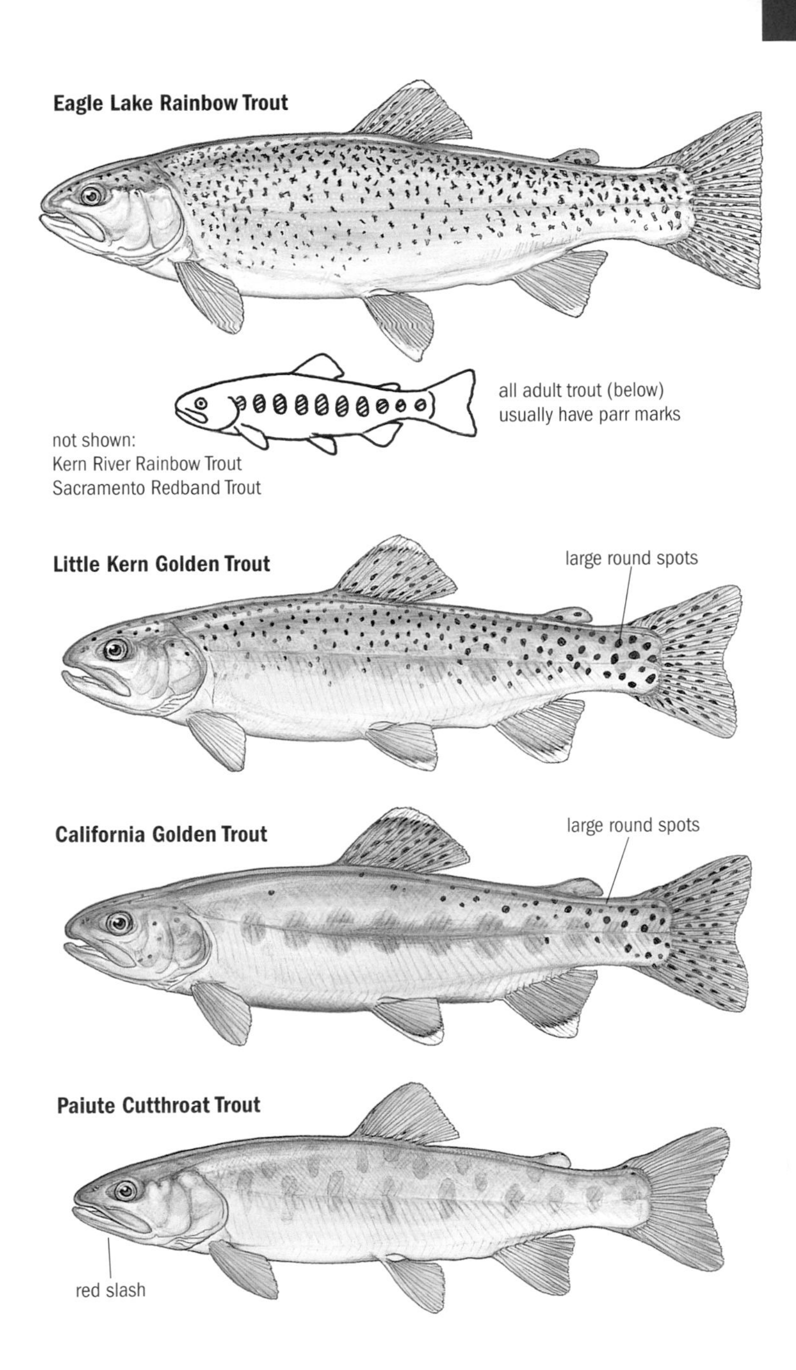
Eagle Lake Rainbow Trout
all adult trout (below)
usually have parr marks
not shown:
Kern River Rainbow Trout
Sacramento Redband Trout
Little Kern Golden Trout
large round spots
California Golden Trout
large round spots
Paiute Cutthroat Trout
red slash

Smelts (Osmeridae)

Delta Smelt *(Hypomesus transpacificus)* PAGE 197

STATUS: Native, ST, FT. **DESCRIPTION:** Small mouth, edge of jaw not extending past middle of eye; no strong concentric striations on operculum; LL incomplete; live fish has a faint cucumber odor; DF = 8 to 11, PelF = 8, AF = 15 to 19; ad. TL 6 to 8 cm (2.4 to 3 in.), max. TL 12 cm (4.7 in.). **DISTRIBUTION:** Middle and lower Delta region.

Wakasagi *(Hypomesus nipponensis)* PAGE 200

STATUS: Introduced. **DESCRIPTION:** Similar in appearance to the Delta Smelt *(H. transpacificus)* but with the following different fin ray counts: DF = 7 to 9, PelF = 8, AF = 13 to 15; LL incomplete; SL 7 to 9 cm (2.8 to 3.5 in.); max. TL 17 cm (6.8 in.). **DISTRIBUTION:** Many reservoirs and the Delta.

Longfin Smelt *(Spirinchus thaleichthys)* PAGE 201

STATUS: Native, CSSC. **DESCRIPTION:** Large mouth; edge of jaw may extend beyond posterior margin of eye; no strong concentric striations on operculum; PecF about as long as head; LL incomplete; DF = 8 to 10, PelF = 8, AF = 15 to 22; ad. TL 8 to 10 cm (3 to 4 in.), max. TL 15 cm (6 in.). **DISTRIBUTION:** Lower Delta and lower Eel River.

Eulachon *(Thaleichthys pacificus)* PAGE 202

STATUS: Native, CSSC. **DESCRIPTION:** Large mouth; edge of jaw may extend beyond posterior margin of eye; strong concentric striations on operculum; PecF shorter than length of head; LL complete; DF = 10 to 13, PelF = 8, AF = 18 to 23; max. TL 20 to 30 cm (8 to 11.8 in.), ad. TL = 15 to 20 cm (6 to 9 in.). **DISTRIBUTION:** Klamath and Mad Rivers (Humboldt County).

Silversides (Atherinidae)

Inland Silverside *(Menidia beryllina)* PAGE 286

STATUS: Introduced. **DESCRIPTION:** Very small, pincerlike mouth, top of head flattened; light, translucent body yellow green above with pronounced silver band along LL; ad. TL 8 to 10 cm (3.4 in.), max. TL 16 cm (6.4 in.). **DISTRIBUTION:** Clear Lake (Lake County), Sacramento–San Joaquin Delta, tributary rivers of the San Joaquin Valley, associated reservoirs; spreading. A native marine silverside, the Topsmelt *(Atherinops affinis)*, may occur with the Inland Silverside in some estuarine areas of central California. The Topsmelt has forked teeth and more AF rays (AF = 1 spine, 19 to 25 rays) than the Inland Silverside (AF = 1 spine, 16 to 18 rays).

Mullet (Mugilidae)

Striped Mullet *(Mugil cephalus)* PAGE 350

STATUS: Native. **DESCRIPTION:** Broad, flat head; very large eyes partially covered with translucent eyelids; horizontal black stripes dominant on upper half of blue silver body; lacks LL; ad. TL 25 to 35 cm (10 to 13 in.), max. TL 60 cm (23.6 in.). **DISTRIBUTION:** Lower Colorado River below Imperial Dam; occasional southern California coastal streams.

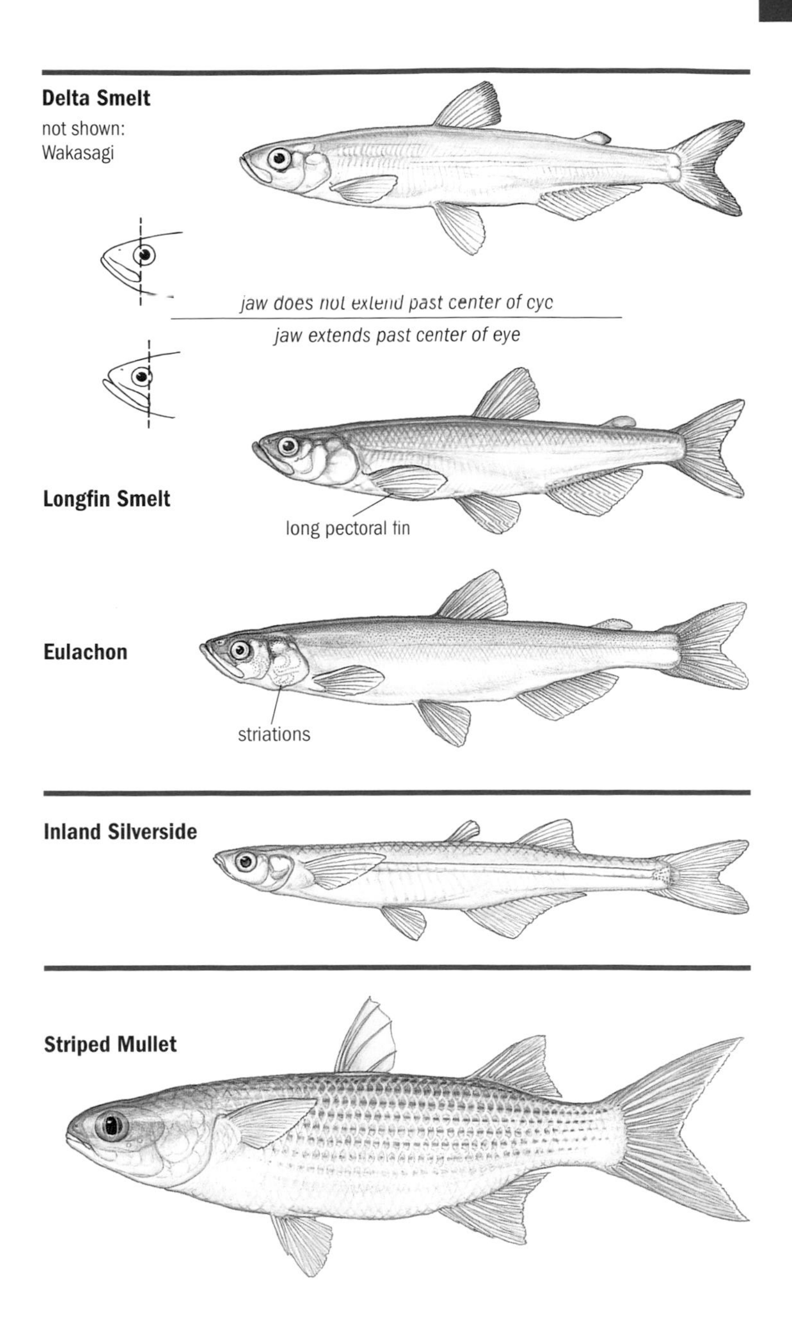
Delta Smelt
not shown:
Wakasagi
jaw does not extend past center of eye
jaw extends past center of eye
Longfin Smelt
long pectoral fin
Eulachon
striations
Inland Silverside
Striped Mullet

Temperate Bass (Moronidae)

Striped Bass *(Morone saxatilis)* PAGE 294

STATUS: Introduced. **DESCRIPTION:** Body depth less than one-third SL; continuous black horizontal stripes on silver body; AF = 3 spines, 9 to 11 rays; ad. TL 50 to 80 cm (19.5 to 31.2 in.), max. TL (California) about 130 cm (4.3 ft), max. wt. (California freshwater record) 41 kg (90 lb). **DISTRIBUTION:** Sacramento–San Joaquin Delta system, Millerton Lake, San Luis and San Antonio Reservoirs, Lake Mendocino, and the lower Colorado River.

White Bass *(Morone chrysops)* PAGE 303

STATUS: Introduced. **DESCRIPTION:** Body depth more than one-third SL; interrupted black horizontal stripes on silver body; AF = 3 spines, 12 to 13 rays; ad. TL 20 to 25 cm (8 to 10 in.), max. TL 45 cm (17.7 in.), max. wt. (California record) 2.4 kg (5.3 lb). **DISTRIBUTION:** Lake Nacimiento, and Ferguson Lake area of lower Colorado River.

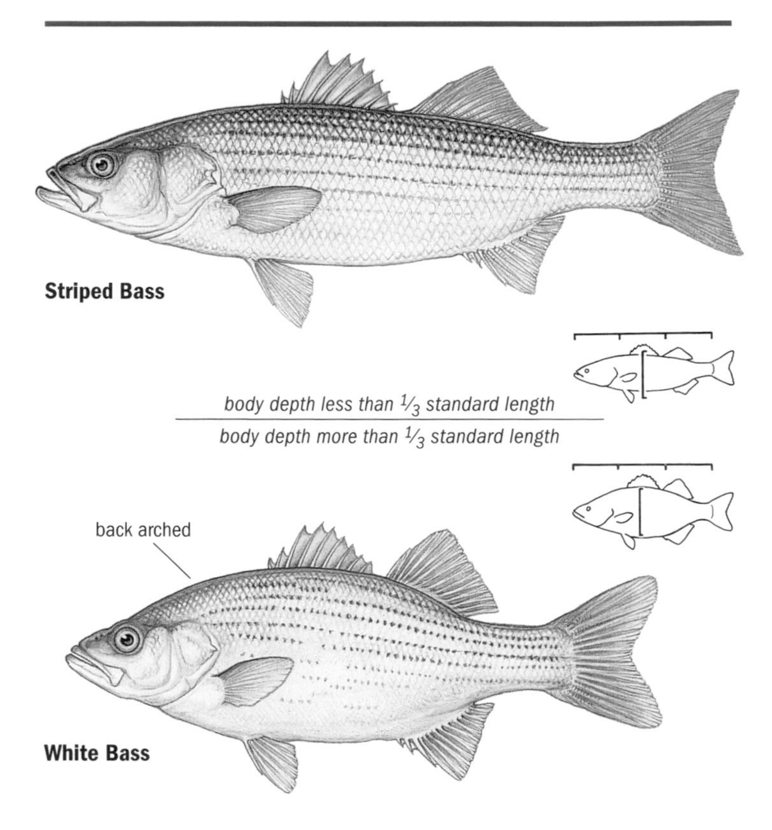

Viviparous Perch (Embiotocidae)

Tuleperch *(Hysterocarpus traski)* PAGE 344

STATUS: Native. **DESCRIPTION:** Sunfishlike build but with a distinct row of large scales at the base of a very long DF; color variable, usually with a dark back and light (white to yellow) belly; bars may be present on the sides; 1st DF = 15 to 19 spines, 2d DF = 9 to 15 rays, AF = 3 spines (which appear fused in male), 20 to 26 rays; ad. SL 10 to 12 cm (4 to 4.7 in.), max. SL about 22 cm (8.8 in.). A subspecies, the Russian River Tuleperch *(Hysterocarpus traski pomo)*, is a CSSC. **DISTRIBUTION:** Lower Delta region and connected reservoirs; Russian River, Clear Lake, and Lake Merced; Sacramento River drainage from Lake Britton to Delta; and O'Neill Forebay.

Shiner Perch *(Cymatogaster aggregata)* PAGE 348

STATUS: Native. **DESCRIPTION:** Sunfishlike body with a distinct row of scales at base of DF; AF with three spines, fused in male; slivery body with three yellow vertical bars, masked in breeding male by dark body color; 1st DF = 8 to 9, 2d DF = 18 to 23, AF = 3 spines, 22 to 25 rays; ad. SL 12 to 14 cm (4.7 to 5.5 in.), max SL 23.5 cm (9.3 in.). **DISTRIBUTION:** In lower reaches of coastal streams.

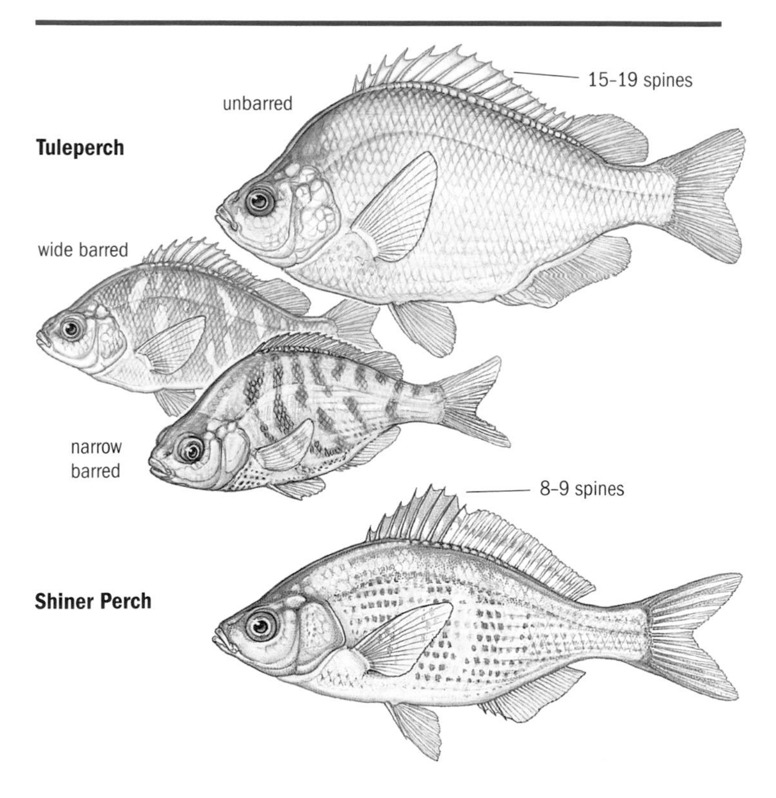

"Black" Basses, Sunfishes, and Crappies (Centrarchidae)

Largemouth Bass *(Micropterus salmoides)* PAGE 322
STATUS: Introduced. **DESCRIPTION:** Jaw extends past posterior margin of eye; continuous black lateral stripe; 1st DF = 9 spines, 2d DF = 12 to 14 rays, AF = 3 spines, 11 to 12 rays; ad. TL 30 to 40 cm (11.8 to 15.7 in.), max. TL about 75 cm (29.5 in.), max. wt. (California record) 9.9 kg (21.8 lb). **DISTRIBUTION:** Most warm, freshwater habitats.

Redeye Bass *(Micropterus coosae)* PAGE 329
STATUS: Introduced. **DESCRIPTION:** Similar to Smallmouth Bass *(M. dolomieu)* but with red eye and reddish bronze body; 1st DF = 9 to 11 spines, 2d DF = 11 to 13 rays, AF = 3 spines, 9 to 11 rays; ad. TL 21 cm (8 in.), max. TL 41 cm (15.7 in.). **DISTRIBUTION:** Sisquoc River (Santa Barbara County), south fork of Stanislaus River, and other Central Valley–foothill sites.

Smallmouth Bass *(Micropterus dolomieu)* PAGE 327
STATUS: Introduced. **DESCRIPTION:** Jaw does not extend to posterior margin of eye; no black lateral stripe; 1st DF = 9 to 10 spines, 2d DF = 13 to 15 rays, AF = 3 spines, 10 to 12 rays; ad. TL 20 to 30 cm (8 to 12 in.), max. TL 69 cm (27.5 in.), max. wt. (California record) 4.1 kg (9 lb). **DISTRIBUTION:** Most rivers and lower- and mid-elevation lakes.

Spotted Bass *(Micropterus punctulatus)* PAGE 329
STATUS: Introduced. **DESCRIPTION:** Similar to Largemouth Bass *(M. salmoides)* but with unconnected dark blotches making up lateral stripe, jaw does not extend beyond eye, has teeth on tongue; 1st DF = 9 to 11 spines, 2d DF = 9 to 11 rays, AF = 3 spines, 9 to 11 rays; ad. TL 20 to 25 cm (8 to 10 in.), max. TL 50 cm (20 in.), max. wt. (California record) 4.5 kg (9.9 lb). **DISTRIBUTION:** Cosumnes and Feather Rivers; Oroville, Perris, Kaweah, and other reservoir lakes.

Pumpkinseed *(Lepomis gibbosus)* PAGE 315
STATUS: Introduced. **DESCRIPTION:** Red spot at tip of dark opercular flap; spots on rayed DF; spotting on body; 1st DF = 10 spines, 2d DF = 10 to 12 rays, AF = 3 to 4 spines, 9 to 11 rays; ad. TL 15 to 20 cm (6 to 8 in.), max. TL 30 cm (11.8 in.), max. wt. 0.5 kg (1 lb). **DISTRIBUTION:** Reservoirs and backwaters of Klamath, Lost, and Susan Rivers; Honey Lake and Big Bear Lake; and in Santa Clara County.

Bluegill *(Lepomis macrochirus)* PAGE 311
STATUS: Introduced. **DESCRIPTION:** Blue black opercular flap, vertical bars on side, large black spot on posterior base of rayed DF; 1st DF = 10 spines, 2d DF = 11 to 12 rays, AF = 3 spines, 11 to 12 rays; ad. FL 15 to 20 cm (6 to 8 in.), max. TL 41 cm (16 in.), max. wt. (California record) 1.6 kg (3.5 lb). **DISTRIBUTION:** Warm, freshwater habitats.

Redear Sunfish *(Lepomis microlophus)* PAGE 314
STATUS: Introduced. **DESCRIPTION:** Opercular flap dark with orange red edge; no heavy spotting on DF; spotting pattern on body; 1st DF = 10 spines, 2d DF = 11 to 12, AF = 3 spines, 10 to 11 rays; ad. TL 20 to 25 cm (8 to 10 in.), max. TL 31 cm (12 in.), max. wt. (California record) 2.4 kg (5.3 lb). **DISTRIBUTION:** Warm freshwater habitats of southern and central California.

3 ANAL SPINES

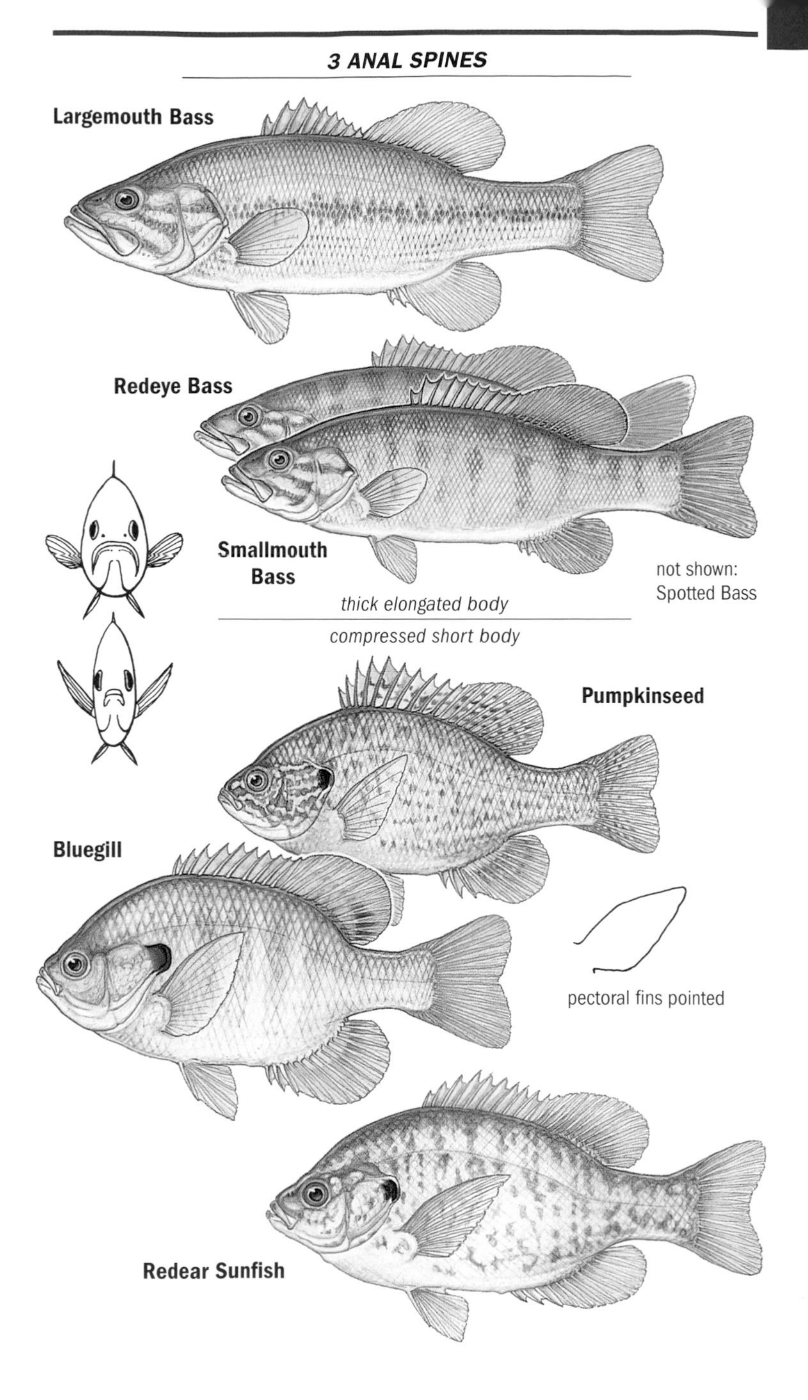

Warmouth *(Lepomis gulosus)* PAGE 318

STATUS: Introduced. **DESCRIPTION:** Similar to Green Sunfish *(L. cyanellus)* but no dark opercular flap, no iridescent green coloration, has teeth on tongue; 1st DF = 10 to 11 spines, 2d DF = 9 to 11 rays, AF = 3 spines, 9 to 10 rays; ad. TL 20 to 28 cm (8 to 11 in.), max. TL 31 cm (12 in.), max. wt. 0.5 kg (1 lb). **DISTRIBUTION:** Throughout Central Valley, selected middle-elevation reservoirs, and the Colorado River.

Green Sunfish *(Lepomis cyanellus)* PAGE 317

STATUS: Introduced. **DESCRIPTION:** Rounded PecF; 8 to 9 AF rays; jaw extends to middle of eye; dark opercular flap with white edge; iridescent blue green markings on body and head; 1st DF = 9 to 11 spines, 2d DF = 10 to 12 rays, AF = 3 spines, 8 to 10 rays; ad. TL 15 to 20 cm (6 to 8 in.), max. TL 30 cm (12 in.), max. wt. 1 kg (2.2 lb). **DISTRIBUTION:** Low- and middle-elevation freshwater habitats, except Klamath River drainage.

White Crappie *(Pomoxis annularis)* PAGE 320

STATUS: Introduced. **DESCRIPTION:** Five to 6 dorsal spines; silver body with spotted vertical bands; 1st DF = 5 to 6 spines, 2d DF = 13 to 15 rays, AF = 6 to 7 spines, 16 to 18 rays; ad. TL 20 to 25 cm (8 to 10 in.), max. TL 35 cm (13.8 in.), max. wt. (California record) 2.05 kg (4.6 lb). **DISTRIBUTION:** Throughout southern California, Clear Lake, and Central Valley.

Black Crappie *(Pomoxis nigromaculatus)* PAGE 320

STATUS: Introduced. **DESCRIPTION:** Seven to 8 dorsal spines; dark, random spots on body; 1st DF = 7 to 8 spines, 2d DF = 15 to 16 rays, AF = 6 spines, 17 to 19 rays; ad. TL 25 to 35 cm (10 to 13.8 in.), max. TL 47 cm (18.5 in.), max. wt. (California record) 1.9 kg (4 lb). **DISTRIBUTION:** Low- and middle-elevation reservoirs, and slow streams.

Sacramento Perch *(Archoplites interruptus)* PAGE 308

STATUS: Native, CSSC Clear Lake and Delta region. **DESCRIPTION:** Twelve to 14 dorsal spines; brown body with dark vertical stripes; 1st DF = 12 to 14, 2d DF = 10 rays, AF = 6 to 8 spines, 10 to 11 rays; ad. TL 25 to 35 cm (10 to 13.8 in.), max. TL 61 cm (23.6 in.), max. wt. (California record) 1.6 kg (3.5 lb) (larger elsewhere). **DISTRIBUTION:** Redistribution program for this species has placed it in over 36 new locations, mostly in central and southern California.

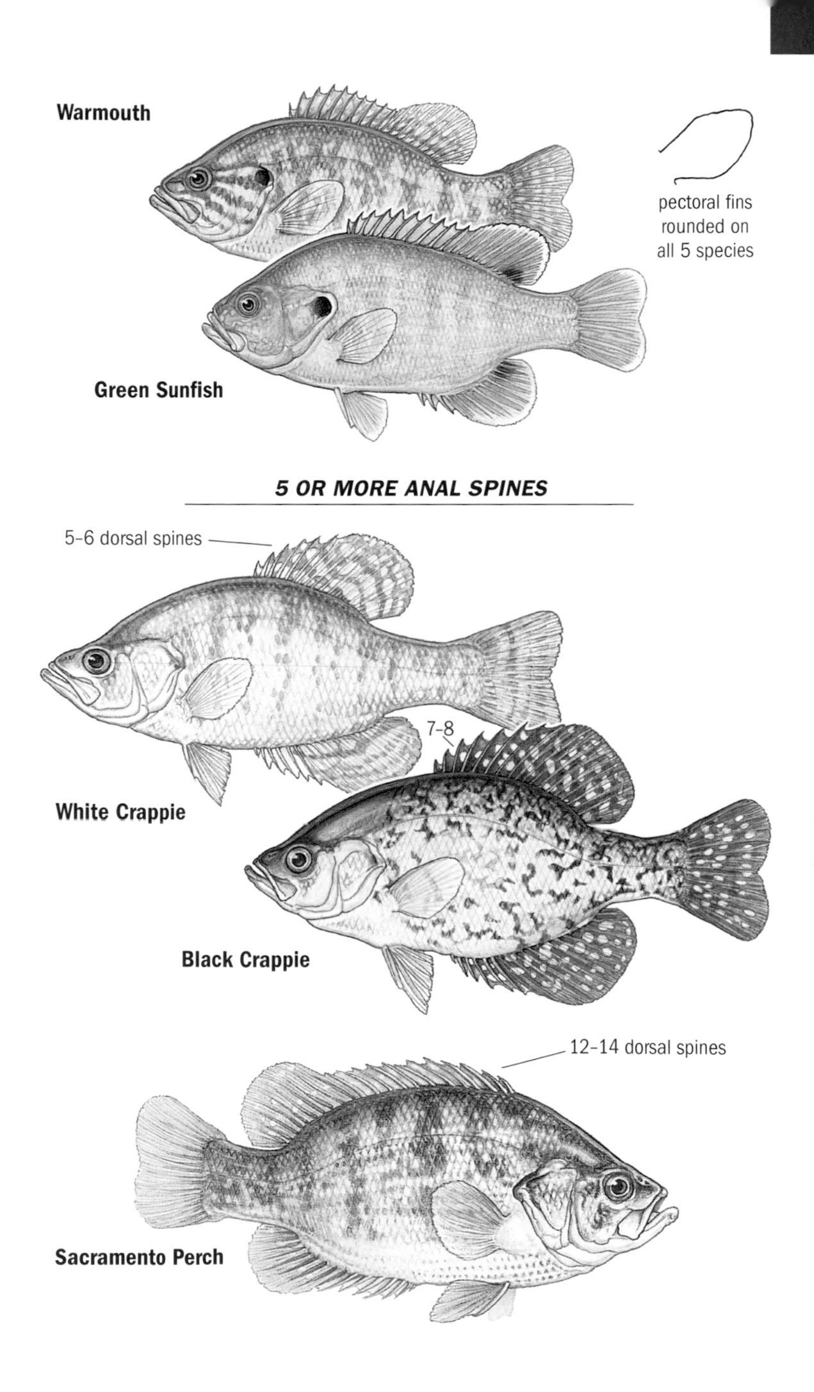
Warmouth
pectoral fins
rounded on
all 5 species
Green Sunfish
5 OR MORE ANAL SPINES
5–6 dorsal spines
7–8
White Crappie
Black Crappie
12–14 dorsal spines
Sacramento Perch

Suckers (Catostomidae)

When identifying species in this family, consult the written descriptions first and then go to the illustrations for final species determination. The basic question to ask when identifying a sucker species in the field is "Where am I?" When ranges overlap, species are identified by scale size, lip structure, papillae number, or prominent features such as a hump on the back.

Santa Ana Sucker *(Catostomus santaanae)* PAGE 254

STATUS: Native, FT. **DESCRIPTION:** Small sucker with lips separated by a lateral notch; shallow notch in lower lip; papillae weak in upper lip; pelvic axillary process like a fold, poorly developed; LL scales 67 to 86; dark back with irregular dorsal blotches, silvery belly, dark membrane between caudal fin rays; max. TL about 18 cm (7 in.). **DISTRIBUTION:** Found only in the Santa Ana, Santa Clara, San Gabriel, and Los Angeles Rivers. Owen aqueduct has brought it into contact with the Owens Sucker *(C. fumeiventris);* separate by lateral lip notch.

Mountain Sucker *(Catostomus platyrhynchus)* PAGE 253

STATUS: Native, CSSC. **DESCRIPTION:** Small sucker with lips separated by lateral notch; well-developed axillary process at the base of each PelF; LL scales 60 to 108; light membrane between caudal fin rays; dark back, gold to white underbelly, sides dusky with dark lateral band (red orange during breeding season); max. TL 25 cm (10 in.). **DISTRIBUTION:** Streams and lakes on west side of the Sierra from Lassen County to El Dorado County; range overlaps with that of Tahoe Sucker *(C. tahoensis)* (see lip detail).

Shortnose Sucker *(Chasmistes brevirostris)* PAGE 256

STATUS: Native, SE, FE, FPS. **DESCRIPTION:** Medium to large sucker with large head, mouth almost terminal, often with small dorsal hump; eye usually closer to snout than to operculum; thin lips, wide notch on lower lip; LL scales 67 to 92; dark back, silver white belly, reddish cast to scales during breeding season; max. TL over 50 cm (20 in.). **DISTRIBUTION:** Copco Reservoir on Klamath River and Clear Lake Reservoir on Lost River. This species hybridizes with the Lost River Sucker *(Catostomus luxatus)* and may no longer be present in California in its pure genetic form. If it is, it may be distinguished from the Lost River Sucker by having a bulkier head and more terminal mouth.

Lost River Sucker *(Catostomus luxatus)* PAGE 250

STATUS: Native, SE, FE, FPS. **DESCRIPTION:** Large sucker with a long slender head, inferior mouth, snout with dorsal hump; eye usually midhead or closer to operculum; deep notch on lower lip; relatively small scales, LL scales 82 to 113; dark back and sides, yellow white belly; max. TL 1 m (3.3 ft). **DISTRIBUTION:** Lost River, Tule Lake, and Upper and Lower Klamath Lakes; range overlaps with that of Klamath Largescale Sucker *(C. snyderi)* (distinguish by degree of lower lip notch and head shape) and with Shortnose Sucker *(Chasmistes brevirostris)* (distinguish by position of mouth and head size).

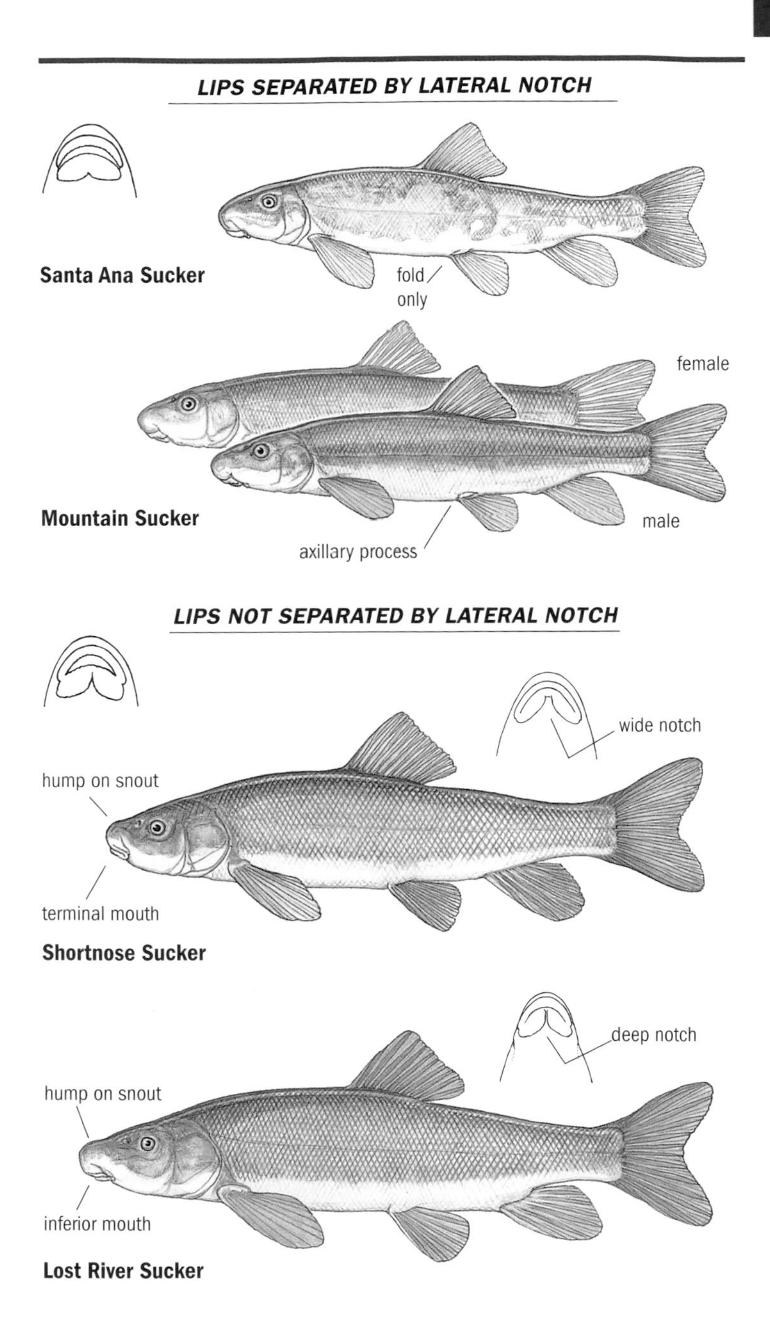
LIPS SEPARATED BY LATERAL NOTCH
Santa Ana Sucker
fold
only
female
Mountain Sucker
male
axillary process
LIPS NOT SEPARATED BY LATERAL NOTCH
wide notch
hump on snout
terminal mouth
Shortnose Sucker
deep notch
hump on snout
inferior mouth
Lost River Sucker

Razorback Sucker *(Xyrauchen texanus)* PAGE 255
STATUS: Native, SE, FE, FPS. **DESCRIPTION:** Large sucker with pronounced hump or keel on back just behind head; LL scales 68 to 87; deep notch separates lower lip; dusky olive back, white to yellow orange belly (bright orange in breeding male); max. TL about 1 m (3.3 ft). **DISTRIBUTION:** Colorado River.

Klamath Smallscale Sucker *(Catostomus rimiculus)* PAGE 252
STATUS: Native. **DESCRIPTION:** Medium to large sucker; relatively small scales, LL scales 81 to 93; 5 to 6 rows of papillae on upper lip, 2 or more rows on lower lip; dusky back, light belly; max. TL 50 cm (19.8 in.). **DISTRIBUTION:** Trinity River and Klamath River below Klamath Falls. Range overlap in Klamath Reservoir with Lost River Sucker *(C. luxatus)* and Shortnose Sucker *(Chasmistes brevirostris)* (distinguish by lack of snout or back hump) and with Klamath Largescale Sucker *(Catostomus snyderi)* (distinguish by number of lip papillae rows and LL scales).

Sacramento Sucker *(Catostomus occidentalis)* PAGE 247
STATUS: Native. **DESCRIPTION:** Medium to large sucker; relatively large scales, LL scales 56 to 75; 4 to 6 rows of papillae on upper lip, 1 row completely across lower lip; lower lip deeply notched; greenish back, dusky yellow belly, faint lateral red stripe, brighter in breeding male; max. TL about 56 cm (22 in.). **DISTRIBUTION:** Sacramento and San Joaquin Rivers and tributaries; tributaries of San Francisco Bay; Russian River; Clear Lake; reservoirs of the Delta system; Pajaro and Salinas Rivers; Eel, Bear, and Mad Rivers (Humboldt County); and Goose Lake (Modoc County). Range overlaps with that of Modoc Sucker *(C. microps)* (distinguish by scale size and number of papillae on lips).

Klamath Largescale Sucker *(Catostomus snyderi)* PAGE 251
STATUS: Native, CSSC. **DESCRIPTION:** Medium to large sucker; large scales, LL scales 67 to 81; 4 to 5 rows of papillae on upper lip, 1 row across lower lip; dark back, dusky to light belly; max. TL 55 cm (21.8 in.). **DISTRIBUTION:** Klamath and Lost Rivers; range overlaps with Lost River Sucker *(C. luxatus)* (distinguish by scale size and snout hump), with Shortnose Sucker *(Chasmistes brevirostris)* or hybrids (distinguish by snout hump and head width), and with Sacramento Sucker *(C. occidentalis)* (distinguish by DF size and rays).

Tahoe Sucker *(Catostomus tahoensis)* PAGE 249
STATUS: Native. **DESCRIPTION:** Medium to large sucker; relatively small scales, LL scales 82 to 95; 2 to 4 rows of papillae on upper lip, 1 row on lower lip; dark back, yellow white belly, dark lateral band, red band on breeding male; max. SL over 60 cm (23.6 in.). **DISTRIBUTION:** Eastern side of the Sierra from Eagle Lake to El Dorado County; abundant in Lake Tahoe, Fallen Leaf Lake, and Truckee River and in upper Feather and Rubicon Rivers; range overlaps with that of Mountain Sucker *(C. platyrhynchus)* (distinguish by lateral lip notch).

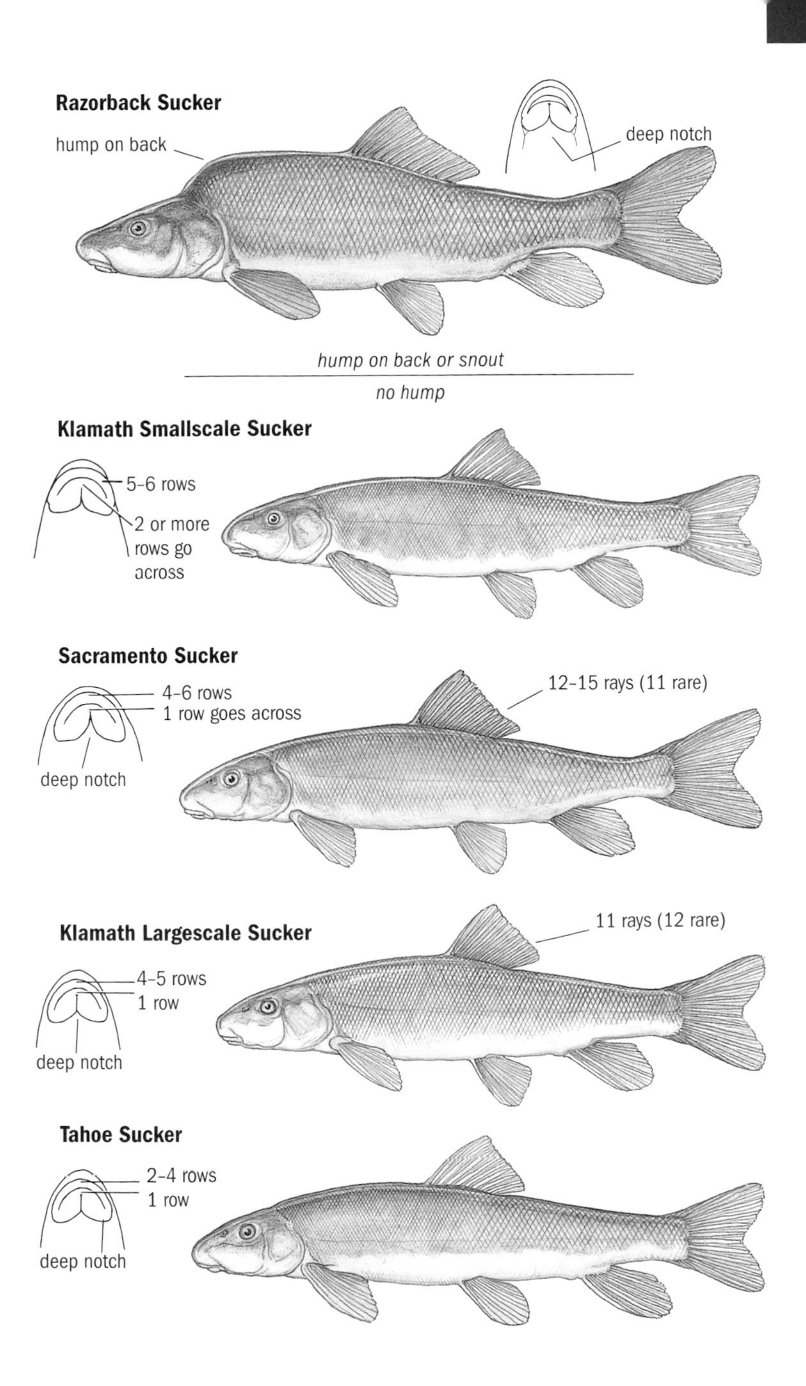
Razorback Sucker
hump on back
deep notch
hump on back or snout
no hump
Klamath Smallscale Sucker
5-6 rows
2 or more rows go across
Sacramento Sucker
4-6 rows
1 row goes across
deep notch
12-15 rays (11 rare)
Klamath Largescale Sucker
11 rays (12 rare)
4-5 rows
1 row
deep notch
Tahoe Sucker
2-4 rows
1 row
deep notch

Owens Sucker *(Catostomus fumeiventris)* PAGE 250
STATUS: Native, CSSC. **DESCRIPTION:** Similar in size and color to Tahoe Sucker *(C. tahoensis)*; LL scales 66 to 85; max. SL 50 cm (20 in.). **DISTRIBUTION:** Owens River and tributaries, and June Lake. Owens Aqueduct has brought it into contact with Santa Ana Sucker *(C. santaanae)* with which it has hybridized.

Modoc Sucker *(Catostomus microps)* PAGE 252
STATUS: Native, SE, FE, FPS. **DESCRIPTION:** Small sucker; relatively small scales, LL scales 79 to 89; similar to Tahoe Sucker *(C. tahoensis)* but with slightly smaller head (distinguish by range); SL 15 cm (6 in.), max. SL 28 cm (11 in.). **DISTRIBUTION:** Southern Modoc County and northern Lassen County, and in Johnson, Rush, Ash, Washington, Turner, and Hulbert Creeks.

Minnows (Cyprinidae)

Cyprinids represent a challenge to the first-time key user because of their generally small size and similar appearance. The species are separated here into five groups based on similarities in scale size, fin characteristics, body cross section, and lateral body form. Before attempting to key a minnow, page through the illustrations to see how these characteristics differ.

Slightly Indented to Unforked Tail

Tench *(Tinca tinca)* PAGE 240
STATUS: Introduced. **DESCRIPTION:** Tail only slightly indented; small barbel at the end of jaw; tiny scales with 90 to 115 in LL; ad. male has thickened ray at leading edge of PelF; fins dark; max. TL 80 cm (31.5 in.), max. wt. 3 kg (6.6 lb). **DISTRIBUTION:** Ponds and reservoirs in San Mateo and Santa Cruz Counties.

Carp-Type Minnows (One Dorsal Spine, Very Large Scales, Thick Body)

Common Carp *(Cyprinus carpio)* PAGE 227
STATUS: Introduced. **DESCRIPTION:** Deep-bodied, large minnow; long DF, with 1 spine, 17 to 21 rays; two barbels on posterior edge of upper jaw; serrated spine on leading edges of both DF and AF; large scales up to 38 in LL; greenish bronze color; some carp have only patches of scales (Mirror Carp) or completely lack scales; max. TL about 75 cm (29.5 in.), max. wt. about 40 kg (88 lb), California record 26.3 kg (58 lb). **DISTRIBUTION:** Most freshwater habitats except those of the Klamath River drainage.

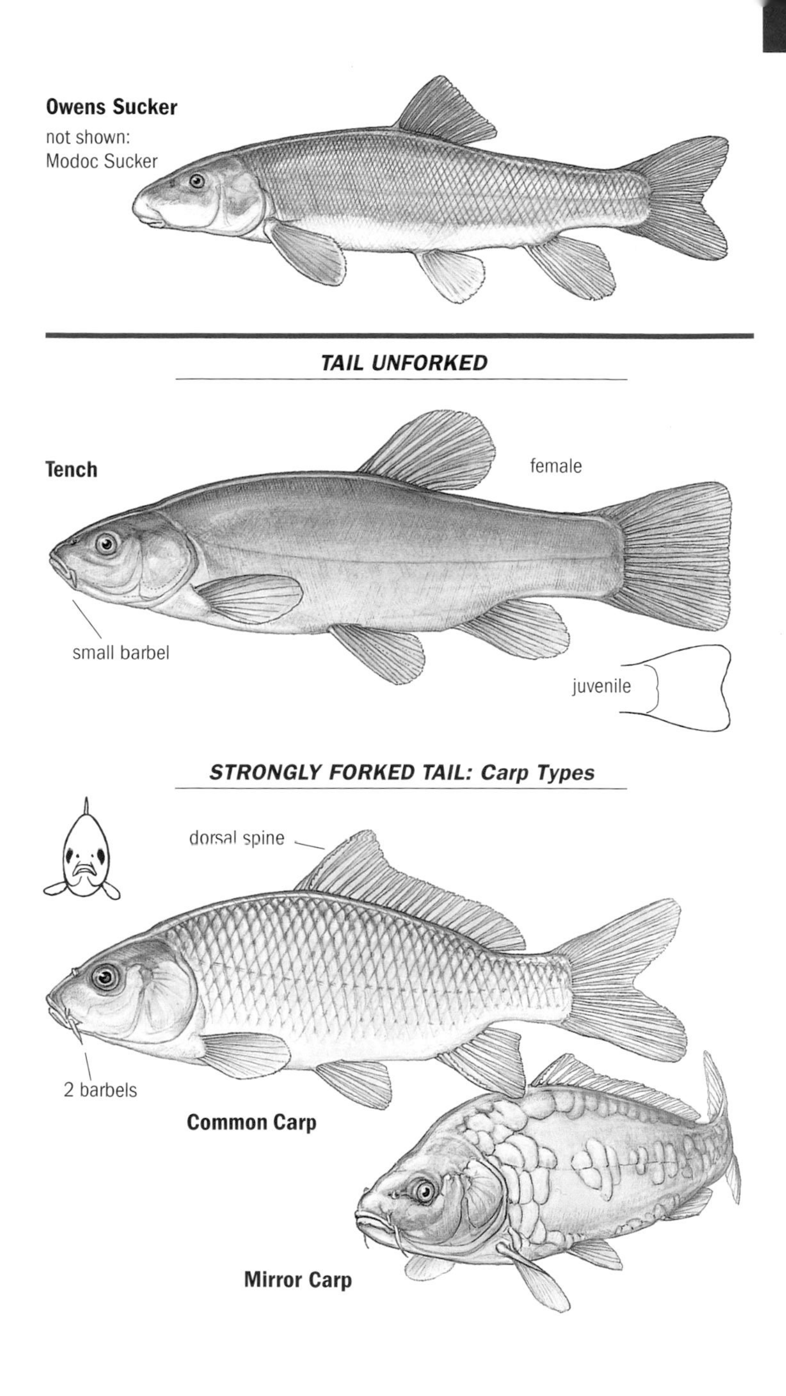
Owens Sucker
not shown:
Modoc Sucker
TAIL UNFORKED
Tench
female
small barbel
juvenile
STRONGLY FORKED TAIL: Carp Types
dorsal spine
2 barbels
Common Carp
Mirror Carp

Chub-Type Minnows (Shortened, Thick Body)

Large Scales

Grass Carp *(Ctenopharyngodon idella)* PAGE 242
STATUS: Introduced. **DESCRIPTION:** Wide scaleless head, body with thick caudal peduncle, very large scales outlined in black on body, terminal mouth, no barbel, short upright DF with no spine; LL scales 34 to 35; max. TL 1.5 m (5 ft). **DISTRIBUTION:** Imperial Valley canals; golf-course ponds.

Fathead Minnow *(Pimephales promelas)* PAGE 239
STATUS: Introduced. **DESCRIPTION:** Head wide and blunt; LL only half length of body; scales behind head more tightly spaced than on rest of body; first DF ray unbranched and separated by membrane; back olive brown, sides dusky; breeding male very dark, with tubercles on snout and PecF; DF = 8, AF = 7; max. TL about 11 cm (4.3 in.). **DISTRIBUTION:** Confined to small streams of Central Valley.

Tui Chub *(Siphateles bicolor)* PAGE 212
STATUS: Native. **DESCRIPTION:** Complete LL; jaw does not reach anterior edge of eye; scales larger than those of closely related Blue Chub *(Gila coerulea)* (LL scales 41to 64); color brassy brown back, silver belly; young more silvery overall; DF = 7 to 9, AF = 7 to 9; max. SL about 40 cm (15.7 in.). **DISTRIBUTION:** Throughout the Lahontan and Klamath fish provinces, the Owens, Mojave, and upper Pit Rivers and Goose Lake. Two subspecies, the Owens Tui Chub *(S. b. snyderi)* and the Mojave Chub *(S. b. mohavensis)*, both endangered (SE, FE, FPS), survive in spring pools at the California State University Desert Studies Center, Mojave Desert.

Blue Chub *(Gila coerulea)* PAGE 214
STATUS: Native. **DESCRIPTION:** Similar to Tui Chub *(Siphateles bicolor)*, but larger mouth extends to anterior margin of eye; smaller scales than Tui Chub, LL scales 58 to 71; dusky back, silvery blue sides; DF = 9, AF = 8 to 9; max. FL about 38 cm (15 in.); ad. SL 12 to 15 cm (4.7 to 5.9 in.). **DISTRIBUTION:** Klamath River and Lost River drainages.

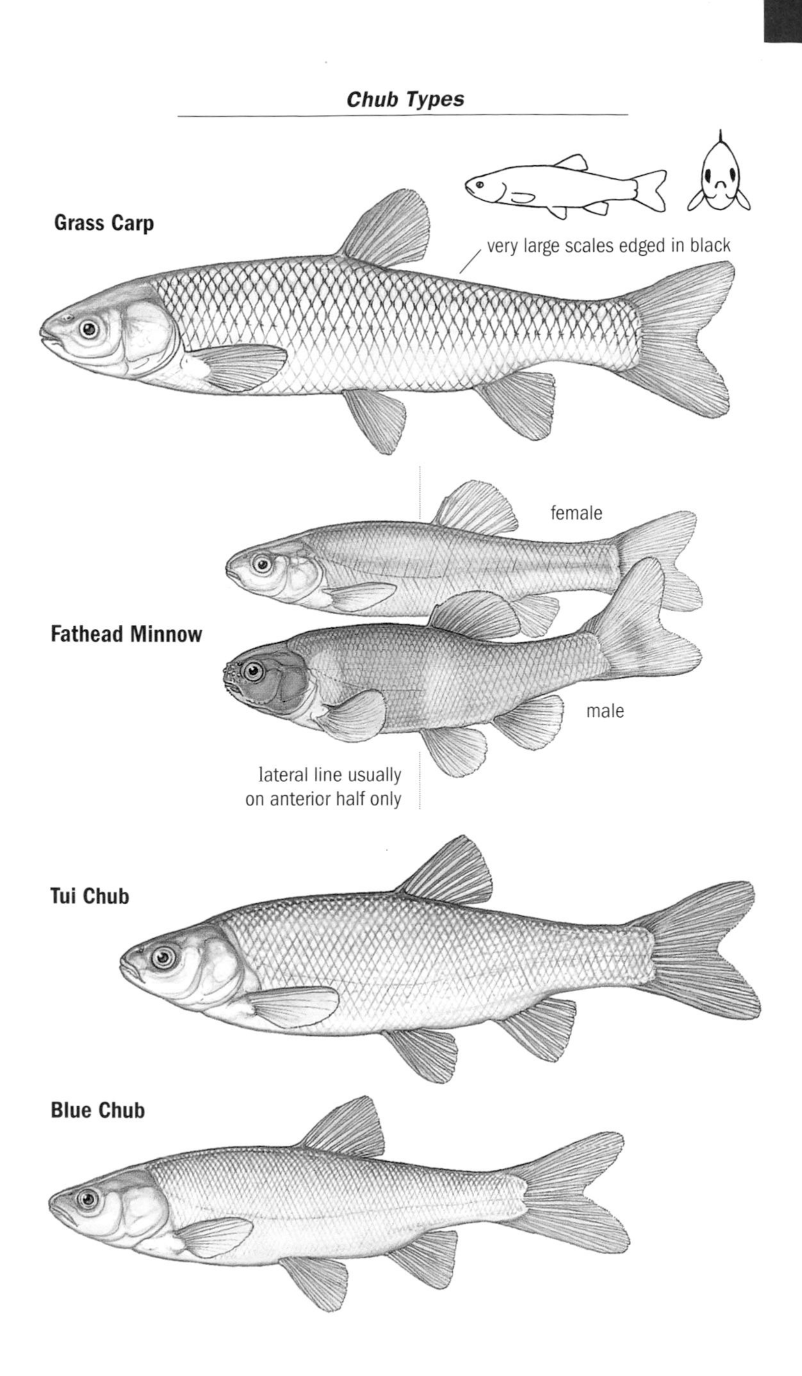
Chub Types
Grass Carp
very large scales edged in black
female
Fathead Minnow
male
lateral line usually
on anterior half only
Tui Chub
Blue Chub

Arroyo Chub *(Gila orcutti)* PAGE 214

STATUS: Native, CSSC. **DESCRIPTION:** Only large-scaled chub in southern California streams; LL scales 48 to 62; gray green on back, white belly; DF = 8, AF = 7; max. FL about 9 cm (3.3 in.). **DISTRIBUTION:** Creek and river systems of southwestern California; introduced into Mojave River, where it now hybridizes with Tui Chub.

Thicktail Chub *(Gila crassicauda)* PAGE 215

STATUS: Native, extinct. **DESCRIPTION:** Caudal peduncle thicker than in other California chubs; large scales, LL scales 49 to 60; dark back, yellowish sides and belly; DF = 8 to 9, AF = 8 to 9; max. TL about 25 cm (10 in.). **DISTRIBUTION:** Once numerous in backwater areas of the pristine Central Valley.

Small Scales

California Roach *(Lavinia symmetricus)* PAGE 206

STATUS: Native, CSSC. **DESCRIPTION:** Dark on back, silvery on belly, usually with a dark lateral stripe; DF = 7 to 9, AF = 6 to 8; LL scales 47 to 63; small size, max. TL about 10 cm (4 in.). **DISTRIBUTION:** Sacramento–San Joaquin drainage, Russian River, Gualala River (Sonoma County), Navarro River (Mendocino County), tributaries of San Francisco Bay, Monterey Bay, and Goose Lake, and the Pit and Eel Rivers.

Lahontan Redside *(Richardsonius egregius)* PAGE 210

STATUS: Native. **DESCRIPTION:** Dark back, silvery belly; breeding male and female have bright red lateral stripe, persisting as dark band during nonbreeding season; DF = 7 to 8, AF = 8 to 10; LL scales 52 to 63; max. SL 17 cm (6.4 in.). **DISTRIBUTION:** Confined to the eastern side of Sierra Nevada; possible introduction into scattered sites in the upper Sacramento River drainage.

Speckled Dace *(Rhinichthys osculus)* PAGE 211

STATUS: Native. **DESCRIPTION:** Color highly variable, normally with speckles or blotches that may form an irregular midlateral band; fins may appear reddish during spawning; mouth subterminal but not inferior as in suckers (Catostomidae); tiny barbel may be present on each maxilla; DF = 6 to 9, AF = 6 to 9; LL scales 47 to 89; max. SL about 11 cm (4.4 in.). **DISTRIBUTION:** Only native freshwater fish to occur in most habitats in all fish provinces; least abundant in the Santa Ana and Coastal Provinces.

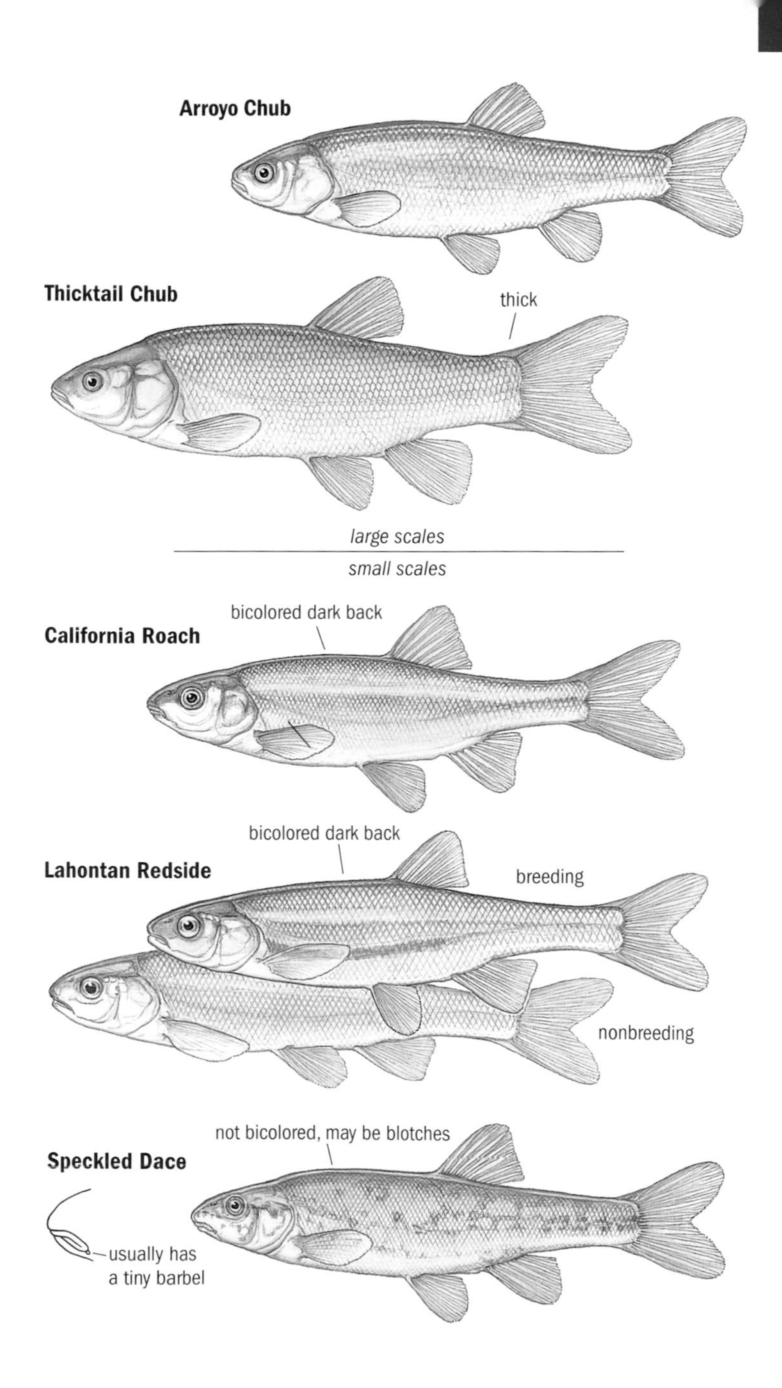
Arroyo Chub
Thicktail Chub
thick
large scales
small scales
bicolored dark back
California Roach
bicolored dark back
Lahontan Redside
breeding
nonbreeding
not bicolored, may be blotches
Speckled Dace
usually has
a tiny barbel

Pikeminnow-Type Minnows (Elongate, Thick Body)

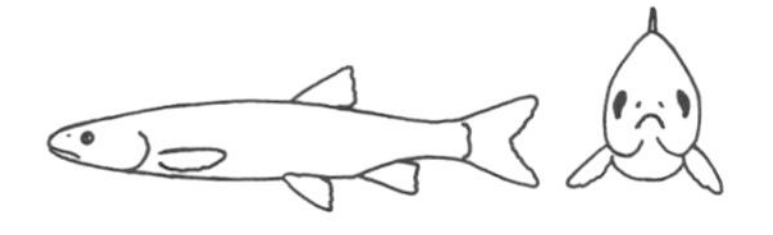

Small Scales

Sacramento Pikeminnow *(Ptychocheilus grandis)* PAGE 222
STATUS: Native. **DESCRIPTION:** Long, pointed snout; jaw extends beyond anterior margin of eye; jaw without teeth; olive brown back, yellowish belly; during breeding season fins are orange; young, smaller individuals more silvery; DF = 8, AF = 8; LL scales 65 to 78; very large minnow, max. TL about 1.4 m (4.7 ft), max. wt. 14.5 kg (32 lb). **DISTRIBUTION:** Sacramento–San Joaquin River system; Pajaro, Salinas, Russian, Eel, and Pit Rivers.

Colorado Pikeminnow *(Ptychocheilus lucius)* PAGE 225
STATUS: Native, SE, FE, FPS. **DESCRIPTION:** Long, pointed snout; jaw extends beyond anterior margin of eye; jaw without teeth; color similar to Sacramento Pikeminnow *(P. grandis)*; DF = 9, AF = 9; LL scales 76 to 97; largest of all North American minnows, max. TL 2 m (6.6 ft), max. wt. 45 kg (99 lb). **DISTRIBUTION:** Sites in the Colorado River, though presently rare in, or absent from, the California portion; no range overlap with the Sacramento Pikeminnow.

Hardhead *(Mylopharodon conocephalus)* PAGE 219
STATUS: Native, CSSC. **DESCRIPTION:** Long, pointed snout; jaw does not extend beyond anterior margin of eye; a fold of skin (frenum) connects upper jaw to side of head; bronze brown on back, silvery on belly; DF = 8, AF = 8 to 9; LL scales 69 to 81; max. SL about 30 cm (11.8 in.). **DISTRIBUTION:** Sacramento–San Joaquin drainage, and Napa and Russian Rivers.

Sacramento Splittail *(Pogonichthys macrolepidotus)* PAGE 216
STATUS: Native, FT, CSSC. **DESCRIPTION:** Upper half of caudal fin longer than bottom half; small barbel usually present at posterior edge of jaw; back olive gray, sides silver gold; orange PecF, PelF, and caudal fin in breeding specimens; DF = 9 to 10, AF = 7 to 9; LL scales 57 to 64; max. TL about 45 cm (17.7 in.). **DISTRIBUTION:** Delta region and lower Sacramento and San Joaquin Rivers.

Clear Lake Splittail *(Pogonichthys ciscoides)* PAGE 218
STATUS: Native, extinct. **DESCRIPTION:** Similar to Sacramento Splittail *(P. macrolepidotus)* but with no range overlap. **DISTRIBUTION:** Clear Lake and tributary streams.

Bonytail *(Gila elegans)* PAGE 215
STATUS: Native, SE, FE, possibly extinct in California. **DESCRIPTION:** Very long, narrow caudal peduncle; deeply forked caudal fin; small head with moderate dorsal hump behind; DF = 10 to 11, AF = 10 to 11; LL scales 75 to 99; max. TL about 64 cm (25.6 in.). **DISTRIBUTION:** Colorado River drainage, but scarce in, or absent from, California portion.

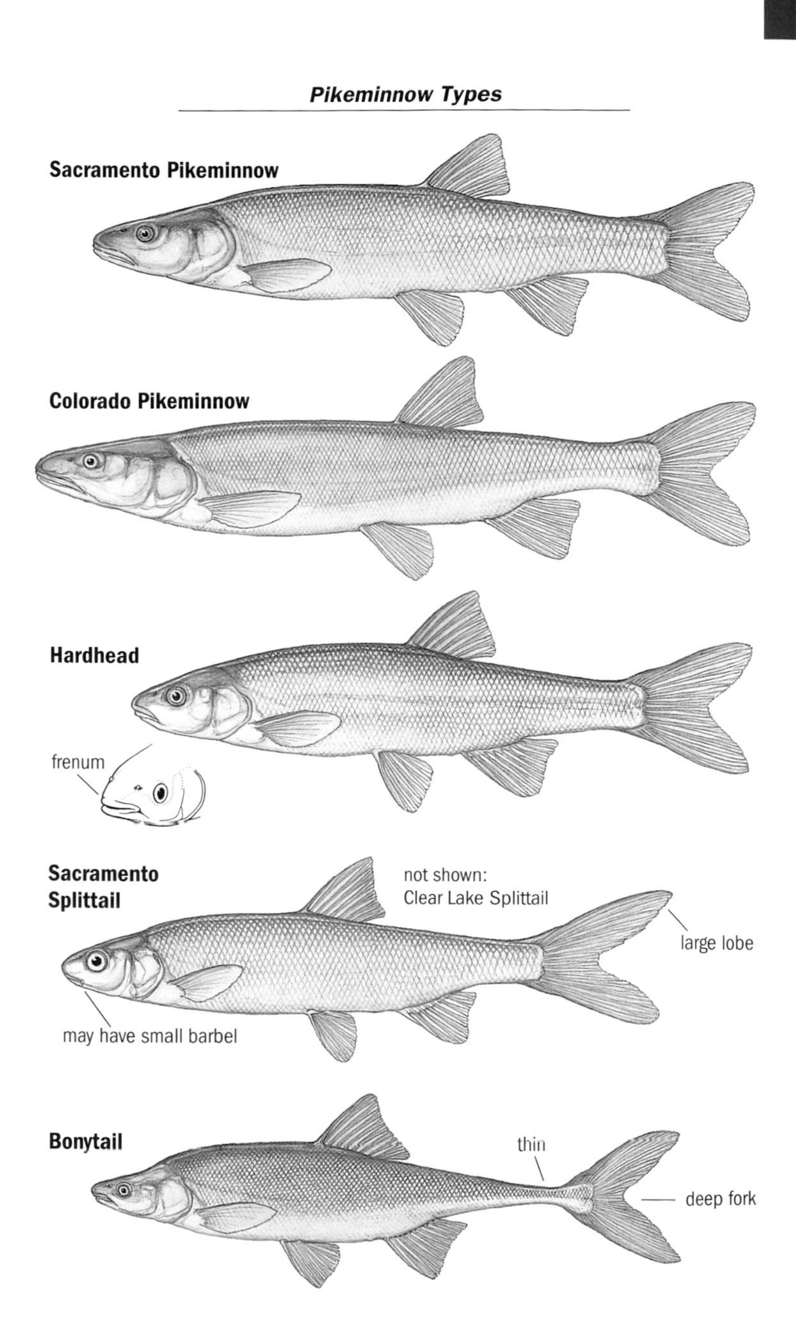
Pikeminnow Types
Sacramento Pikeminnow
Colorado Pikeminnow
Hardhead
frenum
Sacramento Splittail
not shown:
Clear Lake Splittail
large lobe
may have small barbel
Bonytail
thin
deep fork

Very Small Scales

Sacramento Blackfish *(Orthodon microlepidotus)* PAGE 220
STATUS: Native. **DESCRIPTION:** Very small scales (LL 90 to 114), but without the unusual body form of the Bonytail *(Gila elegans);* wide flat forehead; DF origin above or slightly in front of PF origin, similar to the Sacramento Splittail *(Pogonichthys macrolepidotus);* young silvery; ad. has dark dorsal bicolor pattern; DF = 9 to 11, AF = 8 to 9; max. TL about 55 cm (21.7 in.), max. wt. 1.5 kg (3.3 lb). **DISTRIBUTION:** Sacramento–San Joaquin drainage; Pajaro, Salinas, and Russian Rivers; Clear Lake; Delta area; Coyote, Alameda, and Walnut Creeks; California Aqueduct; and San Luis Reservoir.

Shad (Clupeidae)

both shad have sawlike keel along belly

Threadfin Shad *(Dorosoma petenense)* PAGE 144
STATUS: Introduced. **DESCRIPTION:** Small, silvery fish with thin, deep body; sharp, sawlike keel along length of belly; long, threadlike projection from posterior of DF; single black spot just behind head; ad. TL about 8 to 10 cm (3 to 4 in.), max. TL 33 cm (13 in.). **DISTRIBUTION:** Most lower- and middle-elevation freshwater habitats in California.

American Shad *(Alosa sapidissima)* PAGE 141
STATUS: Introduced. **DESCRIPTION:** Large, silvery fish with thin, deep body; sharp, sawlike keel along length of belly; fry may be confused with Threadfin Shad *(Dorosoma petenense)* but lack threadlike projection from DF; several small spots along LL behind head; max. FL about 48 cm (19 in.), max. wt. 5.2 kg (11.4 lb). **DISTRIBUTION:** Spawning runs and fry are found in the Sacramento, San Joaquin, American, Feather, Yuba, Mokelumne, Stanislaus, Russian, Eel, and Klamath Rivers; populations persist in Millerton Lake (Fresno County), San Luis Reservoir, and other waters of Central Valley irrigation system.

Pike (Esocidae)

Northern Pike *(Esox lucius)* PAGE 370
STATUS: Introduced. **DESCRIPTION:** Very elongate body, DF and AF located far back on body at anterior edge of caudal peduncle; large, duck-bill-shaped mouth; color varies from dark gray to green back, sides with pale oval spots (wavy lines in young), and white or yellow belly; large ad. TL 70 to 80 cm (27.5 to 31.5 in.), max TL in North America 110 cm (43 in.), max. wt. in North America 14.2 kg (31.2 lb). **DISTRIBUTION:** Davis Reservoir (Plumas County) and possibly Lake Oroville (Butte County).

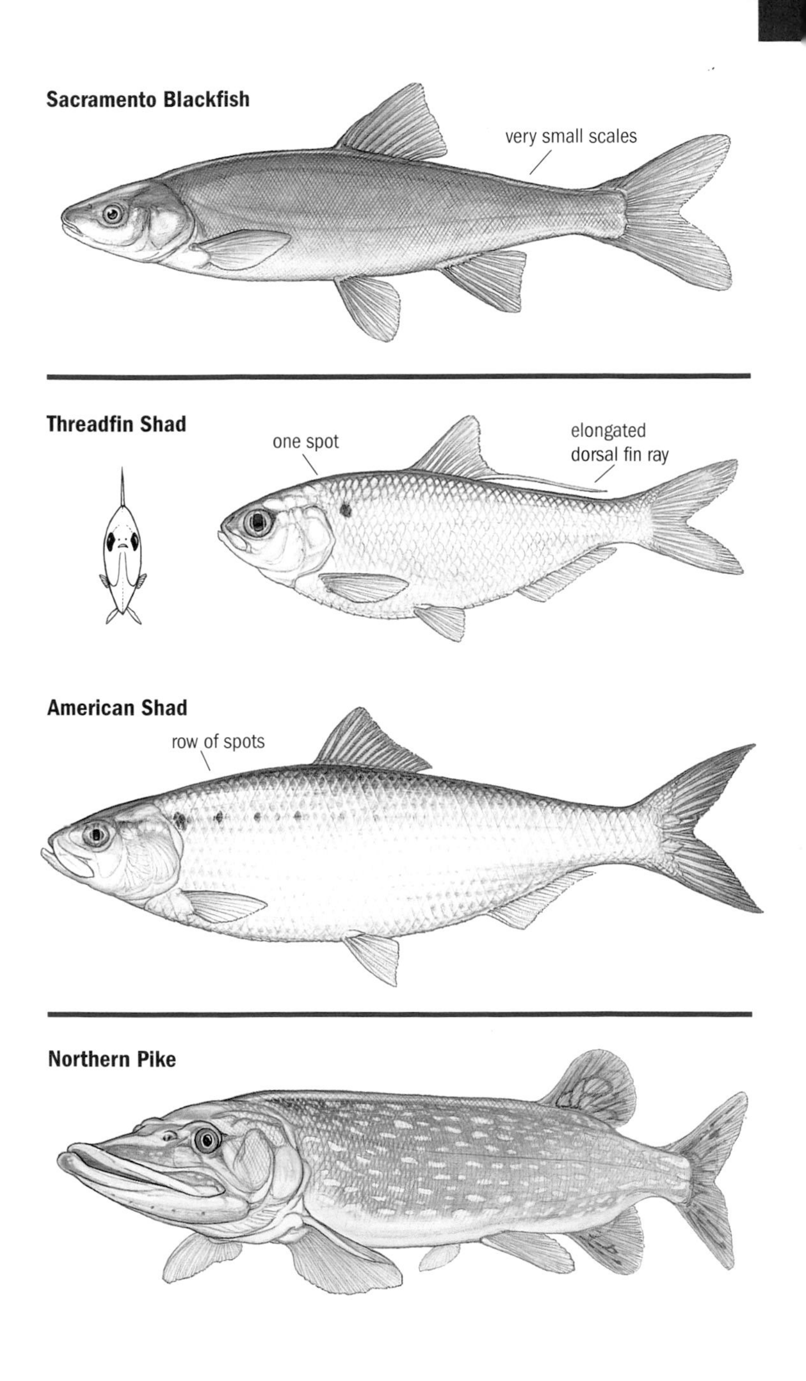
Sacramento Blackfish
very small scales
Threadfin Shad
one spot
elongated
dorsal fin ray
American Shad
row of spots
Northern Pike

SPECIES ACCOUNTS AND ANGLING NOTES

LAMPREYS (Petromyzontidae)

Lampreys are vastly different from all other freshwater fishes, and well they should be. Along with their marine cousins, the hagfish, they are all that remain today of the first vertebrate animals. The evolution of this ancestral group of all present-day vertebrates, including, of course, ourselves, began about 430 million years ago and is represented in the fossil record by small, fishlike animals called ostracoderms. This name means "bony skin" because their bodies were covered with bony plates, not scales. The lack of any jaw structures in their fossils implies that they were filter feeders.

The real importance of these first fishes was not in their external body but instead what was inside. Running down the length of this elongate body was a stiff, cartilaginous rod, the notochord, and surrounding it were additional small blocks of cartilage to which muscle groups could easily attach. This initial combination of vertebral column and muscles made possible a sculling action in the ostracoderm body that is the basis of modern fish movement. The ostracoderms and their allies eventually gave rise to new fish groups that possessed jaws, and it was this latter structure that, in combination with the vertebral column, set the anatomical stage for the evolution of land vertebrates. Ironically, the body plates of the ostracoderms, which were initially effective protection against invertebrate predators, were apparently unable to withstand the increasing strength and size of the evolving fish jaw, and by the middle of the Devonian, some 375 million years ago, all but one family of freshwater jawless fishes became extinct. This one group, the lampreys, is still alive and well today, especially in California and has more species and subspecies than any other geographic unit in the world. However, because lampreys spend most of their life sequestered in soft-bottom areas of creeks and streams, most Californians are unaware of their existence.

The lampreys have persisted to the present apparently by having turned the evolutionary table on the newly evolved jawed fishes and becoming, as adults, efficient external fish parasites. It is not so much the adult lamprey that attracts the evolutionist's attention, but rather its larval form. This larva (ammocete) represents a beautiful connecting link between the vertebrates and their prevertebrate ancestors. The ammocete is also the most long-lived stage of the lamprey's life cycle. In some species all growth takes place in this stage for up to seven years before individuals metamorphose into a relatively short-lived adult form.

Ammocetes are filter feeders like some of their ostracoderm ancestors, and they spend most of their lives buried tail first in the bottom substrate of streams with just their mouths exposed. At metamorphosis, the filtering mouth is reshaped into a formidable rasping and sucking disc. In the nonparasitic species this mouth is never used for feeding, and adults never migrate from their home stream, but remain to reproduce and then die, usually within six months of their metamorphosis. This type of life history is identical to that of several insect groups in which the only function of the adult form is reproduction.

In contrast, adults of parasitic species migrate from the spawning streams to either the sea or a lake, where they may remain for up to two years. During their pelagic life they feed on large, soft-scaled fishes such as salmon. Feeding is accomplished by first attaching to the side of their prey by means of their round, sucking mouth. Next, the strong, rasping tongue and surrounding tooth plates penetrate the outer layers of the host fish, after which the body fluids are extracted with a strong sucking action. When the flow of body fluids lessens, the lamprey releases its hold and begins to search for another host. This presumably is easily accomplished with shoaling species such as salmon because another potential host is always nearby. Some parasitized fishes may die either from the loss of blood and other fluids or from the wound, which is left open to infection. It is very detrimental, however, for any parasite to kill its hosts en masse, for such behavior can only lead to the eventual demise of the parasite as well. Thus, species such as our Pacific Lamprey *(Lampetra tridentata)* and the anadromous Pacific salmonid host species with which they evolved have apparently struck a workable balance in which many attacks are not lethal and numerous hosts survive. When examining whole, wild-caught Chinook Salmon *(Oncorhynchus tshawytscha)* in a fish market, you may occasionally see quarter-size lamprey scars, indications of some past nonfatal encounter.

This equitable relationship between parasite and host can be possible only if the latter has ample time to slowly adapt to the presence of the former. This was not the case when the Sea Lamprey *(Petromyzon marinus)*, an East Coast species, arrived at Lake Michigan in the early 1930s after beginning its journey in the late 1800s from the Atlantic Ocean to Lake Ontario and then on to Lake Erie via Welland Canal in the early 1920s. Up to that time Lake Michigan supported a lucrative Lake Trout *(Salvelinus namaycush)* fishery, with many small harbors along both the Michigan and Wisconsin shores that displayed prosperous fleets of small, family-owned fishing boats. As a boy growing up in Wisconsin in the late 1940s, however, my only remembrance of these harbors is of boats in dry dock with peeling paint and rotting hulls. In less than a decade the Sea Lamprey had decimated the once massive Lake Michigan Lake Trout population. It took over two decades of trial-and-error field experimentation to finally discover an ammocete-specific poison that could effectively control this invader and start Lake Michigan Lake Trout on a partial road to recovery.

A comment on classification is perhaps necessary before we leave this most unique group of "first fishes." The current state of lamprey taxonomy and systematics continues to exhibit an evolution of its own. For instance, since the first edition of this book in 1984, one species has been downgraded to subspecies status, another new species has been recognized by most authorities, and the common and Latin names for a third have been changed. Given the elusiveness of these animals, especially the nonparasitic forms, and the recent upsurge of interest in them, we may expect more changes before some future ichthyologist writes the final chapter of the Petromyzontidae of California.

PACIFIC LAMPREY *Lampetra tridentata*

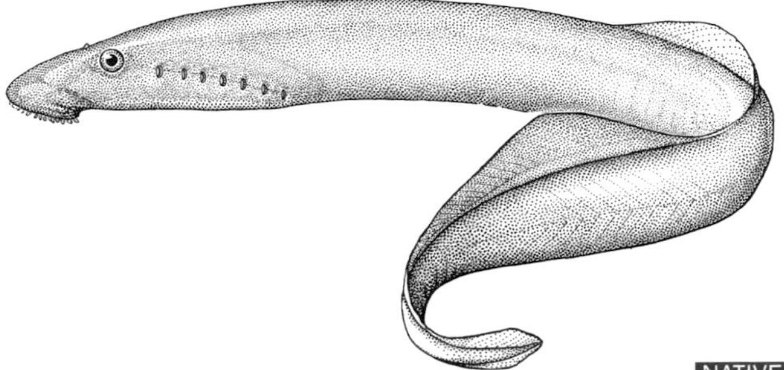

NATIVE

This is the largest species of California lamprey, with adults attaining lengths up to 76 cm (30 in.) (pl. 17). The adult stage lives up to two years and inhabits estuaries and nearby ocean areas where it parasitizes fishes such as salmon (Salmonidae) and larger "flatfish." In several landlocked populations of this species, the main hosts are suckers (Catostomidae) and larger minnows (Cyprinidae). Most spawning takes place between late spring and the end of summer, as schools of Pacific Lampreys move slowly inland to upstream spawning beds of fine gravel. Such movements usually take place at night, another feature of lamprey life that makes these fishes invisible to most people. During present-day spawning migrations, anadromous lamprey have been able to surmount vertical stream barriers better than their fellow salmonid migrants. This is possible through the use of their strong mouth suction, which allows them to hold fast to the face of barriers as temporary periods of exhaustion set in (see "Anatomy and Physiology of Freshwater Fishes"). Once recovery occurs, the Pacific Lamprey can resume its upward trek, while a salmon experiencing the same exhaustion/recovery syndrome has to start all over again at the bottom of the obstacle. Spawning is surprisingly similar to that of the salmonids, with both sexes constructing nests by removing stones and gravel from circular plots about the diameter of their bodies. After the minute eggs are laid and fertilized, the spawner loosens gravel from the upstream side of the nest so that it then washes down and buries the eggs. No egg counts have as yet been reported for the Pacific Lamprey, but estimates range as high as 200,000 eggs per female. Like the salmon

Plate 17. Pacific Lamprey, adult and larva.

that they parasitize, most adult lampreys die soon after spawning, although some occasionally survive this initial effort and make a second round-trip to the sea.

Newly hatched ammocete larvae move downstream with the current to areas with good organic bottom debris. Here they bury tail first with only the large oral hood protruding just above the substrate surface and take up a filter-feeding existence. They remain buried throughout most of their five to seven years of larval life. This is an important survival strategy because their small, wormlike bodies are potential food for fishes. They undergo a rapid metamorphosis at about 15 cm (6 in.) and then move downstream to the estuary and sea to begin the parasitic phase of their life cycle.

Although presently bypassed by anglers, the Pacific Lamprey was a major food source of several coastal California Native American tribes and was caught in a most unusual manner. The angler stood knee deep in water, bent over with his open hands between his lower legs. As a migrating lamprey brushed by, it was grasped and flung onto the shore in one continuous motion, the slender, mucus-coated body being almost impossible to hold for more than a brief moment. Most of the meat was dried, to be eaten at a later time. Fresh lamprey is reported to be quite tasty and is a popular item in some Asian recipes.

The Eel River in Mendocino and Humboldt Counties is named for the massive Pacific Lamprey spawning runs that once occurred there. It was named the "Eel River" rather than the "Lamprey River" because the old popular name for the lamprey was "Lamprey Eel," which incorrectly implied that this ancient group is closely related to the modern American Eel *(Anguilla rostrata)*, a catadromous bony fish species of the Atlantic coast whose numbers have been drastically reduced by overfishing.

RIVER LAMPREY — *Lampetra ayresi*

NATIVE

PROTECTIVE STATUS: CSSC.

The River Lamprey is the other parasitic lamprey species in California. It occurs from San Francisco Bay northward but appears to be most numerous in the Sacramento and San Joaquin Rivers. Like its larger relative, the Pacific Lamprey *(L. tridentata)*, the adult feeds on soft-scaled species of marine and estuarine fishes. In addition, it has been found on salmon in freshwater. Its smaller size (up to about 30 cm [12 in.]) appears to favor attacks on smaller prey such as Pacific Herring *(Clupea pallasii)*. The life history of this species has not been fully studied, but it is assumed to follow the general pattern of that of the Pacific Lamprey. However, the metamorphosis period is much longer (up to 10 months), and the adult stay at sea is much shorter (3 to 4 months). This relatively short parasitic period may compensate for the fact that it feeds on the dorsal lateral muscles of its prey and will continue to do so up to and even after its host dies. The exceptionally well-developed pair of sharp cusps on the supraoral plate appear to promote this carnivorous type of feeding. Given such behavior, the term "predator" may be more appropriate than "parasite" for this species. Perhaps the very short adult feeding period ensures that a given breeding population of River Lamprey does not completely eliminate breeding populations of salmonids that occupy the same area.

PIT-KLAMATH BROOK LAMPREY — *Lampetra lethophaga*

NATIVE

This is a nonparasitic species of lamprey. The adults do not feed but instead remain in the streams, where they soon spawn and die. The lamprey larvae filter feed in the bottom mud of streams for at least four years before metamorphosis. They grow to 21 cm (8 in.), which is also the maximum adult size, because with no adult feeding, growth is not possible after metamorphosis. Spawning is presumed to follow the pattern of the Pacific

Lamprey *(L. tridentata)*. As its name implies, this species is limited to the Pit and Klamath Rivers and Goose Lake.

KLAMATH RIVER LAMPREY *Lampetra similis*

NATIVE

PROTECTIVE STATUS: CSSC.

This small parasitic species occurs in the Klamath River drainage. Adults are under 27 cm (10.6 in.) total length and do not go to sea but instead feed on local sucker species (Catostomidae) and large cyprinids in the Klamath River and associated lakes and reservoirs. The Klamath River Lamprey was recently recognized as a distinct species based on both biochemical and anatomical features. As is often the case, taxonomic studies precede ecological investigations, and what is now needed is life history information for this nonanadromous parasitic form.

KERN BROOK LAMPREY *Lampetra hubbsi*

NATIVE

PROTECTIVE STATUS: CSSC.

This species of nonparasitic lamprey was originally described from 11 newly transformed adults and one ammocete from the Friant-Kern Canal, east of Delano, in the southern San Joaquin Valley. Since then it has also been found in several tributaries of the San Joaquin River. Like the other two California nonparasitic species, the adult, nonfeeding form has poorly developed tooth plates. As with other nonparasitic species, the average adult length (11 cm [4.5 in.]) is slightly less than that of the ammocete (13 cm [5.1 in.]). The Kern Brook Lamprey presumably has a life cycle similar to those of other nonparasitic species and is likewise in need of in-depth ecological study.

WESTERN BROOK LAMPREY *Lampetra richardsoni*

NATIVE

This is the only nonparasitic lamprey that occurs in creeks of Mendocino, Lake, and Sonoma Counties and is also found in the Sacramento–San Joaquin drainage. The species is similar in size to the Kern Brook Lamprey *(L. hubbsi)*. Little is known of its life history, though it is presumed to be similar to that of the Pit-Klamath Brook Lamprey *(L. lethophaga)*.

STURGEONS (Acipenseridae)

"What kind'a fish is that?!" is often the initial utterance upon first seeing a sturgeon. At such a moment, flashbacks to some past biology class or nature film may occur as you recall the two basic fish groups: the Chondrichthyes class (sharks, rays, and skates) and the Osteichthyes class (the bony fish taxon to which almost all California freshwater fishes belong). The former has a cartilaginous skeleton, externally opening gill slits, a spiracle, and in sharks, a heterocercal caudal fin. In contrast, the latter has a skeleton of bone, no spiracle, an operculum covering the gill slits, and a homocercal caudal fin. Here, however, is a fish with the skeleton, caudal fin, and spiracle of a shark, but also with bony plates protruding through its skin and with the operculum of a bony fish (pls. 18, 19). At this moment the suggestion that it could be some sort of macrohybrid may not seem too far-fetched.

Plate 18. Green Sturgeon (above), White Sturgeon (below).

The answer to this puzzle lies not in these external features but deep within the sturgeon body. Passing through the middle of each cartilaginous vertebra is a flexible rod, the notochord. This structure can be clearly seen in cross section in the sturgeon steaks occasionally displayed in fish markets or a sturgeon

Plate 19. Heterocercal caudal fin, White Sturgeon.

angler's kitchen. This key structure in vertebrate evolution has long since disappeared in the adult form of all other California freshwater fishes except for the lampreys (Petromyzontidae). Its persistence in the sturgeons signifies that these are "living fossils," members of a taxonomic group known collectively as the Chondrostei, which gave rise to all present-day bony fishes, or teleosts. Sturgeons are as highly evolved and adapted to their habitats as any other present-day fish species, but because basic sturgeon biology has so many broadly adaptive features, there has apparently been little need for modification during the long evolutionary journey from the Devonian to the present.

Heading the list of adaptive features is the sturgeon feeding niche, in other words, the type of food a species eats, the habitat where it finds it, and the method by which it obtains it. The downward protruding, tubular lips act like a suction hose, enabling sturgeons to capture most small prey species that sequester within the rich, soft substrate of river estuaries (pl. 20). This ability plus a wide range of food preferences allowed ancestral sturgeons to maintain their broad-spectrum bottom-feeding niche as the various prey species continued to change over past millennia. Such animals are often referred to as "feeding generalists" and normally have a much greater evolutionary longevity than "feeding specialists," which are highly adapted to feed on one or two specific prey items. When its prey go extinct, the feeding specialist soon follows.

Plate 20. Protruding tubular lips, White Sturgeon.

Sturgeons have also acquired two rather universal forms of protection against predation: size and a tough body covering. If a species is bigger by far than any other with which it shares a habitat, it is normally free from all nonparasitic predation. This is certainly the case with sturgeons, which are the largest freshwater fishes in the world. The North American White Sturgeon *(Acipenser transmontanus)* can grow to at least 6 m total length (about 20 ft) and weighs up to 630 kg (almost 1,400 lb). Even at this impressive size it is dwarfed by the Beluga Sturgeon *(Huso huso)* of Eurasia, which grows to 8.5 m (26 ft) total length and weighs up to 1,297 kg (2,850 lb). Coupled with this enormous adult size is fast growth during the first year of life in most species. The White Sturgeon can attain a length of 30 cm (about 1 ft) in this time, which places it beyond the normal prey size for the average piscivore in the San Francisco Estuary. A formidable covering has been added to these large bodies in the form of several rows of bony plates or shields that emerge at an oblique angle from the underlying dermal skin layer. Like the hard body covering of another group of living fossils, the tortoises, these shields have been able to buffer the bites of a long succession of predators over the past several hundred million years.

A final requisite for persistence through time is a reproductive biology that can cope with long- and short-term habitat changes. Species that become specialized for spawning on a specific substrate or at certain depths usually become extinct when geological or climatic changes alter those conditions. In contrast, sturgeons spawn by scattering very large numbers of adhesive eggs over a variety of river rock and gravel substrates, riverine features that presumably have persisted over much of geological time. In this form of "caviar roulette," even if a minute fraction of the eggs hatch and grow to maturity, the adult spawners will be replaced within their lifetime. Such replacement can occur because of the great longevity of sturgeons—up to 100 years in some species. Even though most species do not reach sexual maturity for 10 years or more, they still have many decades to successfully produce their replacements, even if natural drought periods or unnatural human water diversions negate several successive spawning attempts.

Given this array of long-term survival features, sturgeons should persist for many eons to come, barring gross interference by our own species. We are indeed fortunate to have alive today in California two representatives of this ancient fish type, the Green Sturgeon *(A. medirostris)* and the White Sturgeon, with the latter providing one of the major freshwater sport fisheries in this state.

WHITE STURGEON *Acipenser transmontanus*

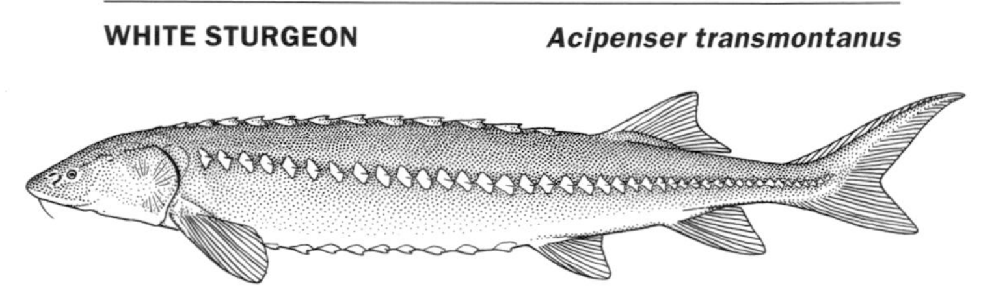

NATIVE

This is the largest freshwater fish in North America. In California it does not attain the maximum size and weight previously mentioned for other areas but still grows to almost 3 m (9.75 ft) and 225 kg (495 lb). It is primarily an estuarine species that moves up large rivers to spawn. Most spawning appears to take place in the segment of the Sacramento River between the Sacramento International Airport and the Sutter Buttes. Spawning has also been observed in portions of the San Joaquin River and may also occur in one or more tributary streams. Several years ago I was walking along the bank of the lower Stanislaus River near my small farm in southern San Joaquin County and came upon the severed caudal fin of what must have been a very large White Sturgeon. Pieces of intestine the size of a car radiator pipe and fragments of sticky eggs were also scattered around, indicating that this had been a female that may have been moving to a spawning area.

The White Sturgeon feeds on a wide variety of bottom-dwelling invertebrates and small, bottom-oriented fishes. It is also an avid egg eater and each year takes large numbers of eggs of the Pacific Herring *(Clupea pallasii),* which spawns at the mouth of the Delta. Its ability to readily shift to new prey species has been recently confirmed from stomach samples containing large numbers of the recently introduced Overbite Clam *(Potamocorbula amurensis),* whose populations are now erupting throughout the San Francisco Estuary. Indeed, sturgeon predation is currently one of the few efficient controls for this unwanted mollusk.

The White Sturgeon breeds for the first time at a more advanced age than any other freshwater fish. A typical female reaches sexual maturity between 12 and 16 years of age, at which time she is about 95 to 135 cm (37 to 49 in.) total length. The spawning migration up the Sacramento and San Joaquin Rivers begins in late winter and extends through early spring. A spawning sturgeon seeks out deep, hard-bottom sites in the main channels of the middle portions of these rivers and the lower reaches of some of their tributaries. The eggs are quite small but very numerous. On her first spawning run a female may deposit only a modest 100,000 or so, but in later years this number may reach 5 million. Individuals do not spawn every year, however, and from two to four years may elapse between each spawning.

Eggs are broadcast over rock- or gravel-bottom areas where the current at midstream spreads them over considerable distances. The sticky eggs eventually attach individually to these substrates, and they hatch seven days later. After hatching, the fry travel downstream with the current and begin feeding on zooplankton and small bottom invertebrates. A young sturgeon grows rapidly and can reach over 30 cm (about 1 ft) total length in its first year. At an annual growth rate of up to 6 cm (2.4 in.) per year, most sturgeons reach the current legal minimum catchable size of 117 cm (46 in.) in about 15 years. This permits a sportfishing harvest of females at the size and age when they become sexually mature, which by itself would not be a good management practice. However, the maximum size limit is 183 cm (6 ft), which hopefully ensures that a supply of middle- to older-age females that can produce well in excess of a million eggs per spawning is always on hand.

This combination of long life and slow sexual maturation makes sturgeons more vulnerable to commercial fishing than most fishes. In California before 1870, the White Sturgeon was considered a nuisance by local commercial fishermen because it became entangled in and then tore up salmon nets. Essentially, no market existed for the species, except in the Chinese community, where the gelatinous notochord was considered a delicacy. Then the sudden decline of the Atlantic sturgeon fishery (presumably due to overfishing) and the influx of easterners to the Pacific Coast created such a demand for White Sturgeon that by 1909 the species was feared near extinction in the Delta, and all fishing was stopped. Commercial fishing was reopened several years later, but catches were so low that it was closed again in 1917. It stayed closed until 1954, when a year-round season was again established, but only for sportfishing. At that time, sportfishing included the method of snagging by dragging large numbers of hooks along the bottom, a technique to which the large, slow-moving sturgeon is very susceptible. Alarmingly large catches resulted from this method, and in 1956 the regulations were modified to restrict fishing to hook and bait only. For the next decade very small annual catches were recorded, mainly as a result of the California angler's inexperience in fishing for this species. Then, in the mid-1960s, it was discovered that Grass Shrimp *(Hippolyte californiensis)* are an excellent bait when fished just off the bottom, and from that time on the sport fishery for this species has steadily increased. The present annual exploitation rate by anglers is estimated at about 10 percent of the population. However, the illegal take by poachers of "black gold," or sturgeon caviar, may double this rate in a given year. Up to the time of the dramatic rise in real gold prices in the latter part of the twentieth century, an ounce of good-quality caviar commanded a restaurant price that approached its weight in gold. Caviar is still worth enough today for poachers to risk their lives in night netting operations and for game wardens to risk theirs trying to stop them. The retail price in 2005 was $2,500 for 1,000g (35 oz.).

Given this history of legal and illegal White Sturgeon harvest in California, most fishery biologists believe that this species should never again be open to commercial fishing. Recent achievements in the area of sturgeon aquaculture, however, may soon modify this outlook. With its reputed national craving for caviar, it is not surprising that Russia was the first country to culture this sturgeon successfully. Prompted by a decline in populations due to overfishing, it began a sturgeon aquaculture program in the early 1960s. Today it produces millions of fingerlings annually to stock sturgeon-depleted rivers. Hatcheries also raise sturgeons in confined freshwater habitats for meat production. By intensive feeding methods, it is possible to produce a marketable fish three times faster than it can grow to the same size in nature.

Encouraged by the success of the Russian program, aquaculturists at the University of California at Davis have developed a program for the replenishment and maintenance of White Sturgeon populations in California. Artificial spawning and fingerling-rearing techniques have been perfected, and excellent growth rates of fishes reared in controlled habitats have been obtained. In addition to providing fish market sturgeon, some look to an expanded hatchery program to solve any future problems of declining sturgeon numbers. As with the salmonid hatchery program, selection is favoring adaptation to hatchery conditions instead of to the natural habitat, and some critical traits for success in the natural habitat could be diminished over time. Until more long-term information can be compiled on the survival and adjustment of hatchery-released sturgeon, the prime focus of management should be on protecting and enhancing the two key habitat areas for this species: its river-bottom feeding and spawning grounds.

ANGLING NOTES: The use of Grass Shrimp as bait in the lower Sacramento–San Joaquin Delta appears to have been the major factor in reviving the White Sturgeon sport fishery in California, although why this one bait seems to have made such a difference is unclear. Stomach analyses have shown that Grass Shrimp is not the only food of this species, but it is probably one of the mainstays of the sturgeon's diet. It may also exude an abundance of attractant odor molecules when placed on a hook, which would then result in a far greater number of strikes than would conventional baits such as clams or herring. Whatever the reason, the initial success of Grass Shrimp, and more recently of Ghost Shrimp *(Callianassa californiensis)*, has greatly expanded sturgeon fishing.

The bottom-fishing technique is the same as that for most catfish. A lead sinker large enough to resist the current is tied at the end of the line, and the bait hook on its own small swivel-attached leader trails out with the current from the sinker. Very heavy test line is considered standard. The rod should be the best possible compromise between strength and sensitivity, because the actual bite is amazingly light given the great size of this fish. Failure to recognize and react to one of these subtle bites is perhaps the most common failing of the novice sturgeon angler. Once a sturgeon has been hooked and successfully played, a large, heavy-duty landing net is needed to successfully bring this large fish aboard. The use of gaffs for this purpose is no longer permitted. Unless you are wearing heavy work gloves, it is not advisable to attempt pulling a sturgeon aboard a boat by hand. The sharp-edged bony plates that have protected this group of fishes against many predators of the evolutionary past work equally well against the tender skin of the human piscivore.

As a reminder, the legal size range is from 46 to 72 in., and the limit is one per day. If a sturgeon measures outside these limits, release is mandatory. Removal of the hook while the fish is held alongside the boat with a landing net or gloves is preferred. With large sturgeons this is sometimes very hard, and temporary restraint within a net on deck may be necessary. Sturgeons endure short periods out of water better than most fishes. If the hook is embedded in the fleshy protruding lips, its best to cut the shank with a wire cutter and then pull the hook out backward. Sturgeons have no teeth, so they cannot bite the hand that tries to free them. For hooks embedded well within the mouth or swallowed, cutting the line as short as possible is the only recourse. Most fishes have an amazing ability to cope with foreign objects in their flesh by encapsulating them in a heavy, scar-type tissue.

White Sturgeon fishing in California is still in its early stage of development, and there are new techniques yet to be learned. Recently, good catches have been obtained in midwinter in the northern portion of San Francisco Bay in conjunction with the large herring spawning runs that occur there at this time. Sturgeons move into herring spawning areas in large numbers to feed on the billions of eggs deposited on algal growths. A soft, mesh bag filled with herring eggs makes an excellent bait, but other offerings, such as fresh herring chunks, also produce results. The better catches are usually obtained early in the herring run when eggs are still somewhat scarce, so that a bait offering has a far better chance to be selected by a feeding sturgeon. Farther up the Delta in the Rio Vista area, night fishing is often preferred over a day outing. Whatever the chosen time period, sturgeon fishing is conducted at a leisurely pace, much like that followed by the quarry itself. As in many other types of seasonal fishing, the key to success resides in knowing just when and where consistent catches are being made, and weekly contact with local bait shops and weekly fishing publications or Internet fishing hotlines can save a lot of wasted time and disappointment.

GREEN STURGEON *Acipenser medirostris*

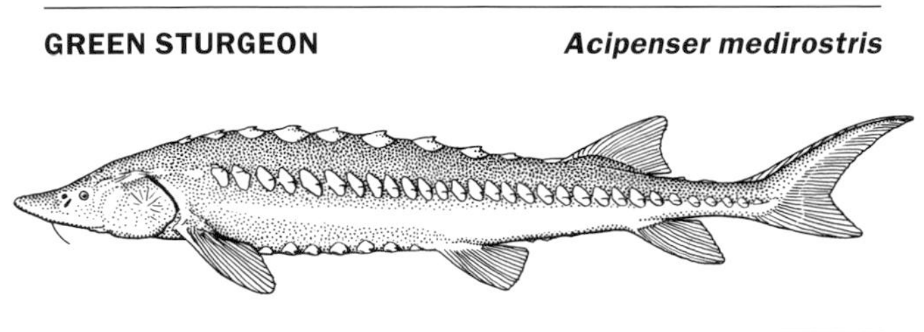

NATIVE

PROTECTIVE STATUS: CSSC.

The Green Sturgeon is the most marine oriented of all sturgeon species. Adults come into the estuary and lower reach of rivers to spawn after spending about 12 of their 14 years of preadult life at sea. Spawning is similar to that of the White Sturgeon *(A. transmontanus),* with small adhesive eggs scattered over rock or cobble substrate in river areas of good water flow. Another similarity is that the fry grow fast to a length of about 30 cm (1 ft) during their first year. Here the similarity ends, because they do not remain to grow up in freshwater but instead migrate to the sea during their second or third year. Two species with very similar feeding niches would be in direct competition with each other if they grubbed for the same bottom prey source in the same area throughout life. Instead, the Green Sturgeon juveniles feed mainly in the river estuaries and then abandon freshwater altogether except for spawning.

A pronounced spatial separation also exists between the two species. Only about one out of every 80 sturgeons caught in the Sacramento River is a Green Sturgeon, and that lopsided ratio is reversed for sturgeons taken in the Klamath River. Their maximum adult size also differs. The recent record for the Green Sturgeon of 270 cm (8.8 ft) and 175 kg (385 lb) falls well short of that for the White Sturgeon in California and is only about one-third of the White Sturgeon world record. Yet another difference between the two concerns our knowledge of their respective natural histories. We have a sound information base for the White Sturgeon, gathered over time by the early commercial fishery and supplemented by the present-day sport fishery and aquaculture research and development. Interest in the Green Sturgeon, on the other hand, has been minimal until very recently. Even the angler, usually a reliable source of information about most fish species, has not been so in this case because the Green Sturgeon does not support a major sport fishery as does the White Sturgeon but instead is taken incidentally, especially in the San Francisco Estuary. Even when caught, the Green Sturgeon is often turned loose by anglers wanting their one "keeper" of the day to be a White Sturgeon. Such actions stem partially from the rumored bad taste of not only Green Sturgeon flesh but also its eggs.

One of the few reliable inputs on population stability comes from several Native American tribes that retain their practice of netting Green Sturgeon during spawning runs on the Klamath River. A noticeable decline in a portion of the Native American fishery in the last two decades of the twentieth century may indicate that troubled waters lie ahead for this species. In response to this possibility, the California Department of Fish and Game has listed it as a species of special concern and has also closed to sturgeon fishing the north coast counties through which the Klamath River flows. This does not apply to the Native American fishery, but because the 2005 regulations simply use the generic term "sturgeon," it does apparently apply to both species.

SHAD (Clupeidae)

This family is perhaps the most economically important fish group in the world and is usually associated with the marine environment. Two groups within this family, the herrings and sardines, along with the anchovy family, account for the greatest part of the world's annual fish catch. The local stock of Pacific Herring *(Clupea pallasii)* still appears to be in good shape and supports an annual commercial harvest of both fishes and eggs (roe) in San Pablo Bay, the northern extension of San Francisco Bay. In addition, a number of anadromous and freshwater clupeid species exist, two of which have been introduced into California freshwater habitats. Species of this family are highly specialized for filter feeding on zooplankton. They therefore compete directly with the young of most other fish species for the zooplankton resource and can have a pronounced effect on the food chain in a given habitat.

AMERICAN SHAD ***Alosa sapidissima***

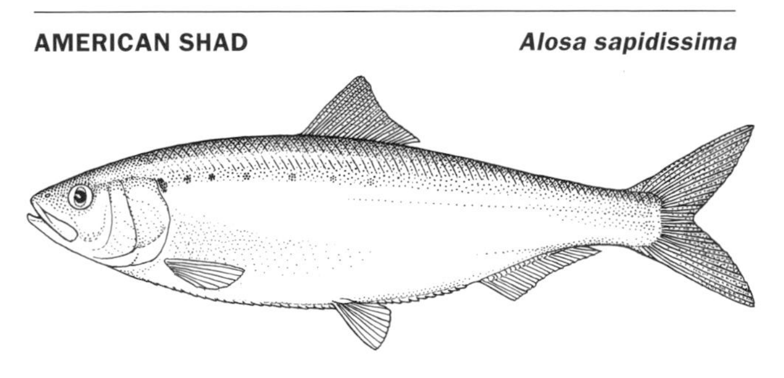

INTRODUCED

The early history of the American Shad in California has already been reviewed in the "Native versus Introduced Species" section. To briefly summarize here, over 800,000 fry were transported from New York State and released in the Sacramento River during the decade following the completion of the first transcontinental railroad in 1869. The establishment of this anadromous species in the San Francisco Estuary was so rapid and successful that a lucrative commercial fishery was established in the early 1880s and persisted through 1956. Then, after a consistent decline of annual catches, commercial fishing for this species was closed in 1958, and only sportfishing has been permitted since. The success of this initial California introduction is vividly seen today with established spawning populations ranging from Baja California, Mexico, to Alaska and on across the Bering Sea to Russia. Spawning populations have also spread from the Sacramento River proper to other Delta tributaries and to the Russian, Eel, and Klamath Rivers. Just a few years ago I collected my first adult American Shad in the lower Stanislaus River at a site where I had sampled annually, with negative results, for many years before.

It is interesting to speculate as to what might have been the California Fish Commission's early policy on introductions if the initial, costly, and well-publicized 1870s shad releases had failed. It is doubtful that a preintroduction study or analysis of the ecological niche requirements of this species was made other than to note that central California has a large river estuary and bay system similar to that seen in the Hudson Bay area. As we look back now, with a moderate grasp of California freshwater fish biology, it is easy to see that this was a case of an empty ecological niche just waiting to be filled. California has several species of native anadromous salmonids, but none are zooplankton filter feeders like the American Shad adult. Furthermore, the lower river sand- and gravel-bed spawning sites of this newcomer did not infringe upon the creek gravel egg deposition sites of salmon (Salmoninae) and Steelhead *(Oncorhynchus mykiss irideus)*. Finally, juvenile shad do not remain long in the Sacramento River system but instead go to sea after only a year or at the most two in the Delta area. Indeed, during fall beach seine sampling with my annual ichthyology class at the confluence of the Sacramento and San Joaquin Rivers over the past several decades, we have captured juvenile American Shad only every third or fourth year. This relatively short juvenile stay in the Delta greatly reduces potential competition with most other fry species for the zooplankton food resource. Given this basic natural history, which appears to have had little overlap with the then completely native California freshwater fish fauna, it is little wonder that this fish, like so many human eastern immigrants, found California very much to its liking and successfully established here.

The American Shad lives for seven years, during which time it grows from about 10 cm (4 in.) as a yearling to a maximum adult size of 48 cm (19 in.). Even with the cessation of the commercial fishery over a half century ago and a relatively light sport fishery since, the San Francisco Estuary population has noticeably declined during the last third of the twentieth century. The "usual suspects" such as water diversions, pesticides, ocean warming, and the like have been proposed as causes. I would also add to the list the introduction of highly competitive small zooplankton feeders to the Delta waters over that time period. It seems unlikely that the introduction of species such as the Threadfin Shad *(Dorosoma petenense)*, Inland Silverside *(Menidia beryllina)*, and Wakasagi *(Hypomesus nipponensis)* can occur without some negative effect on the growth and survival of juvenile American Shad habitat and the fry of other species.

ANGLING NOTES: The American Shad continues to support a sport fishery because of its large adult size and robust activity when hooked. Unlike most of its more elusive river migrant neighbors—the anadromous salmonids—the American Shad strikes at special lures called darts during its spawning run. These upstream movements begin in March as waters begin to warm and increase in intensity through May. An added dimension to shad fishing is available in the Central Valley sportfishing district, where anglers are allowed to use a net, as well, and a hook and line to take their limit of 25 fishes each. A long-handled, wide-diameter landing net is held in the river facing upstream from either a small boat or while wading near shore. When the netter feels a shad in the net, he or she quickly twists it and lifts it out of the water before the fish escapes. The neighborhood bait shop can usually provide up-to-date information as to "where they're running."

Many anglers catch American Shad just for sport, then release them or discard them on the bank without realizing that they are very good to eat. Indeed, it was the easterner's preference for shad and shad roe that was apparently responsible for this first of many California freshwater fish introductions. All larger members of the herring and shad family are also preferred for both smoking and pickling; and the culinary section of this book includes a discussion of American Shad because of its availability in large numbers during spring spawning runs.

THREADFIN SHAD *Dorosoma petenense*

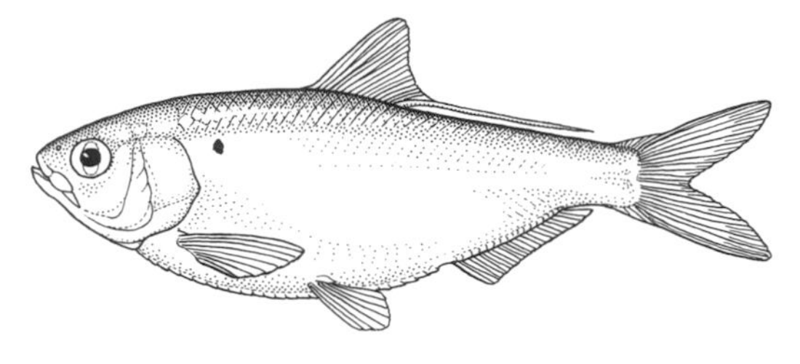

INTRODUCED

This is a small, trim, silvery fish with a black spot just above the apex of the opercular flap. It rarely exceeds 12 cm (about 5 in.) but can grow to more than twice that size when food is abundant. The Threadfin Shad is named for the long, threadlike projection at the posterior edge of the dorsal fin (pl. 21). This species spawns en masse in late spring and summer when surface water temperature rises above 20 degrees C (68 degrees F). The small adhesive eggs are broadcast over almost any submerged or floating object, including aquatic vegetation, and thus successful reproduction can be achieved in a wide variety of habitats. Most spawners are two years old, and they die soon after this event. Animal species that feed on a common and abundant food source grow fast, produce large numbers of young, and die early as the next generation matures are very much like the weeds of the plant world. They are able to quickly colonize a new habitat into which they are introduced, often at the expense of the natives or even earlier introductions residing there. Herein lies the problem with the Threadfin Shad in California.

The California Department of Fish and Game introduced this fish into isolated sites in San Diego County in 1953 and then into Lake Havasu, a reservoir on the lower Colorado River, in 1954. Before the end of 1956 it had spread through the entire lower portion of this river. This should have been a warning sign of just how much of an "ichthyological weed" this little fish can be. Over the next several years the Threadfin Shad was planted in numerous California reservoirs, and finally in the Delta in 1959. Given the many human-made connections between the Delta and numerous reservoirs in both northern and southern California, the Threadfin Shad is now one of the most widespread fishes in the lower elevation waterways of the state. Because of its ability to withstand salinities near that of seawater, it is also found throughout San Francisco and San Pablo Bays.

I shall attempt to be an unbiased reporter on the subject of Threadfin Shad introduction by stating first that in a number of reservoirs the large hatchery Rainbow Trout *(Oncorhynchus mykiss)* did indeed exhibit a size increase after the introduction of Threadfin Shad. Similar reports have been filed for reservoirs con-

Plate 21. Threadfin Shad.

taining several other large piscivorous species. If scored by the trophy angler, these introductions would probably receive an "A". The California freshwater fish conservationist, however, may not be so generous in his or her grading for the following reasons.

Like the American Shad, the Threadfin is a filter feeder on zooplankton, but unlike the former, it lives its entire life in freshwater where it continuously competes with native zooplankton feeders such as the Delta Smelt *(Hypomesus transpacificus)* and the juveniles of nearly all other species for this vital food resource. If we view this situation as if the San Francisco Estuary is an experimental aquarium with a controlled food supply and a set number of consumers, we would expect that the survival of the original complement of fishes would decrease with the addition of a new and highly efficient additional consumer. This has been the case reported by one study of Threadfin Shad–Largemouth Bass

(Micropterus salmoides) interaction in California. During my own two-year study of this and two other zooplankton feeders in the middle San Joaquin River, both otter trawl and beach seine samples produced Threadfin Shad to Striped Bass *(Morone saxatilis)* fry ratios as high as 10 to 1. Such data suggest that the decline of Striped Bass during the latter third of the twentieth century and the establishment of the Threadfin Shad throughout the Delta region during that same period may not be independent events. Instead, the Threadfin introduction could be one of a number of reasons for the Striped Bass decline.

Perhaps the most dramatic example of how such an introduction of a biologically aggressive species can reverberate through an entire aquatic ecosystem is the unsanctioned planting of Threadfin Shad in Clear Lake. By the late 1980s this fish was present in great numbers throughout our one surviving lower-elevation, natural California lake. On my annual weekend ichthyology class trip to Clear Lake State Park in 1989, we collected so many Threadfin in our first beach seine haul that we had to lift the lead line and allow most to escape for fear of ripping the net. Even before 1989, threadfin at the lake had become so numerous that large numbers of migrating piscivorous water birds that would have otherwise made a brief "refueling" stop and then continued to their southern wintering grounds instead stayed on. Indeed, our fall field trips there during the late 1980s often turned out to be more ornithological than ichthyological. The large aggregation of water birds did not discriminate between the threadfin and the juveniles of other fishes, and the latter have suffered from both food scarcity and increased predation.

Depending from which side of the creek one views this situation, threadfin introduction has been either a needed punch to boost sportfishing in this state or a low blow for California freshwater fishes. Regardless of your personal viewpoint, one thing is certain: the Threadfin Shad is here to stay.

TROUT AND SALMON (Salmonidae)

The Trout and Salmon family is the most important sportfishing family in California. In addition, the anadromous members of this group are of great commercial importance. To maintain both fisheries, a multimillion-dollar hatchery program has been created in this state. In fact, the hatchery program is built almost entirely around this family. Several introductions have been added to California's rich native salmonid population, and at present the state has eight native and four introduced species.

The family contains the trout, char, and salmon (all in the subfamily Salmoninae), along with other species such as the whitefishes (subfamily Coregoninae) and the grayling (subfamily Thymallinae). Because of this wide variety of types, the term "salmonid" is often used to refer to the entire group. In general, the salmon species are anadromous, spawning in coastal streams and rivers and then migrating to and maturing in the sea. The trout and other groups normally remain in freshwater all their lives.

Upon closer investigation, however, the story is not quite that simple. A few salmon populations are nonmigratory, and many trout populations exhibit anadromy. In addition, during the past few decades numerous populations of Kokanee Salmon *(Oncorhynchus nerka)* have been established successfully in California lakes and elsewhere. Indeed, the Coho Salmon *(O. kisutch)*, Chinook Salmon *(O. tshawytscha)*, and Pink Salmon *(O. gorbuscha)* have created a whole new sport and commercial fishery in the Great Lakes. It now appears that all salmon are capable of the fresh to saltwater adjustments outlined in "Anatomy and Physiology of Freshwater Fishes," and continued manipulation, particularly the further establishment of inland salmon populations, may be expected throughout the country.

The Salmonidae family is one of the most primitive of the teleost or modern-day bony fish groups. It may have arisen as far back as the Cretaceous, which would place these ancient trout ancestors in streams where dinosaurs waded. The earliest fossil salmonid is from 40- to 50-million-year-old Eocene deposits in British Columbia. A genetic event took place in the salmonid

Plate 22. Parrs: Coho Salmon (above), Steelhead (below).

ancestor(s) around that time that has had a major influence on the versatility of this family: the chromosomal number doubled. Unlike most other fish groups, which are diploid, the trout and salmon family is tetraploid, and thus have had a double genetic deck with which to play the evolutionary game called natural selection.

Many aspects of salmonid biology are strikingly different from those of other families. Most species are finely scaled, and all possess an adipose fin. They are strong-swimming fishes that require high levels of oxygen and relatively cool water temperatures. Many species spawn in fall and winter as well as in spring. Gravel beds in streams or lakes are the preferred site for the construction of the redd, a depression in which the eggs are laid, fertilized, and then lightly buried. The eggs normally take a month or more to hatch, and the alevins (newly hatched fry with large yolk sacs) spend several more weeks within the gravel layer before emerging. The free-swimming fry soon develop vertical, dark ovals on their sides. These are called "parr marks," and these young are referred to as "parr." Parr marks are very useful in identifying the young of most species (pl. 22). They function as a living bar code and can be easily "read" in clear streams with the aid of short-focal-length binoculars. This technique is especially helpful to ichthyologists conducting trout and salmon spawning-creek surveys, particularly in those areas where species have been designated as endangered or threatened. The parr marks are usually concealed during maturation by silvery pigment (guanine). This change is especially noticeable in anadromous species and occurs as the parr begin their downstream journey to the sea. With this color change comes yet another name change, and these new silver juveniles are now known as "smolts." Perhaps the strangest feature in salmon life history is the well-known death-after-spawning syndrome. It is not related to the return to freshwater, because some individuals are known to spend months in rivers and streams before spawning. Instead, there seems to be a massive and irreversible enzymatic change associated with the spawning process that soon leads to a complete breakdown of normal tissue maintenance and to the development of a number of pathologic symptoms. Whatever the cause, it is a beautiful example of natural recycling, because by dying and decomposing at the site of their and their offspring's birth, the nutrient load gained mostly at sea is passed on to each successive generation through the home-creek food chain.

Salmon

COHO SALMON or SILVER SALMON — ***Oncorhynchus kisutch***

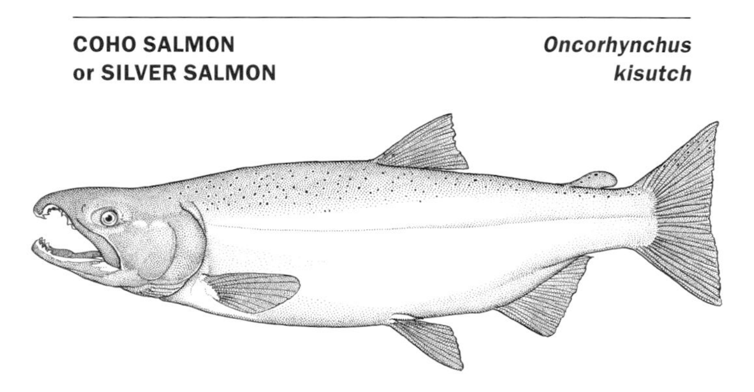

NATIVE

PROTECTIVE STATUS: SE, FT in the Central California Coast ESU; FT in the Southern Oregon/Northern California ESU.

The Coho Salmon and Chinook Salmon *(O. tshawytscha)* are the only abundant anadromous salmon species in California. The North American breeding range for the Coho is from the San Lorenzo River in Santa Cruz, California, northward to Point Hope,

Alaska. Of the nearly 600 California coastal creeks and streams that once supported Coho Salmon spawning, only about half remain as natural hatcheries today. The damming of many northern coastal rivers has been a major factor in this decline. In the southern part of the range, however, the less-visible impact of stream bed siltation has also taken its toll on small spawning populations. During numerous surveys of San Mateo and Santa Cruz County coastal creeks, I found an alarming number of sites where once-viable parr-rearing pools had been turned into sterile silt traps by unregulated logging, quarrying, heavy livestock grazing, and hillside development in the upper watershed. An important first step in halting such critical habitat destruction was the listing of all Coho populations south of San Francisco Bay as Endangered by the California Department of Fish and Game under the state of California's Endangered Species Act. The National Marine Fisheries Service then followed suit by defining two Evolutionarily Significant Units (ESUs) in 1996 and 1997 that designated the Coho within these as a Federal Threatened species. The importance of such listings lies in the fact that any impact to the coastal creek habitat is viewed in the same light as the catching or killing of the protected fish. This in turn provides a legal basis for stopping uncontrolled watershed destruction and seeking mitigation from anyone found guilty of such action.

Coho Salmon reproduction follows the anadromous salmonid pattern described in the family introduction, except that the parr spend up to a year or so in the breeding streams before going out to sea, whereas other Pacific Coast species usually spend only a few months. Coho parr are highly territorial, voracious feeders, and during the latter part of their creek stay may feed on small fishes such as the fry of the California Roach *(Lavinia symmetricus)*, a common companion species in such streams. This longer growth period in freshwater is mirrored in hatchery propagation and results in the release of much larger parr that may have a better rate of survival and apparently a better homing ability to their hatchery stream.

After spending over a year in freshwater, Coho smolts go to sea at an average size of about 12 cm (4.7 in.). They feed on small fishes and large krill for up to 18 months, growing up to 80 cm (31.5 in.) and 6 kg (about 13 lb) before beginning their spawning run to the home creek. The California Coho record is 10 kg (22 lb), taken in Lagunitas Creek, Marin County, in 1959. This creek

Plate 23. Coho Salmon.

remains the only viable Coho and Steelhead *(O. mykiss irideus)* creek in the greater San Francisco Bay area today, mainly because of good watershed preservation and protection efforts and adequate sustained releases from small upstream reservoirs.

Once it attains mature adult size, the Coho begins the return journey to the creek of its origin (pl. 23). It recognizes these sites by means of an olfactory memory of a creek's organic molecular components, as described in the section "Anatomy and Physiology of Freshwater Fishes." This is acquired through the behavioral process of imprinting during the parr stage. The actual movement into the home creek is triggered by the first heavy, late-fall rains. Indeed, fall heavy-stream flows are a necessity in many of the smaller southern range creeks, because during late spring and summer, wide sand berms build at the creek mouths and cut off all direct water flow to the sea. People strolling such beaches at that time often are not aware that they are walking over what will be a fast-flowing creek mouth come winter.

Another area where the Coho Salmon has excelled in recent years is its successful adaptation to lake life. Along with Chinook and Sockeye Salmon *(O. nerka),* the Coho Salmon has made the greatest gains as a landlocked species in a number of California reservoirs, although it does not breed there and must be maintained by hatchery stocking. In the Great Lakes, however, where the introduction of this species has revived both sport- and commercial fishing in Lake Michigan, several breeding populations have been established in tributary streams. The finest Coho catches I have seen were at the small sportfishing harbor at Port Washington, a small town about 30 miles north of my home city of Milwaukee, Wisconsin.

The hatchery rearing success for the Coho appears to ensure the genetic survival of this species, but like all such programs there is a chance that the end product may be fishes ill suited to the wild state. The hatchery megaproduction cannot maintain the interspecific adaptive diversity found in local populations because they usually rely on a genetic stock of mixed ancestry and thus these fishes are not necessarily well adapted to the sites where they are stocked. Hopefully, that will not be the case, and instead the restocking of the many currently degraded coastal creeks and streams will eventually restore the Coho to its former status among California freshwater fishes.

ANGLING NOTES: The voracious feeding behavior of the Coho Salmon parr carries over to the adults and makes them a favorite with salmon anglers. However, because of the recent protective measures for this species, it is vital that you carefully review the current freshwater sportfishing regulations supplied with the purchase of a fishing license. This in itself can be a challenge because in recent years the regulations booklet reads more and more like an IRS Form 1040 instruction book. As you might expect from the preceding comments, many coastal creeks and streams are closed to all fishing. However, there are some where you can still catch a Coho Salmon on a barbless hook as long as you release it. In a number of cases, these catch-and-release creeks are situated between completely closed creeks, and because many small creeks are not labeled with signs, a good set of county or topographic maps can often save the day.

As with all stream angling for salmon, it is most important to know when large numbers of spawning fishes are entering the streams. For the Coho Salmon, this event may take place from October to February, and annual changes in rainfall and water temperatures can cause significant year-to-year variation. Serious anglers must keep at least a weekly check on their favorite site if they hope to have a good season. The most important factor to keep in mind is that when the water moves well, so do the anadromous fishes. A seemingly fishless stretch of stream near the end of a short winter drought period will come alive again within a day or so after the return of substantial rain. The exact lure choice is usually not critical as long as it shows good action in normal stream current.

Reservoir fishing for planted Coho Salmon is still in its infancy in California, and many questions have yet to be answered. Like most salmonids, Cohos respond negatively to warm water and spend the summer months in the lower reaches of the thermocline, where they may be taken by trolling either a minnow or a lively spoon or lure. One of the major problems in this kind of fishing is, of course, locating the thermocline. This area of rapidly changing water temperature may be quickly detected by a portable electric thermometer with a long, weighted sensor cord. Such units are available at electronics stores for a modest price and add a scientific dimension to fishing. Another excellent comparative temperature sensor is human skin, especially on the fingertips and lips. If a series of large lead sinkers is spaced at 1- or 2-m (3- or 6-ft) intervals on a sounding line and allowed to remain in the water for about five minutes, the first distinctively cooler sinker can easily be detected upon retrieval, thus determining the approximate trolling depth. The fall turnover signals the end of the summer thermocline life for this species, and surface trolling at that time may produce some excellent fishing in Coho-stocked reservoirs.

CHINOOK SALMON or KING SALMON — *Oncorhynchus tshawytscha*

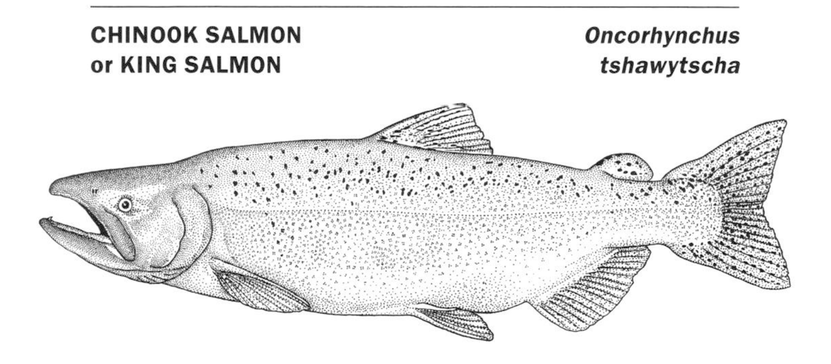

NATIVE

PROTECTIVE STATUS: ST, FT, spring run; SE, FE, Sacramento River winter run; FT, Redwood Creek (Humboldt County) to Russian River (Sonoma County) fall run.

The Chinook Salmon has always been the most abundant salmon species in California, although it is actually the least numerous of

Plate 24. Chinook Salmon.

all Pacific Coast salmon (pl. 24). The relatively new and now official name "Chinook" is attributed to a Native American tribe of the Columbia River drainage whose livelihood was centered around these salmon with which they shared the great river. The old name "King" is still in moderate use today by both anglers and the general public because it remains the common fish-market name. In the category of size, it certainly is a king among salmon species. The current sportfishing record specimen taken in Alaska weighed in at 44.3 kg (97.4 lb), three times the records for Coho Salmon *(O. kisutch)* and Chum Salmon *(O. keta)*. It is also the most familiar salmon species because of its prevalence in fish markets and its starring role at major fish ladder and hatchery facilities in the state.

The many populations of Chinook are defined by their own seasonal spawning run within a specific drainage. Populations whose spawning migration occurs in September and October are referred to as "fall run," and those with a migration peak in December are "late-fall run." A March movement peak is a "winter run," and a fourth, the "spring run," culminates in May and June. Some small streams may have only one run, but large rivers such as the Sacramento exhibit all four.

The fall run is the most common and familiar because the large hatcheries with public viewing facilities operate on this spawning schedule. Under natural conditions, returning adults spawn soon after arriving at upstream gravel-bed sites. Spawning behavior follows closely that described for other salmonids with

Plate 25. Chinook parr.

occasionally one added twist: small yearling males, referred to as "jacks," achieve early sexual maturity during a short time at sea and move in on spawning adult pairs to add their milt to the redd. Selection for such early sexual development and behavior may serve as a way to promote high egg fertility in years when the adult breeding population is low.

Each female lays up to 8,000 eggs, which remain anchored within the gravel-pebble lattice of the redd and supplied with a continuous flow of cool, well-oxygenated water until late winter (January through March). The emerging alevins spend several more weeks within the gravel and then begin moving downstream as yolk sac absorption is completed. The newly developed parr remain in the lower reaches of the stream where they feed on a variety of aquatic insect larvae, adult insect drift, and large zooplankton (pl. 25). They begin the color transformation to smolts in fall of their first year at a size range of 4 to 8 cm (1.6 to 3 in.). This is about half the size at which Coho Salmon smolts go to sea, and it reflects the shorter parr life of the Chinook Salmon. The new sea arrivals grow quickly, and at two years of age they are over 0.5 m (20 in.) total length. Fully mature adults return between the ages of three and five, and the latter age group can be up to 1 m (40 in.) in length. These fully mature adults now begin their final journey back to the bay or estuary of their home stream through the use of sun-compass orientation, possibly subsurface landmark recognition, and orientation to Earth's magnetic field, using iron-bearing capsules in their nasal sacs that act like compass needles. Final location of the parr-rearing site occurs through olfactory memory, but because parr life of the Chinook is considerably shorter than of the Coho and Steelhead *(O. mykiss irideus),* occasional wanderings to other spawning streams occur.

This potential flaw in the homing ability of the Chinook may actually serve as an evolutionary advantage because over time it has probably promoted the reestablishment of runs in creeks that are adequate for spawning only during climatic wet cycles.

In contrast to the fall-run pattern, spring-run Chinooks enter the main river systems in late spring and exhibit peak movement in May through June. Adults summer over in the rivers and spawn in early fall. Under pristine conditions they migrate inland as far as possible into the small creek tributaries of their home drainage. This is possible only in late spring when snow melt runoff swells creeks to depths needed for large-fish movement. Inland journeys of over 630 km (378 mi)have been recorded for these runs. That is comparable to driving from Sacramento to Los Angeles on Interstate 5, but instead of sitting, you are swimming. Once at these distant sites, they remain until September when they spawn and then die. The eggs hatch throughout the winter months. The length of parr life is variable, depending on how well a creek retains water in a given year. When adequate water levels are maintained throughout the year, parr residency may be a year or more, thus producing significantly larger smolts to cope with life at sea.

The late-fall run and the winter run spawning patterns fall between the two just described. The entire run complex represents an intricate natural design for partitioning spawning and juvenile development over the time span of one year and the riverine habitat span of the Pacific Northwest. Unfortunately, past dam construction in addition to ongoing watershed destruction and hatchery manipulation of populations continue to "disimprove" on this design.

The Chinook Salmon has adapted well to hatchery propagation because it has a shorter parr life than either the Coho Salmon or the Steelhead, and it is the only salmon species that has been established outside of North America (in New Zealand). Despite this less than perfect adaptability, hatchery propagation remains the only current solution for run preservation in those situations where the major pathways to ancient spawning grounds have been blocked (pls. 26, 27). Besides the primary function of perpetuating the Chinook populations of the drainages where they are constructed, hatcheries also provide us with one of the few close-up views of a large, wild California freshwater fish and the spectacular and unique vertebrate life cycle it displays. The

Plate 26. Chinook Salmon eggs in incubation tray.

Plate 27. Chinook egg trays in water flow rack.

section "Freshwater Fish Watching in the Wild" gives the locations and suggested times for visiting these sites, and addresses and directions for California hatcheries are listed at the back of the book. When viewing these facilities bear in mind that they are quick and hopefully temporary fixes to the salmonid habitat-loss problem. Perhaps they can buy enough time to allow future and wiser generations to correctly restore many lost segments of our original salmon habitat complex.

ANGLING NOTES: The Chinook Salmon is still "king" of salmon sportfishing in California despite its slow but steady decline in overall numbers. As with all salmon fishing, success lies in a knowledge of what the fishes are doing on a particular day or week. For instance, fall-run Chinook Salmon are noted for their tidal runs into the mouths of prime spawning rivers such as the Klamath, starting in July, whereas the main run in the Feather River west of Oroville is in full swing by late September. Anglers fish from either boat or shore, and when the fishes are there, it is a crowded affair, often with boats and anglers positioned literally side by side. In such situations, the type of lure or bait used is not as important as the type of rod and the weight of line. This is not the place for light tackle. Contrary to the sound sporting outlook that most anglers have, in these circumstances a hooked Chinook should be brought in as soon as possible using a stout rod and heavy test line. The reason is obvious: an attempted long fight with light tackle will certainly result in tangling your line with the lines of your all-too-close neighbors, resulting in not only the loss of the fish, but most certainly incurring the wrath of your fellow anglers.

For light tackle enthusiasts, there are many miles of good upstream water in many spawning streams where you can usually find a stretch all to yourself. Here the technique is much the same as that described for fishing for Coho Salmon. Regarding preferred lures, ask for tips from the local bait store or fellow anglers. Lure preference varies greatly from year to year, or even month to month, and despite much research, scientists are still unsure why fish lure preferences are erratic. Because the homing ability of this species is sometimes less than perfect, some urban creeks have reestablished a small fall-run population. One example is Walnut Creek, for which the city in the East Bay is named. Despite ongoing alterations to the creek by the local flood control district and a devastating aviation fuel spill in the early 1990s, meter-long chinooks still occasionally push their way as far as the first major creek barrier in the city's downtown. If you take the time to search the few remaining gravel substrate sites, you may occasionally find a lone redd, a poignant reminder that the king has not yet conceded the loss of his former realm.

I devote a fair amount of space in this book to fishing because for decades fishing has been the number one sport in the United States on the basis of participation and total dollars spent. However, some people who prefer not to fish nevertheless like to see fishes in their natural habitat. A similar difference exists between bird watchers and bird hunters. Unfortunately, freshwater habitats provide relatively few good opportunities for fish viewing because of less-than-favorable water clarity. However, migrating salmon in shallow creek water are a very visible and even spectacular sight. Besides the great size of these fishes, their ability to move through riffles only a few centimeters deep is amazing.

Small spawning streams immediately below dams on coastal rivers are excellent venues for salmon watching. At sites such as the mouth of Bogus Creek, just below Iron Gate Dam and Hatchery on the Klamath River a few miles east of Interstate 5, you can sit on the river's north bank opposite the creek entrance and watch as the salmon leave the river and thrash into the creek. September and October are perfect months to enjoy such an experience. For other sites, consult the section "Freshwater Fish Watching in the Wild." However, anglers and fish watchers alike should not take Chinook Salmon on the Klamath River for granted. At the end of the unseasonably dry spring and summer of 2002, the Secretary of the Interior decided farmers should have more than their allotted share of the river's water and personally opened the headgates of a dam above Klamath Falls, diverting about 146 billion gallons into irrigation ditches. This resulted in the thermally induced death of 35,000 early migrating chinooks, one of the largest salmon kills on record. A second wave of approximately 65,000 migrants faced a similar fate due to the low, warm river water, but an eleventh-hour FedEx delivery of 227 kg (500 lb) of rotting salmon, courtesy of the local Yurok and Hoopa Native American tribes, to the Interior Department in Washington, D.C., finally got the government's attention. Within days "spare water" was suddenly discovered and used to flush the river, thus saving these fish.

KOKANEE SALMON or SOCKEYE SALMON *Oncorhynchus nerka*

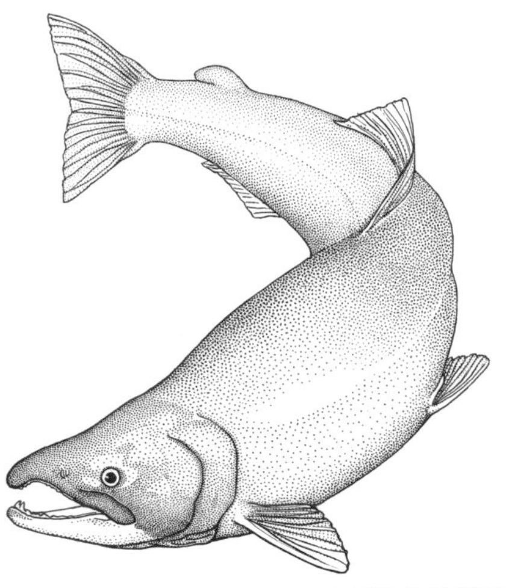

NATIVE/INTRODUCED

Although the Sockeye Salmon is the second-most abundant salmon species on the west coast of North America, only the occasional stray finds its way up a north coast stream in California. Such wanderers are apparently experiencing the same lack of homing ability exhibited by a small percentage of Chinook Salmon *(O. tshawytscha);* they never spawn here. On this basis alone, the anadromous Sockeye deserves only slight mention in a book on California fishes. However, in 1941, Kokanee Salmon from Idaho were introduced into a small northern California reservoir. The Kokanee is a landlocked race of Sockeye Salmon, so it never goes to sea. In 1951, Lake Shasta was stocked with Kokanee from Kootenay Lake in British Columbia, and since that time a hatchery program for Kokanee Salmon has developed in California. Today, numerous reservoirs and natural Sierra lakes contain this fish. Lake Tahoe is the best known of these, and the annual Kokanee spawning run from this lake into Taylor Creek has garnered much public attention in recent years. The Kokanee Salmon has even taken up an urban existence with its introduction into Lake Merced in the middle of "the city" in San Francisco County.

Unlike lake-stocked Coho Salmon, Kokanee Salmon have adapted to spawning both in feeder streams and in the inshore gravel beds of lakes. The spawning process closely follows that of other salmon species, except that both parents vigorously defend the redd against intrusion. An average of 800 eggs are deposited and fertilized, after which both partners die. Eggs laid in midfall hatch in late winter or early spring, depending on water temperature, and three weeks later alevins emerge from the redd and float down the spawning creek to the lake, where they transform into parr.

A mature Kokanee Salmon falls far short in adult size compared to its genetic twin, the seagoing Sockeye Salmon. A one-year-old Kokanee Salmon averages about 18 cm (7 in.) and when food is plentiful may mature at a little over 50 cm (20 in.) and 1.5 kg (3.3 lb). The California record Kokanee, taken in Lake Tahoe, was 2.2 kg (4.8 lb). Although this may seem impressive, it is less than half the average weight and length attained by the Sockeye: 65 to 80 cm (25 to 31 in.) and 4 to 7 kg (9 to 15 lb). The reason for this marked difference is food availability, and it vividly demonstrates the advantage of the anadromous lifestyle. Sockeye Salmon and Kokanee Salmon are essentially created equal, especially with respect to food preference and gill rakers. The former dictates that they eat mostly zooplankton, and the latter, which are long and closely spaced, allow them to filter feed on this food source. Whereas the Pacific continental shelf is noted for its rich zooplankton supply, the reservoirs and mountain lakes of California are not. Coupled with this environmental reality is the fact that fishes are indeterminate growers, allowing individuals within a species to sexually mature and breed at sizes considerably smaller than normal adult size. This syndrome is often seen in farm ponds stocked with highly fecund sunfish species such as the Bluegill *(Lepomis macrochirus)*. This direct link between size and food supply makes it easy to see that anadromy accrues a significant selective advantage to the evolving salmon species. With the acquisition of a homing ability, these prehistoric seafarers became able to return to spawn with a far greater egg capacity and vigor than those who stayed behind.

Unfortunately, the scientific worth of comparing two lifestyles within the same species is offset by the impact to the ecology of most lakes caused by Kokanee Salmon introductions. This species is a planktivore throughout life and, as seen in fig. 3, occupies the same secondary consumer position in lakes as the fry of nearly all other species. Most of these vacate the zooplankton feeding niche as they mature, but not the Kokanee. California reservoirs, and especially Sierra lakes, are not noted for their abundance of zooplankton, especially when compared to the eutrophic lakes of the middle and eastern United States. In Kokanee Salmon–dominated lakes, the basic rule of zooplankton food supply and planktivore demand is a major force affecting the welfare of all species. Several native California fishes in these lakes, primarily members of the minnow family (Cyprinidae), have been dealt yet another blow by the introduction of the Kokanee Salmon into their habitat.

On the other hand, one highlight of Kokanee Salmon introduction is the opportunity it affords the public to view the salmonid reproductive phenomenon in a small, easily accessed area without the need for a salmon hatchery. An excellent place to witness this is in Taylor Creek, a tributary of Lake Tahoe. In early fall, Kokanee emerge from the lower thermocline where they have spent much of the summer and congregate at the mouth of Taylor Creek and other spawning creeks in the same manner that seagoing salmon aggregate in estuaries before swimming inland. At this time males undergo a striking color change, acquiring a deep red body hue and a black head. You can view these fishes both in the creek and at a U.S. Forest Service underwater stream observation facility (see the section "Freshwater Fish Watching in the Wild"). The photo in pl. 28 was taken by a professorial colleague in a shallow segment of Taylor Creek using a partially submerged aquarium as an underwater camera housing (see "Freshwater Fish Photography," pl. 98). Patches of deteriorating skin can be seen on the otherwise brilliant scarlet breeding body coloration of one Kokanee Salmon male, whereas another in the background has just begun the nutrient recycling process.

ANGLING NOTES: Kokanee Salmon fishing continues to attract California anglers, and fishing derbies for this species are held at some foothill reservoirs. These can be challenging events because the reservoir-dwelling Kokanee presents a dual problem to anglers: determining the depth of the thermocline where it is usually found, and choosing an appropriate bait. Like the Coho Salmon, the Kokanee Salmon remains in deep water throughout the warm months. In Lake Shasta, it has been known to go so deeply into the hypolimnion, where both temperature and oxygen levels are low, that occasionally, small numbers of fishes suffocate (see fig. 2). Therefore, determining the depth range of the thermocline by methods suggested for Coho fishing is a must.

Plate 28. Kokanee Salmon male after spawning.

Because it is a filter-feeder, the Kokanee is not a voracious lure grabber like its piscivorous Coho cousin and therefore requires a different bait. Trolling in the lower thermocline with small lures may occasionally produce moderate to good catches during summer, but the Kokanee's preferred food is zooplankton, especially the species group known collectively as *Daphnia*. Individual *Daphnia* are only about half a millimeter long, and even the most skilled fly tier would most likely fail to successfully reproduce it (although I know some who would try!). What kind of bait, then, will a planktivore consistently take?

I am sorry to report that I do not have the answer, but I do know some foothill-reservoir anglers who achieve fairly consistent results with corn—not on the cob but instead presented by trolling a kernel or two on a small hook with spinner or on the hook tips of a favorite lure. Some Kokanee anglers prefer a long gang of large spinners and flashers ahead of their corn-baited hook. Still others, not content with the natural product, dip the corn kernels in a commercial bait scent or even brew up their own. As a scientist, I find it interesting that corn pops up as a recommended bait for a wide variety of fishes. I can only speculate that something in its molecular composition closely resembles the olfactory signals that certain foods, possibly zooplankton, emit. If you do try corn kernels, keep in mind that you are trying to catch a filter-feeding planktivore, which normally scoops its small invertebrate prey, as opposed to a piscivore, which vigorously grabs its small-fish prey. You must therefore resist the impulse to quickly "strike back" and instead let the Kokanee run with the bait, giving the bait time to move well into the mouth and past the long, filamentous gill rakers. Then you can gently but firmly set the hook. At any rate, it may be worth picking up a dozen ears of freshly picked corn at a roadside stand on the way to your next summer Kokanee fishing trip. Even if the salmon are not biting, the corn will provide pretty good vegetarian fare in camp that night.

PINK SALMON and CHUM SALMON

Oncorhynchus gorbuscha* and *Oncorhynchus keta

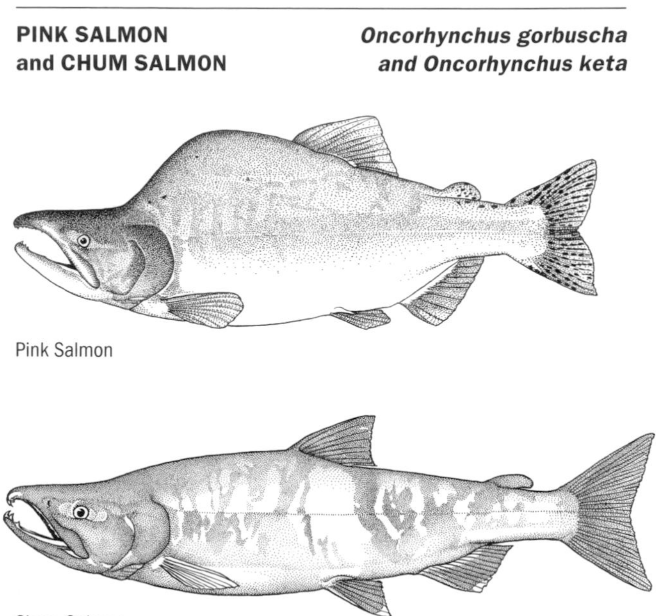

Pink Salmon

Chum Salmon

NATIVE

Like the Sockeye Salmon *(O. nerka)*, these two species are rare in California rivers, although historically a number of small runs may have occurred in several northern California rivers. Today, the occurrence of occasional specimens in these drainages can be attributed to the same sort of "off course" wandering seen in the Sockeye Salmon and nonhoming Chinook *(O. tshawytscha)*. In addition, small annual runs of the Chum Salmon persist in the

Klamath, Smith, and Trinity Rivers. Neither species has been successfully introduced into western lakes and reservoirs, but introductions of the Pink Salmon into Lake Superior have spread to all five Great Lakes and now provide a new sport fishery there along with introduced Coho Salmon *(O. kisutch),* Chinook Salmon, and Steelhead *(O. mykiss irideus).* Despite their near-total absence in California, the Pink Salmon and Chum Salmon are the first- and second-most abundant northeastern Pacific salmon species, and for that alone deserve some mention in this book.

The Pink Salmon is the most marine of the Pacific salmon and usually spawns in the lower reaches or near the mouth of coastal streams, sometimes even in intertidal waters. Spawning takes place in fine gravel beds where a female may construct several redds to accommodate all of her 1,500 or so eggs. Between bouts of spawning, the male vigorously guards the redd site, and with his sharply pointed and hooked upper jaw can inflict serious punishment on intruders. Both parents die within a few days of spawning in early fall. Eggs hatch about five months later, but alevins do not emerge from the redds until the yolk sac is fully absorbed. At that time they begin a floating migration downstream where the fry spend several months feeding in the nutrient-rich river estuaries. Notice that I have not used the word "parr" here, because in this one species of western salmonids, parr marks never develop (see parr illustrations, p. 91). Instead, the fry body and fins have the smoltlike silver coloration that is apparently more advantageous for estuary life. They go to sea at a very small size (about 5 cm [2 in.]) and after growing rapidly begin feeding on small fish, squid, and large krill species. They return to their home stream near the end of their second year, the shortest sea life of all salmon species. At this time they range in size up to but rarely beyond 60 cm (about 2 ft) and 2.5 kg (5.5 lb).

The Chum Salmon exhibits a more normal life span of three to five years. The migrating adult is less able to traverse small, natural obstacles in spawning rivers; therefore, most populations spawn in the lower reaches of these streams. When small falls or other barriers are not present, Chums have been known to migrate great distances inland. In the Yukon River, for instance, they have been known to move upstream about 2,400 km (1,500 mi), but this is exceptional. Different runs of this species may spawn from summer to late fall, and hatching time varies from two to six months. When spawning takes place far inland in creeks with few barriers, the resulting parr behave much like those of Chinook Salmon and feed for several months while moving downstream. Once at sea, the diet of the growing adult Chum differs markedly from that of other Pacific coast salmon. It feeds primarily on the medusa forms of jellyfishes and the closely related ctenophores. This is a good example of feeding-niche division in a habitat utilized by several closely related species; however, the Chum Salmon can also adapt to other foods when its preferred jellyfishes are scarce. Mature adults average about 65 cm (25.6 in) in length and 6 to 7 kg (13 to 17 lb).

The Chum Salmon is the most abundant (in biomass) of all salmonid species. It is the principal species used for "ocean salmon ranching." Between four and six million fry are released annually in Alaska, Japan, and Russia. If only 1 percent return to be harvested, the catch would total 40 to 60 million adults (about half a billion pounds). Fishing techniques for these two species are essentially the same as for Coho Salmon and Chinook Salmon, although the Chum seldom takes a bait. Both species are rarely caught in California.

Trout

Because trout are such an ecologically and economically important group of fishes in California, a brief introduction to this segment of the trout and salmon family seems appropriate. The study of trout is a complex discipline, especially the taxonomy. Between 1792 and 1972, some 50 species of western trout were described and named in the ichthyology literature. With a better understanding of western geological history and with vital new information from protein electrophoresis and mitochondrial DNA analysis, Robert J. Behnke, a world authority on trout taxonomy and systematics, was able to reduce this number to just five species by 1991 and, finally, to four just one year later. Dr. Behnke also emphasizes that no one has all of the correct answers to this ichthyological puzzle, but I choose to follow his classification system in this book. I might be accused of personal bias in this decision, because it was Bob who introduced me to California ichthyology when we were both Ph.D. candidates at the University of California at Berkeley in the early 1960s. He has devoted his entire academic life to the study of trout. Even when he takes a rare break from his laboratory work, he goes trout fishing. With that kind of devotion to his subject, his taxonomic views cannot be too far out of line.

Setting the stage for the recent evolution of California trout were geologic events in the late Pliocene, about two to three million years ago. These triggered large-scale extinctions and sweeping changes in the distribution of freshwater fishes west of the Rocky Mountains. In particular, this period saw an extinction of western catfishes and bullheads (Ictaluridae) and all but one member of the sunfish group (Centrarchidae) in California. This left open some very broad feeding niches that were soon filled by the evolving western trout species. About this time, the ancestral western trout gene pool divided into two segments: one led to the Cutthroat Trout *(O. clarki)*, and the other to the Rainbow Trout *(O. mykiss)*.

Regardless of the evolutionary routes that present-day California trout took to achieve their current position in this state, one thing is clear: no other fish has dominated California sportfishing in general, and the California Department of Fish and Game in particular, as has the Rainbow Trout.

RAINBOW TROUT *Oncorhynchus mykiss*

NATIVE

The Rainbow Trout in California is represented by the following six subspecies: Coastal Rainbow Trout or Steelhead *(O. m. irideus)* (Steelhead has Federal Endangered/Threatened status), California Golden Trout *(O. m. aguabonita)*, Little Kern Golden Trout *(O. m. whitei)*, Kern River Rainbow Trout *(O. m. gilberti)*, Sacramento Redband Trout *(O. m. stonei)*, and Eagle Lake Rainbow Trout *(O. m. aquilarum)*.

Because of a variety of behavioral and physiological adaptations to a wide spectrum of habitat conditions, including those of fish hatcheries, the Rainbow Trout has emerged as the most abundant trout species in California and the best-known and most commonly introduced trout in the world (pl. 29). Indeed, when you order "trout" in a restaurant or buy it at a fish market or supermarket, it is "a rainbow." Despite this apparent familiarity, a widespread misunderstanding persists as to just what a Rainbow Trout is.

Plate 29. Coastal Rainbow Trout, juvenile.

A good example is the confusion behind the question: What is a Steelhead? Answer: It is an anadromous **COASTAL RAINBOW TROUT** ***(O. m. irideus)*** that is silver, not "rainbow" in color and that after two years at sea can grow to over 9 kg (20 lb) compared to the stay-at-home Coastal Rainbow Trout, which usually weighs less than a pound (0.45 kg) after the same amount of time. The Steelhead often spawns in the same creeks where the nonanadromous Coastal Rainbow Trout is found. The Steelhead parr spends an average of two years there during which time it grows to 15 to 20 cm (6 to 8 in.). It then smolts and migrates to the sea, where it roams thousands of miles for anywhere from 15 to 30 months before returning to the home creek in either a fall or a spring run. A fall-run adult spawns in early spring and then returns to the sea, but a spring-run fish is usually not sexually mature and must wait in deep portions of the creek habitat until it is able to spawn the following spring. Unlike other anadromous salmonids, the spawning adult does not die but returns once again to the sea; however, tagging studies have shown that only about 10 to 20 percent return for a second spawning. This low return rate probably reflects natural predation at sea rather than a delayed spawning-related death. Despite the rigors of life at sea, some individuals become very skilled at predator avoidance and return up to four times to spawn.

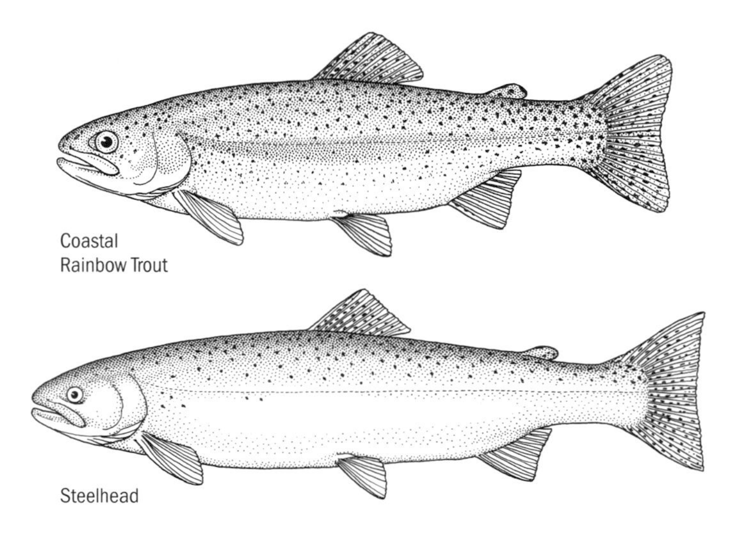
Coastal Rainbow Trout

Steelhead

The Steelhead was the first anadromous salmonid to be established in the eastern half of the United States. The early propagation of Coastal Rainbow Trout in California led to the stocking of streams in the northern Midwest. These trout brought with them the genes for anadromy, and soon a Steelhead population established in Lake Superior and spawned in its feeder creeks.

In contrast to the Steelhead and its dynamic life history, its genetic near-twin, the nonanadromous, naturally occurring Coastal Rainbow Trout, passes through all of its life history stages within the confines of the creek habitat. In this habitat, the parr compete with those of the Steelhead and occasionally with Coho *(O. kisutch)* and Chinook Salmon *(O. tshawytscha)* parr for the invertebrate food resource. As the stream-dwelling Coastal Rainbow Trout approaches sexual maturity, it loses its parr marks but does not acquire the solid silvery color of the young seagoing Steelhead. Instead, the familiar diffuse, reddish, longitudinal band develops along each side and is flanked by black-spotted olive green above and an essentially spotless silver white below. The primary food for the Coastal Rainbow Trout in streams is drift, and individuals set up and defend feeding territories. These territories are often both intraspecific and interspecific, because other salmonids appear to recognize and react to the stiff-swimming and opercular-flaring aggressive displays of the Coastal Rainbow. Such displays are often well developed by the second year of parr life, and along with their larger size these second-year fishes hold their own with the more numerous Coho parr with which they often share a creek.

Between two and four years of age the Steelhead and the Coastal Rainbow Trout become sexually mature and spawn for the first time. Except for the difference in the number of eggs they lay (up to 12,000 for the Steelhead versus about 1,000 for creek-resident Coastal Rainbows), spawning is essentially the same. The female scoops out a redd with her tail in an area of coarse gravel and good water flow, and after the pair spawns, the eggs are covered by loosening gravel from the upstream edge of the nest. In Steelhead, as in Chinook spawning, occasional "sneaker males" also get in on the spawning action. These can be jacks (early returning anadromous males) or even sexually mature parr that have never gone to sea. Hatching occurs within a month, and alevins emerge several weeks later. By late spring the new parr have formed small shoals in creek pools where they begin their initial growth before going to sea as migrating Steelhead smolts or attaining adult size as resident Coastal Rainbows.

Returning to our original question, "What is a Steelhead," perhaps an alternate answer would be that it is a Coastal Rainbow Trout with a midlife pattern different from that of other trout species. Other than that, the genetic differences between the Steelhead and the Coastal Rainbow Trout are less than the differences between other Rainbow Trout subspecies from different major drainages. In technical terms, these are "sympatric" (living together) races of the same subspecies. Despite all such attempts at clarification, I still have to remind myself of this close subspecies affiliation when I see a large, silvery Steelhead from the Olympic Peninsula of Washington and a small, highly colored hatchery Coastal Rainbow Trout displayed together in a fish market.

The expression of this subtle genetic difference was vividly revealed to me some years ago by the following incident. Before Steelhead in California were listed as a Federal Threatened species, I collected a small parr under permission granted by my California Department of Fish and Game permit for display in my ichthyology course lab. It put on a great show, feeding on live brine shrimp until its little belly looked like it would burst. It was doing so well by the end of the fall class that I decided to keep it around for another year of instructional displays. It was housed in a refrigerated aquarium in a windowless room with other tanks containing fishes and invertebrates. The room lighting was on a 12-hour timer, and room air was a constant 20 degrees C (68 degrees F). A student attendant fed it at two-day intervals. The parr grew well and after its second year in captivity was over 6 in. long. When I returned for the fall term that second year, I went to the tank room to see how the parr, which had by now ascended to the level of a pet, was doing. Upon entering the room I saw not my beautiful speckled old friend but instead some sort of flashy, silver fish that someone had apparently put in its tank. What had they done with my Steelhead parr? It was not until I got near enough to the aquarium for my near-sighted vision to better examine this intruder that I realized this indeed was the Steelhead parr—it had simply smolted, as it should have at the end of its second year. And it did so under the ad hoc controlled environment of the tank room. During its nearly two years in captivity, the pages of some internal biologic calendar had been turning independent of external cues.

We cannot leave a discussion of the Coastal Rainbow Trout without a reminder that this subspecies is the foundation of the California salmonid hatchery program. Since the first hatchery rearing of Steelhead in 1887, hatchery-derived varieties of the basic Coastal Rainbow Trout and Steelhead have been introduced around the world, both as an aquaculture and as a sportfish species. Indeed, the first Steelhead I caught was in 1958 on a small creek in the Porcupine Mountains along the south shore of Lake Superior in Upper Michigan.

To fully understand the effects of the California hatchery system, we must keep in mind that the hatchery form of the Coastal Rainbow Trout is a cultigen, a variety cultivated by humans and not found naturally in the wild. The original hatchery Coastal Rainbow is believed to have come from a naturally occurring hybrid resulting from the movement of resident Coastal Rainbows from the Upper McCloud River over the falls to the Lower McCloud River where they crossed with Steelhead spawning in that area. Since the establishment of this first hatchery variety at several sites in the San Francisco Bay Area and the first hatchery on the McCloud River in 1880, there have been numerous additional genetic modifications, and at present seven distinct hatchery cultigens of Coastal Rainbows exist in California. For well over a century these various crosses have been planted into most California streams and lakes, often with devastating repercussions for native trout and other naturally occurring animal species (pl. 30).

Because the introduced cultigens have interbred with naturally occurring populations of Coastal Rainbows, it is unlikely that we will ever know the true nature of many of the original populations. Perhaps the most dramatic example of this has been

Plate 30. Hatchery Coastal Rainbow Trout (above) and Eagle Lake Rainbow Trout cultigens.

the hybridization of the Coastal Rainbow cultigens with the official California state freshwater fish, the Golden Trout. All populations of this subspecies subjected to the introduction of hatchery cultigens have been lost to hybridization. The adverse effects of mountain stocking of hatchery trout do not stop at the fishes. Even in previously fishless Sierra lakes where Coastal Rainbow cultigens have been stocked, there has been a marked reduction or elimination of populations of the Mountain Yellow-legged Frog *(Rana muscosa)*, a species currently proposed for listing as Federal Endangered.

The production of hatchery Rainbow Trout, with their wide range of tolerances to habitat conditions, has culminated in a trout-dominated program for both the California freshwater sport fishery and the California Department of Fish and Game.

As of 2002, California had 28 trout, Steelhead, and salmon hatcheries in California, 20 of these operated by the CDFG. Twenty years ago when the first edition of this book was published, two warm-water (nonsalmonid) hatcheries remained, but these have since fallen by the wayside. We are now an "all salmonid state" (see "State Fish Hatcheries" at the back of this book for a complete list). Each year these hatcheries produce millions of salmonids, both catchable size (19 cm [7.5 in.] or larger) and fingerlings, for

Plate 31. Hatchery feed dispersal truck with trout breaking water.

placement in California lakes and streams at a cost of many millions of dollars (pl. 31). Numerous pros and cons to such a one-sided type of management program can be cited, and only a few of the more apparent features are discussed here.

Trout hatcheries are most defensible for fingerling, or "put and grow," stocking in lakes and reservoirs where natural reproduction does not occur. If conditions are favorable for fingerling growth and maturation in such habitats, a reasonable supply of trout can be sustained for the angler at a comparatively moderate cost. However, of the 40 percent of the total CDFG budget devoted to hatcheries, only a small amount is devoted to put-and-grow stocking. Instead, approximately 97 percent of the trout biomass produced in hatcheries is in the form of catchable trout for put-and-take stocking.

We may easily appreciate the attractiveness of such a massive stocking program both to the CDFG and to much of the general public. At first glance, there is no greater tangible result of a fisheries program than a tank truck backing up to the shore of a lake and dumping 10,000 catchable trout. Unfortunately, the economics of "put-and-take" stocking in California are not so impressive. The state's hatchery program receives over 50 percent of the fishing license fee money, but put-and-take trout support only 7 percent of the total angler days in California. Furthermore, less than 10 percent of the anglers harvest 50 percent or more of all stocked catchable trout. These findings, derived by Dr. Robert J. Behnke from CDFG data, strongly suggest that a new breed of angler, specializing in put-and-take trout, has arisen as a result of this program.

Two additional questions should be asked. First, are we continuously expanding our hatchery trout program at the expense of the ever-increasing warm-water fishery in California? Except for the occasional stocking of species, some of which are not the best choices for a given habitat, many of California's reservoirs remain unstudied and unmanaged. If only a modest portion of the present hatchery efforts in California was redirected toward improving littoral zone conditions of reservoirs, for example, the current warm-water reservoir program could be significantly expanded.

Second, is it possible that the put-and-take stocking program, and the trout angler it attracts, are caught up in a snowballing cycle in which the lure of instant trout attracts more anglers who, in turn, will need more trout, and so on? Left without a put-and-take-program, or with a reduced one, the anglers of California might distribute themselves more equally throughout the wide-ranging warm- and cold-water sport fisheries that this state offers.

And what is the indirect effect of put-and-take stocking on the aesthetic value of the fishing experience? Since my youth, I have been an avid reader of fishing magazines and other popular fishing literature, including newspaper sportfishing columns. In recent years, I have found several of the latter quite disturbing. Instead of describing the beauty and natural challenges of the area to be fished, they instead focus on the techniques of finding out exactly when and where the next planting of catchable trout will take place and how to plan your next trip so as to outmaneuver your fellow put-and-take anglers in acquiring your fair share.

And because fishing is a true family sport, is this the "fishing experience" that future generations will grow up with and carry on? My youthful remembrances of fishing are of the many trips with my father and teenage friends to beautiful Wisconsin lakes and streams where the real challenge was to try and figure out where the fishes were and then how we might catch a few. Once in a while these efforts resulted in impressive catches of Northern Pike *(Esox lucius),* Yellow Perch *(Perca flavescens),* Largemouth Bass *(Micropterus salmoides),* and the ever-present Bluegill *(Lepomis macrochirus),* but more often we returned with empty or near-empty stringers. Many years later, I have vivid memories of the challenge of the natural fishing experience, but little memory of the actual numbers of fishes we caught.

Recently, while conducting a brief fish survey on a Central California reservoir, I recognized a shoreline angler as a staff member at my university. He was accompanied by his two teenage children, and they were bait fishing from the barren, sun-baked shore of the drawn-down reservoir. They occasionally pulled in a plump hatchery Rainbow, most likely from a planting the day before. When I finally approached to say hello, the kids proudly displayed their catch, and there was no doubt that this was a most enjoyable family outing. However, I could not help but feel that a significant part of the fishing experience was lacking for these folks, and I imagined what their experience could be if at least some of these reservoirs were managed to promote productive literal zones and naturally reproducing populations of both native and introduced game-fish species.

If the current put-and-take program was significantly reduced, withdrawal pains of the relatively small number of anglers who rely heavily upon it could be partially alleviated by encouraging the expansion of pay-and-fish trout lakes. For many decades, Type B pheasant hunting clubs in California have operated on a plan whereby hunters pay for hatchery-raised birds that are released just prior to the "hunt." Well-managed pay-and-fish trout lakes could also serve to stimulate the very young angler whose attention span is not yet suited to wild-fish angling.

In recent years, a different type of trout management has appeared, aimed at sustaining wild trout populations rather than continuously supplementing them with expensive hatchery fishes. In designated segments of creeks and streams where wild populations still maintain themselves, the catch limit has been reduced from the usual five per angler to three or two or, at some sites, zero. In addition, only artificial lures with single, barbless hooks may be used. Such regulations usually result in fewer fishes actually being caught and encourage the release of unwanted fishes in good condition, successfully avoiding the hatchery syndrome.

An important step toward preserving and restoring the coastal stream habitats of California was the 1996 Federal Threatened listing of all Steelhead populations in three geographical districts (Central California coast, south-central California coast, and the Sacramento River basin) and the Federal Endangered listing for populations in the southern California district. As with a similar listing for Coho Salmon, this affords direct protection not only to the Steelhead that breed there, but to their breeding habitat as well. The listing mandates that the U.S. Fish and Wildlife Service and National Marine Fisheries Service assume a watchdog role with respect to these creeks and streams and ensure that any proposed land use changes within their watersheds will in no way affect these habitats. Furthermore, in those cases where such activities continue to adversely impact these sites through siltation of spawning gravel and parr pools, creek tree canopy removal, creek course gravel removal, cattle intrusion, and the like, these agencies are empowered to prosecute and seek payment for restoration from offenders. In other words, the hope is that the many practices that have degraded our coastal waterways for well over a century will now cease. If administered correctly, this federal protection should prove a landmark effort in California anadromous salmonid preservation.

Unfortunately, such correct administration is not always forthcoming. For instance, in fall 2002, a massive (35,000) kill of adult Chinook Salmon occurred in the lower Klamath River. The cause was the diversion of Klamath River water to agricultural areas instead of maintaining an adequate flow for migratory salmon as mandated by the federal Endangered Species Act. Hopefully, this will be a wakeup call for all those who control reservoir stream releases throughout the state as well for as the U.S. Department of the Interior, the agency with federal jurisdiction in the matter and the one that made this disastrous decision.

ANGLING NOTES: Trout fishing and Steelhead fishing are sciences unto themselves, and any attempt to cover either subject fully in a broad-spectrum book such as this would be foolish. My best advice is to read one or more of the many good books written on these subjects and, if at all possible, talk at length with a trout angler from the area you intend to fish. Perhaps the best service here would be to remind the reader of a few basic bits of Rainbow Trout behavior that must be the focal point of any fishing strategy.

In stream fishing for Rainbows, you must bear in mind the innate tendency of this fish to secure good cover, either on the downstream side of large rocks in riffle areas or, to a lesser extent, beneath root or log cover in pool areas. The undercut bank is also a prime cover area, and each section of a stream should be checked carefully to see if such a situation exists. A knowledge of what the trout in a particular stream are eating is also important, and there is no better way to find out than to catch those first few fishes by trial and error and look in their stomachs; in California streams, two of the most common finds are mayflies and caddis fly larvae. Many novice trout anglers are easily swayed by the beautiful assortment of dry flies offered by the local bait store. Most of these are actually constructed for trout fishing in the East, where a great variety of drifting adult insects is far more common. Although in many of this state's streams adult insects such as stone flies are more important as trout food than larvae, more attention should be paid to the dull, rather bedraggled little offerings that more closely approximate the most abundant food items. Such nymphs, as they are usually called by fly manufacturers, when drifted around the edges of large riffle rocks, are by far the best bet in most streams.

When lake fishing for Rainbows, review the comments above for fishing the lake-dwelling Coho Salmon. Preference for the cool water in the lower thermocline causes the Rainbow to avoid the surface during the summer months. Thus you must first determine the thermocline depth and then troll with spinners or bright, active spoons or lures. Live-bait minnows are also a good bet in reservoirs that permit live-bait use, because the Rainbow feeds during much of the year on the numerous Threadfin Shad *(Dorosoma petenense)* in these habitats. In summer the threadfins remain in the epilimnion, and you might surmise that a stray small fish in the lower thermocline would be greeted eagerly by these large piscivores. Also bear in mind that the fall turnover changes this entire pattern: in fall, surface techniques ranging from dry flies to lures are the main means for taking Rainbows.

As for Steelhead fishing, a first glance at the many pages of fine print in the current California sportfishing regulations might give you the initial impression that you can no longer legally catch a Steelhead in California. But do not give up! Get out the magnifying glass and keep reading. In a number of coastal creeks within and north of the protected Steelhead districts, any species may be caught on a barbless hook and then immediately released. In other creeks, you may actually keep a Steelhead if it has a healed scar where the adipose fin was clipped before being released as a smolt from a hatchery. In other words, given the magnitude of the California salmonid hatchery system, hatchery Steelhead cannot be designated threatened or endangered.

As for the actual planning of your Steelhead trip, knowledge of when and where the fishes are moving is of prime importance. Calling resort or bait shop owners in a prospective fishing area, reading fishing reports in the daily paper or on the Internet, and watching fishing news programs on radio and TV are advisable for the Steelhead angler, who must drive considerable distances to fish. For decades, the fall run has been the favorite of most anglers, but remember that Steelhead tend to taper off in their feeding as the early spring spawning time approaches. Summer runs have become more popular with the catch-and-release anglers. In rivers such as the Klamath, Trinity, and the middle fork of the Eel, spring-run Steelhead take up temporary residence until early the following spring when they spawn. During the early part of this stay, they continue to feed on both insects and small fishes, presenting a variety of angling possibilities.

REDBAND TROUT (*Oncorhynchus mykiss* subsp.)

NATIVE

Now that we have had our crash course on Steelhead and Coastal Rainbow Trout, we know all about Rainbow Trout, right? Well, not exactly. The real confusion is yet to come, especially when you start reading about Rainbow Trout and begin to see the name "Redband Trout" popping up here and there. This generic term applies to nearly all other subspecies and races of Rainbow Trout that have a very well-defined, broad, lateral red band. The best known subspecies in this group is the California Golden Trout.

Plate 32. Golden Trout.

The **CALIFORNIA GOLDEN TROUT** ***(O. m. aguabonita)*** is the official freshwater fish of California and the most colorful of the western salmonids (pl. 32). It is a splendid example of a species resulting from mountain stream isolation in ancient California. Originally, it occurred in only a few streams in the upper Kern River at elevations of about 2,000 m (7,700 ft). However, when David Starr Jordan, father of California ichthyology, originally de-

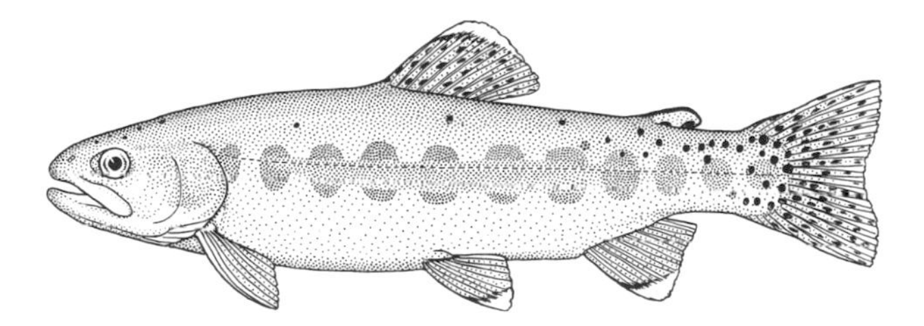

scribed this species in 1892, the specimen with which he worked was taken from Cottonwood Creek in Inyo County. This in itself was not a unique event, for the 1800s was the era of discovering and naming new species. In this case, however, Dr. Jordan's type specimen was the result of an ad hoc introduction in 1876 in which 13 Golden Trout from Mulkey Creek on the west side of the Continental Divide were transported over this famous boundary in a large coffee pot and poured into Cottonwood

Creek. But this fish story does not end here. The Mulkey Creek Golden Trout population was a result of an earlier introduction, in 1872, of the Golden Trout from its native waters, the South Fork of the Kern River. It appears that the urge to move fishes around in California is not new.

Unfortunately, the history of the Golden Trout and introductions that affected it did not end in the nineteenth century. Numerous stockings of hatchery Rainbow Trout cultigens in the Kern River's south fork have resulted in hybrids between these two forms. The widespread introductions of Brook Trout *(Salvelinus fontinalis)* from the eastern United States and the Brown Trout *(Salmo trutta)* from Europe have resulted in the disappearance of the Golden Trout from numerous additional sites. Of historical interest is an early study of Kern River trout, ordered by President Teddy Roosevelt, that concluded that the only serious threat to these fishes was overfishing. It is only fair to bear in mind that at the time of this study, in 1904, Mendel's basic laws of genetics were yet to be "rediscovered," and the concepts of ecological niche and intraspecific competition were not proposed until decades later. Ironically, the state hatchery programs that spawned these near tragedies are now one of the best insurance measures against complete loss of the Golden Trout (albeit Golden Trout cultigens that do not accurately represent the original pure gene pool).

As for Golden Trout natural history, this versatile invertebrate feeder appears to utilize nearly all insect and other invertebrate life, including very small zooplankton normally bypassed by other trout species. It spawns in late June and July and follows the usual trout reproductive pattern. Reproduction takes place in creeks, even when a terminal lake with gravel shores appropriate for redd construction is available. Eggs hatch in three weeks, and alevins emerge from the gravel nests several weeks later, moving to either creek pools or terminal lakes to begin the parr growth stage. With respect to the term "parr" as used to this point in the book, all three subspecies of Redband Trout from the Kern River, along with one from the upper McCloud River, retain parr marks into adulthood. In shallow, clear, creek habitats, the possession of this sort of disruptive coloration can be more advantageous for eluding predators than the uniform sheen of the basic Rainbow Trout. Some accounts of the Golden Trout stress that it evolved in a predator-free environment. This statement only holds true for fish predators, because avian piscivores such as the Belted Kingfisher *(Ceryle alcyon)* and one or more merganser species could put a considerable dent in a population of a trout subspecies that is sexually mature at only 20 cm (8 in.) or so. If food is plentiful and predators are scarce, however, Golden Trout can grow to double that size by their seventh year.

ANGLING NOTES: Most Golden Trout populations occur in high mountain streams and lakes in Tulare and Fresno Counties. Bighorn Lake in Fresno County is a good habitat for this species. When organizing a fishing trip for a restricted species such as this, soliciting local tips on sites where "Goldens" have recently been taken is of prime importance. Mountain lake fishing for this species is much like fishing for Rainbow Trout in mountain streams, except that in lakes, the shallow, inshore waters produce most of the aquatic insect larvae and thus most of the good fishing. Like most fishes, the Golden Trout is often reluctant to move into shallow feeding sites at midday but utilizes these areas at dawn and dusk. The nymph-type wet fly is the best all-around bet, but a careful look in the stomach of that first catch will reveal the most plentiful current food item.

The **LITTLE KERN GOLDEN TROUT** ***(O. m. whitei),*** a Federal Threatened species, is a near twin of the Golden Trout but genetically different enough to merit subspecies status. Like the latter, it evolved in an isolated creek system void of other fish species. Fishes with such a history do not usually fare well when other species are placed in their once-exclusive domain. The Little Kern Golden Trout has hybridized with hatchery Rainbow Trout for many years in the Kern River and has been replaced in several former habitats by introduced Brown and Brook Trout. This barrage of human-caused impacts on such a restricted species resulted in its listing as a Federal Threatened species in 1978, only five years after the Endangered Species Act came into being. Its life history closely parallels that of the Golden Trout, but its Threatened status precludes any fishing for it in many areas.

KERN RIVER RAINBOW TROUT ***(O. m. gilberti),*** a California Species of Special Concern, is the third subspecies of Redband trout native to the Kern River drainage. Its heavy black speckling and more diffuse red band is more characteristic of the Coastal

Rainbow Trout than of its Golden Trout neighbor. Like the latter, it retains a set of light parr marks into adulthood. Over 100 years of hatchery trout introductions into its habitat have resulted in hybridization so extensive that all original, genetically pure populations have most likely been eliminated. With the hope that a few may still persist, the California Department of Fish and Game has designated the Kern River Rainbow Trout a California Species of Special Concern, although fishing with a possession limit of two fish no greater than 25 cm (10 in.) is allowed. It is not the angler who has caused the current problems for this subspecies, but rather the compounded culprits of stream siltation, riparian zone destruction by cattle, and the introduction of beavers, which have wrought a change in the original hydrology of portions of its habitat. Hopefully, there will indeed be some "special concern" for the Kern River Rainbow that will begin to repair the many adverse impacts.

THE SACRAMENTO REDBAND TROUT ***(O. m. stonei),*** is the subspecies assigned to all other races of Redband Trout, at least by the "lumpers" (those favoring fewer subspecies designations). The "splitters," on the other hand, have proposed yet more subspecies for this group. Lumpers are the champions of ichthyology students because their taxonomic views translate into fewer subspecies names on the next exam. However, with the continual advances in molecular biology, the splitters have a far greater arsenal of data with which to defend their viewpoints, and this taxonomic tug-of-war is just beginning. In keeping with the natural history concept of the California Natural History Guide series by the University of California Press, and with my personal goals of introducing readers to and stimulating their interest in the state's freshwater fishes, I have reflected the lumper's taxonomic views in most cases but occasionally inserted the splitter's outlook where it seemed appropriate.

Note that a population of Sacramento Redband Trout in the upper McCloud River is a California Species of Special Concern.

The **EAGLE LAKE RAINBOW TROUT** ***(O. m. aquilarum)*** is one race that does appear to deserve subspecies designation. It evolved in the closed basin of Eagle Lake in northeastern California. Evolution of a subspecies in an isolated lake is in itself not always a qualification for separate taxonomic status, but besides its aquatic isolation, Eagle Lake has a very high alkali level that is lethal to most other fish species. In this sense, the original Eagle Lake Rainbow Trout population existed in a situation much like that of several pupfish (Cyprinodontidae) species: isolated in a habitat made endurable only by its physiological adaptations. Unfortunately, this isolation ended when the lake's only spawning creek for this fish was degraded to the extent that it could no longer support reproduction. As a result, an emergency hatchery program was set up for the Eagle Lake Rainbow in the late 1950s. It saved this subspecies and showed that hatchery propagation can be a successful management tool if stocks are not mingled. In this case, the only source for breeder fishes was Eagle Lake itself, but even then subtle selection for survival of the young in hatchery runways instead of a natural creek can result in a slow drift of a gene pool away from its original composition. The long-term solution is to restore the lake's only spawning creek to its original productive state, a project that has already begun.

Because of this history, the Eagle Lake Rainbow Trout has been accorded California Species of Special Concern status. As a result of the hatchery maintenance program, anglers can still catch this trout in designated parts of the lake during the summer and fall months. Because of its relatively large size, this fish is rapidly becoming a favorite for stocking in reservoirs, where it adapts well to the water's lower alkalinity.

Two other Sacramento Redband Trout populations appear on the 2005 California Species of Special Concern list by common, but not scientific, subspecies names. These are the McCloud River Redband Trout and the Goose Lake Redband Trout. We, along with the CDFG, will just have to wait patiently while the lumpers and splitters battle this one out.

CUTTHROAT TROUT *Oncorynchus clarki*

NATIVE

The present-day Cutthroat Trout group originated over two million years ago, perhaps in the Columbia River basin of that era. Three of the dozen or more current subspecies occur naturally in California. "Cutthroat" is an old fisherman's name that refers to the bright, red orange lines or slashes on the underside of the lower jaw. David Starr Jordan, who was quite fond of this species, and others attempted to have its name changed because they felt it degraded this noble fish. But as with many other species that have been christened by the general public and not the scientist, the name is still with us today. Our three native California subspecies are the Coastal Cutthroat Trout *(O. c. clarki)*, the Lahontan Cutthroat Trout *(O. c. henshawi)*, and the Paiute Cutthroat Trout *(O. c. seleniris)*. The Coastal Cutthroat Trout is a California Species of Special Concern, and the Lahontan Cutthroat Trout and Paiute Cutthroat Trout are Federal Threatened species.

The **COASTAL CUTTHROAT TROUT** ***(O. c. clarki)***, like the Coastal Rainbow Trout *(Oncorhynchus mykiss irideus)*, occurs as both resident and anadromous populations. However, the semi-anadromous populations maintain a far closer tie to freshwater than their Steelhead *(O. m. irideus)* cousins. Most go to sea only during the summer months, and even then they never stray far from the coast or local estuary. Other populations never go to sea

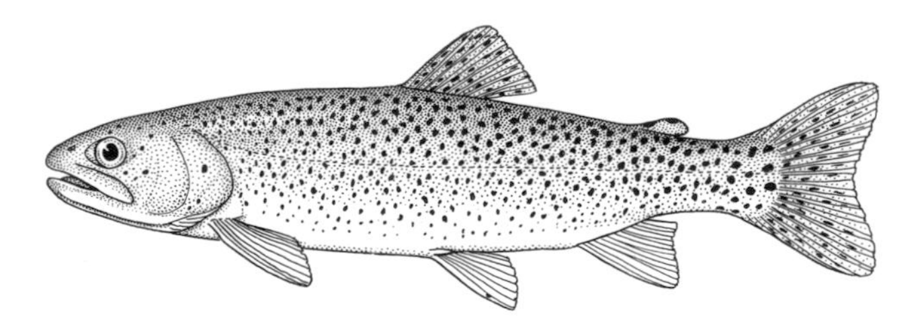

but instead conduct annual migrations between spawning creeks and the major river they feed in. These migratory patterns very likely reflect the manner in which full-scale anadromy as seen in Steelhead and salmon species originally developed.

The basic coloration of the Coastal Cutthroat Trout is similar to that of the Coastal Rainbow Trout but with considerably more spotting over the entire body. In sea-run populations the red throat lines become very faint and the body color more washed out than in the resident creek forms. The full adult length and weight of an anadromous Coastal Cutthroat is considerably less than that of a Steelhead and, once again, demonstrates the direct relationship of the time spent feeding at sea to mature adult size. Despite these morphological and behavioral differences between the Coastal Cutthroat Trout and the Coastal Rainbow Trout, the two species can hybridize and occasionally produce entire hybrid populations. Hatcheries could not resist experimenting with what is apparently an incomplete genetic species separation and have produced "cutbows" that when released, further complicate the taxonomic picture.

The anadromous Cutthroat Trout spawns for the first time between its second and fourth year. The spawning process is similar to that of the Steelhead and takes place from late winter through midspring. Eggs hatch in about six weeks and the alevins emerge a week or two later when the yolk sacs have been absorbed. The new parr find their way to slow-flow areas of the creek where they begin the juvenile growth process. After two to three years of creek life, they make their first journey to the inshore marine habitat at a size of 25 to 30 cm (about 11 in.). The average adult breeding size is 35 cm (14 in.). When at sea or in estuaries the adult feeds primarily on small fishes, including smolts of other salmonid species. This piscivorous pattern continues during the long stay in the creek habitat. Here, the adult is the top piscivore and thus has a variety of prey to choose from.

ANGLING NOTES: The basic fishing methods for anadromous and nonanadromous Coastal Cutthroat Trout mirror closely those for the same two groups of Coastal Rainbow Trout. In-depth study of the current fishing regulations are also necessary because of this subspecies' status as a California Species of Special Concern. In a number of areas, however, such as designated sections of the Smith River, two Coastal Cutthroat Trout may be taken where possession of no other salmonids, especially Steelhead, is allowed. One further point of caution: some populations of anadromous Coastal Cutthroat Trout can look a lot like a Steelhead, especially if you fail to notice the faint throat lines on these sea-run forms. If you have a nice fish in hand and still doubt its species after a visual inspection, then you must perform a quick dental exam. Gently open the fish's mouth and slide your index finger in past the tongue and down to the posterior mouth floor. If you feel numerous small, sharp teeth with your fingertip just behind the base of the tongue, it is a cutthroat. If the posterior mouth floor is smooth, it is a Steelhead.

LAHONTAN CUTTHROAT TROUT FOLLOWS ➤

Both the **LAHONTAN CUTTHROAT TROUT** ***(O. c. henshawi)*** (below) and the **PAIUTE CUTTHROAT TROUT** ***(O. c. seleniris)*** (bottom) have been replaced in most of their original habitats by large-scale introductions of Brook Trout *(Salvelinus fontinalis),* Brown Trout *(Salmo trutta),* and hatchery Rainbow Trout. In Lake Tahoe, the combination of a former commercial Lahontan Cutthroat Trout fishery and the introduction of still another species—the Lake Trout *(Salvelinus namaycush)*—has resulted in the complete disappearance of a population of giant Lahontan Cutthroat Trout. The highly restricted population of Paiute Cutthroat Trout in Alpine County faces a similar danger because it hybridizes with

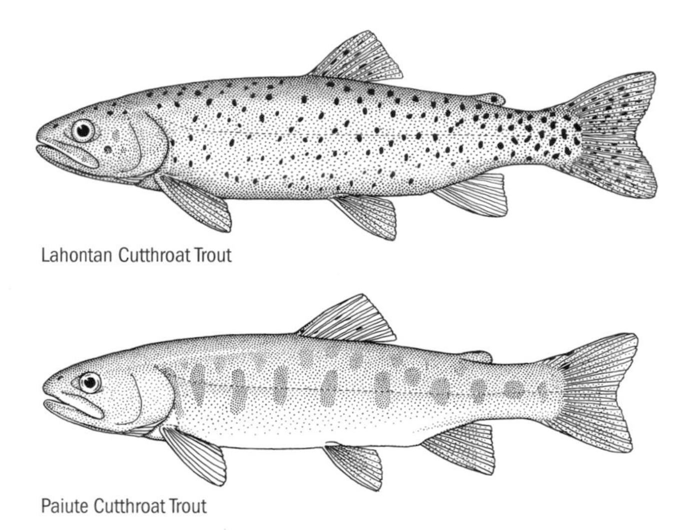

Lahontan Cutthroat Trout

Paiute Cutthroat Trout

introduced Rainbows and Lahontan Cutthroat Trout in Silver King Creek. Luckily, other Paiute Cutthroat populations have been established in Cottonwood Creek, Mono County.

As a result, the Paiute Cutthroat Trout and the Lahontan Cutthroat Trout were listed as Federal Endangered species under the forerunner of the present Endangered Species Act in 1967 and 1970, respectively. In 1975, two years after the final version of the ESA became federal law, this status was downgraded to Federal Threatened species status so that hatcheries could begin recovery stocking programs.

BULL TROUT *Salvelinus confluentus*

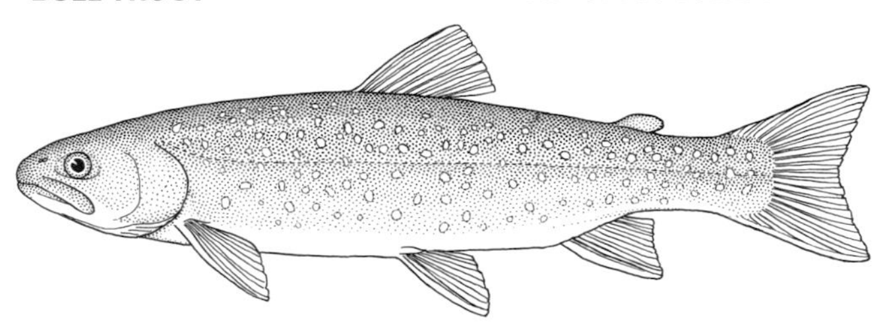

NATIVE

PROTECTIVE STATUS: FT, SE, extinct in California.

For over a century, a controversy has existed as to whether the char genus *Salvelinus* is represented in western North America by one species or by two. Until recently it was assumed that the only member of this genus native to California was the Dolly Varden Trout *(S. malma)* and that this species was found only in the McCloud River. Recently, however, experts in trout taxonomy have reexamined museum specimens of *Salvelinus* from the McCloud and have found that these are the Bull Trout and not the Dolly Varden Trout.

The confusion began in 1872 when anglers began catching a large, salmonlike trout in the McCloud River and named it "Dolly Varden" after a Charles Dickens character who wore a green calico garment with red dots that somewhat resembled the Bull Trout's coloration. The real Dolly Varden trout, which ranges from the California-Oregon border northward to Alaska, was also assigned that name because of its similar coloration.

As you may gather, these two species appear quite similar; However, the Bull Trout's head is more flattened than that of the Dolly Varden. The Bull Trout is also the more piscivorous of the two, relying heavily on small fishes for its major food source. Unfortunately, these taxonomic discussions are purely academic today because it is now believed that the Bull Trout is extinct in California. It was probably never very abundant in the McCloud River, and the construction of two dams on this drainage, the introduction of the nonnative Brown Trout *(Salmo trutta)* (which occupies a similar ecological niche), and heavy fishing pressure apparently have eliminated it completely. I use the word "apparently" because, with any recent presumed extinction, you can never be sure whether a few specimens still linger on. To give such a species no mention in a field guide would almost surely eliminate chance recognition by an angler or naturalist. The Bull Trout still exists as a Federal Threatened species outside of the Golden State, with populations extending from southern Oregon to northern British Columbia.

BROOK TROUT *Salvelinus fontinalis*

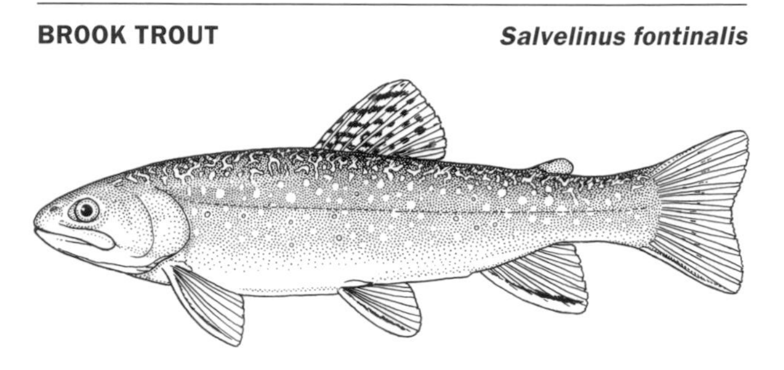

INTRODUCED

The Brook Trout is by far the most numerous introduced trout in California and is second only to the Rainbow Trout *(Oncorhynchus mykiss)* in total numbers (pl. 33). It seldom reaches more than 30 cm (12 in.) and 340 g (0.75 lb) in this state but grows considerably larger within its native range. It belongs to that group of the trout and salmon family known as chars and possesses a number of behavioral traits that are different from those of present native California species. Perhaps the most important of these traits is its ability to spawn successfully in lakes that lack feeder streams. In such lakes the Brook Trout seeks out gravel-bottom areas that contain springs. Spawning occurs in fall, and eggs overwinter in the gravel and hatch in spring.

Brook Trout introduction in California began in 1871. Its in-lake reproductive pattern and tolerance of very cool water have led to the establishment of large populations in more than 1,000 high- and middle-elevation lakes and more than 1,000 miles of Sierra and foothill streams in California. In most of these small headwater streams it has quickly outnumbered native California trout species. Indeed, the indiscriminate state-sponsored stocking of this species has most likely been the single most destructive force in the demise of native mountain trout populations. It has also impacted populations of the Mountain Yellow-legged Frog *(Rana mucosa)* and Yosemite Toad *(Bufo canorus)*, two amphibian special status species (species protected by state and/or federal law).

This biologic price that California has paid for Brook Trout greatly exceeds any returns to the state's anglers. Its high reproductive rate and dominance in small mountain lakes has led to the same overpopulation and stunting witnessed in warm-water species such as the Bluegill *(Lepomis macrochirus)*. A number of mountain lakes have been literally overrun with stunted Brook Trout. Years ago I accompanied my old friend Bob Behnke on a

Plate 33. Brook Trout.

fish survey of a small, private lake in the Trinity Alps range. The lake owner was upset because he could not catch big trout anymore. Everywhere we cast our lines we caught Brook Trout, and every fish was less than 6 in. long. Once the word gets out that a particular lake holds only stunted trout, control by fishing is no longer possible, and the situation gets progressively worse. The best remedy for such a problem is the removal of most of the fish, either by sustained gill netting or by rotenone poisoning, followed by heavy, sustained fishing pressure.

ANGLING NOTES: Given all of the above "bad press" courtesy of this author, you might get the impression that the Brook Trout is a pretty bad customer, which is certainly not the case in its native range in the northern Midwest and eastern United States.

Although "brookies," as anglers call them, feed on essentially the same drift items and larval bottom food as Rainbow Trout, they are usually more abundant in the slower, more sheltered water areas, where they vigorously defend feeding territories against individuals of their own and other species. They are the masters of the undercut bank, and where such banks exist in streams and lakes, the angler should fish them carefully while remaining well back from the water's edge.

One of my first trout fishing trips was with a high school friend and his trout fisherman father who drove us to a Wisconsin dairy farm pasture one spring morning and instructed us to rig up our poles and then follow him. There was no stream in sight, and us wise guys started concocting a contest for who was going to catch the biggest cow pie. After a little while, the father ordered us down on all fours, and in this manner we proceeded toward what now appeared to be a slight linear break in the meadow. As soon as we recognized it as a creek, we were halted and told to start casting while still kneeling. Soon, from this prayerful position, we each caught a couple of nice brookies.

Suggestions for artificial lures fall along the same lines as those for Rainbow Trout. The brookie, however, usually seems more eager than other trout species to hit the ubiquitous angleworm. Perhaps this is a more common food in its preferred backwater and undercut-bank areas, but whatever the case, a can of worms is a sound investment for any Brook Trout expedition.

The last tip for this species is presented as a "survival aid." For some as yet unexplained reason, the Brook Trout is extremely susceptible to the old-time method of fish capture known as "tickling," which works well only in undercut-stream-bank situations. First, you locate the trout by feel while lying at streamside with one hand and arm under the bank. The trick is to move your hand quietly and slowly along until it gently contacts some part of the fish. By continuing to move slowly, you should be able to work your hand up or down the fish with the ultimate goal of positioning the thumb and forefinger on either side of the operculum. A good sense of touch and knowledge of fish anatomy will lead you to this area. If a "grab" is attempted short of this goal, the fish usually wriggles free, but the gill chamber provides the one deep-purchase area on the fish and should be carefully sought out. You may now be wondering, "Why doesn't the fish swim away upon first contact?" Perhaps it is because the fish is already in a dark, cloistered subbank area and the added presence of a hand around its body does not elicit alarm signals from the lateral line organ. Whatever the reason, the method has the potential of saving you from starvation, or at least from an empty creel and a good deal of ribbing back at the fishing camp.

BROWN TROUT *Salmo trutta*

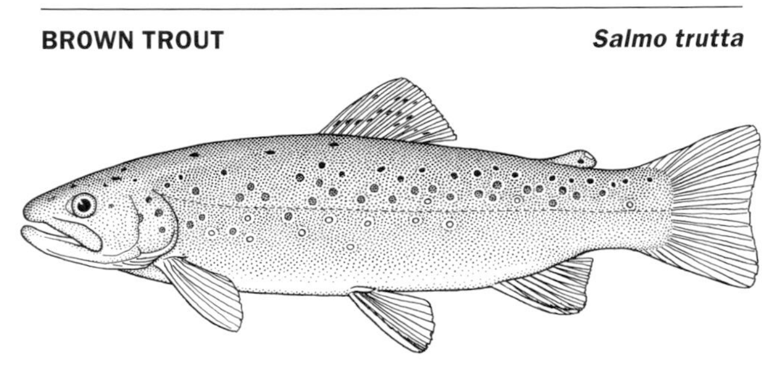

INTRODUCED

This is an introduced species not only in California but also in the Western Hemisphere. It is the trout of Europe and the British Isles, and well before introduction into the eastern United States in 1883 it had adapted to a wide range of habitat conditions. It has the ability to live in streams with summer thermal conditions well above the preferred range, or even the tolerance, of other trout species. The Brown Trout, or "brownie" as it is sometimes called, can hybridize with the Brook Trout *(Salvelinus fontinalis)* and produce a sterile but otherwise vigorous hybrid called the Tiger Trout. This hybrid is especially famous for its voracious feeding behavior, which makes it a favorite of anglers.

Brown Trout have been known to grow to over 1 m (39 in.) in length and to live up to 16 years in large lakes, but about nine years is the maximum longevity in California. It is one of the most territorial of all trout and literally rules the still-water pools of streams where it resides. Like the Brook Trout, it prefers deep-water areas but is also found in riffles. In many streams, it is quite sedentary, and a particular undercut bank section or root tangle may harbor the same brownie for years. It is a drift and bottom-larvae feeder like most other trout, and the adult is also piscivorous. The Brown Trout is a true trout, not a char, and thus requires a stream habitat for spawning, which occurs in fall and early winter when the female digs redds in gravel areas in the normal trout manner. Eggs hatch in about two months, and alevins emerge a month later.

The Brown Trout is well established in many California streams and lakes. In some cases it has added an extra dimension to trout fishing, especially in wild or nonhatchery trout streams such as Hat Creek, Shasta County. Once again, this success has not been without its price, and the introduction of the Brown Trout into the McCloud River is perhaps the main reason for the disappearance of the Bull Trout from these waters and the elimination of the Paiute Cutthroat Trout *(Oncorhynchus clarki seleniris)* from most of its range. The Brown Trout's competitive and territorial nature has also all but excluded the Rainbow Trout *(O. mykiss)* from slower sections of streams in which the two now exist.

ANGLING NOTES: Despite the brownie's wide distribution in California habitats, it is one of the trout found least often in the creel. This is because of its extremely cautious feeding behavior. A large Brown Trout may take more than 15 minutes to "decide" to leave a deep lair and come to the surface to strike. Even when it does respond, it takes the lure slowly into its mouth and produces only the slightest sensation at the rod tip. If an angler is not alert to such soft strikes, the fly or bait will be rejected before he or she realizes just what has happened. After a certain amount of Brown Trout fishing, you get the strong impression that large adult fishes may not engage extensively in surface feeding, but instead rely heavily on subsurface food that drifts or swims. This way, they can dash at the food horizontally and then quickly return to their lair. For this reason, the best first tries in a brownie stream are often spinners, minnows, or worms drifted skillfully past the door of the hideout. True, this method will cost you a few entangled lures or hooks by the end of the day's fishing, but it is also the most consistent method of latching onto this cautious feeder.

LAKE TROUT or MACKINAW TROUT *Salvelinus namaycush*

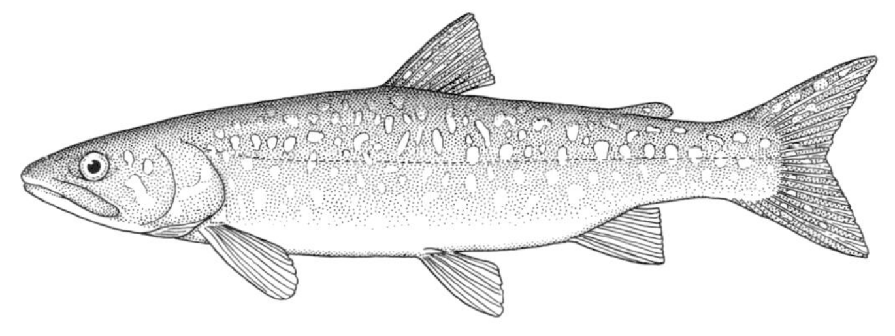

INTRODUCED

This is the largest trout in California, with specimens reaching well over 1 m (39 in.) in length. It was introduced into Lake Tahoe from the Great Lakes region in 1885. As its name implies, it is strictly a lake species and does not even utilize feeder streams for spawning. As an adult it is exclusively piscivorous, a habit also unique among California trout. It is very pelagic and restricts its activities to the thermocline during summer but, like lake-dwelling Kokanee Salmon *(Oncorhynchus nerka),* moves to the upper levels with the fall turnover. It is found only in the deep lakes of the Tahoe basin that have thermal characteristics similar to the natural Lake Trout habitats in the northern parts of the Midwest and eastern United States.

Like other introduced eastern species, the Lake Trout is a fall spawner. It is the only trout that does not dig a nest in the gravel substrate but instead spawns in rock rubble areas, where the fertilized eggs fall between the stones and boulders. Eggs may take up to six months to hatch, and the fry live among the rocks for several more months before becoming pelagic. Zooplankton provides the majority of food during the first couple of years of life, after which the Lake Trout becomes piscivorous. This adult piscivorous habit may well have been responsible for the disappearance of the Lahontan Cutthroat Trout *(Oncorhynchus clarki henshawi)* in Lake Tahoe, once it had been reduced in numbers by commercial fishing and the blockage of spawning tributaries. The Lake Trout now preys upon other native fishes, such as the Mountain Whitefish *(Prosopium williamsoni),* Tahoe Sucker *(Catostomus tahoensis),* and Tui Chub *(Siphateles bicolor)* and on the introduced Kokanee Salmon *(O. nerka),* all of which appear to fare well under such predation pressure.

ANGLING NOTES: The Lake Trout continues to support a major sport fishery in Lake Tahoe, where trolling with spoons has long been the tried and proven method of fishing. Native Americans and European pioneers in the northern Great Lakes states caught Lake Trout by using bright clam shells with hooks attached that wobbled through the water. The depth of the thermocline is all-important in summer, so do not forget your trusty electric thermometer, as discussed in the section on lake fishing for Coho Salmon *(O. kisutch)*. The heavy lines and rods and large boats used to for Lake Trout fishing in Lake Tahoe constitute the closest thing to deep-sea fishing in inland California. Also, do not overlook the smaller, ice-covered lakes of the Tahoe Basin: Fallen Leaf and Stony Ridge. In natural Lake Trout lakes, such as Green Lake in Wisconsin, excellent catches are taken through the ice each year using live minnows. When snow conditions and season permit, this technique is certainly worth a try in California. As a former Wisconsin ice fisherman, I must add a strong note of caution: always become thoroughly familiar with both ice conditions and ice-fishing techniques before attempting this demanding type of winter angling.

MOUNTAIN WHITEFISH *Prosopium williamsoni*

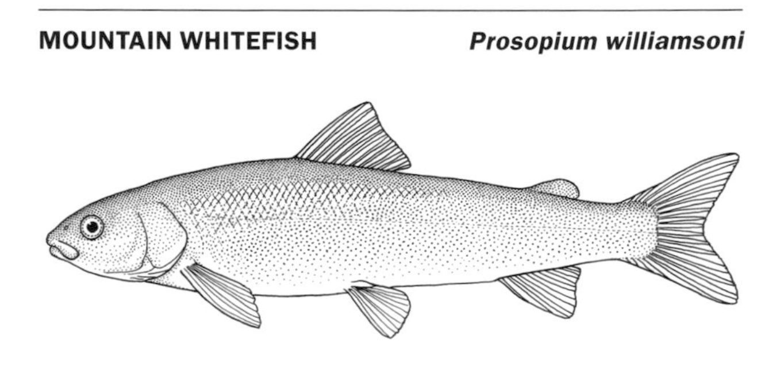

NATIVE

This native species, which occurs only in the Lahontan Fish Province in California, is perhaps the most secretive and little-noticed salmonid in California. Its inferior mouth and large, un-troutlike scales present a marked deviation from the salmonid body plan. It is an opportunistic bottom feeder and takes a variety of aquatic insect larvae, especially those of the mayfly and caddis fly. It also feeds on zooplankton and will eat small crayfish when available. The Mountain Whitefish is a slow-growing species and attains an adult size of 25 to 50 cm (10 to 20 in.) at between five and 10 years of age. In fall, spawning takes place in stream gravel beds, where eggs are scattered rather than deposited in nests as with most other salmonids. This behavior may have developed in response to this species' habit of eating eggs, including its own. Fry hatch in spring and filter feed for the first year or so. Lake populations inhabit the thermocline area during the summer months, and from here they may dive briefly through the hypolimnion to grab chironomid larvae that reside in the near anaerobic bottom layer. Less is known about the ecology of this species than of any other California salmonid, presenting a rich area for future investigation.

ANGLING NOTES: The first task for the novice angler is to learn to recognize this species. The Mountain Whitefish is often hurriedly misidentified and discarded under the assumption that it is some sort of sucker or minnow. But these folks are throwing away what many feel to be the best eating of all the salmonids! The second point to keep in mind is that this is mainly a bottom feeder, so the bait should be placed there. Use individual salmon eggs or small pieces of worm, and bounce them along the bottom for greatest attraction. Stream populations also respond well to hellgrammites and crayfish and occasionally rise to flies, both wet and dry. Finally, when planning a Mountain Whitefish trip, remember that it is found only in streams and lakes on the eastern slopes of the Sierra Nevada and north of the Walker Lake area in western Nevada. Lake Tahoe is more or less the center of its north-south distribution.

SMELTS (Osmeridae)

Smelt are small, trim, silvery fishes that may be found in both marine and freshwater habitats. In California we have three freshwater species, the Delta Smelt *(Hypomesus transpacificus)*, the Longfin Smelt *(Spirinchus thaleichthys)*, and the introduced Wakasagi *(H. nipponensis)*, and one anadromous form, the Eulachon *(Thaleichthys pacificus)*. Like most small pelagic fishes, smelt live in large shoals or schools. They are short-lived fishes and most die after spawning in their second or third year. Smelts in California, especially those inhabiting Delta waters, are important forage fishes for a number of large piscivorous species. In recent years, this role has been partially taken over by introduced species such as the Threadfin Shad *(Dorosoma petenense)* and the Inland Silverside *(Menidia beryllina)*. Indeed, you must now carefully examine any small, trim, silvery fish taken in the Delta waters to correctly differentiate between a smelt and an Inland Silverside. The presence of a small adipose fin as the second dorsal fin remains the single best distinguishing characteristic of smelts. Smelts also have much larger mouths than the introduced Inland Silverside. This feature, along with numerous, long gill rakers on each gill arch, allows smelts to filter feed on a variety of zooplankton species. In this respect, their feeding niche more closely resembles that of the Threadfin Shad than that of the Inland Silverside. All three groups are zooplankton feeders throughout life, and this similarity again raises the question of what effect the rapidly increasing populations of Inland Silverside, Threadfin Shad, and most recently, the Wakasagi will have on our most freshwater-oriented native species, the Delta Smelt.

DELTA SMELT *Hypomesus transpacificus*

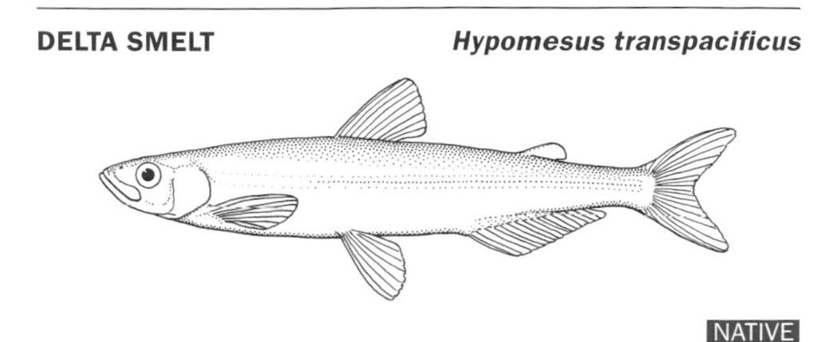

NATIVE

PROTECTIVE STATUS: ST, FT.

The Delta Smelt is a native of the San Francisco Estuary. It is moderately euryhaline and although it can tolerate salinities up to 19 ppt, it prefers the much lower range of 2 to 7 ppt. Even for a smelt, the Delta Smelt is quite small, with fully mature adults attaining a maximum standard length of only 12 cm (4.7 in.) (pl. 34). They eat a variety of zooplankton but also take small aquatic insect larvae when available.

Plate 34. Delta Smelt.

The Delta Smelt life history follows the "one year plan," meaning that individuals spawn at the end of their first year and then die. A small percentage of each year's class survives a second year at the end of which they spawn for a first or possibly second time before dying. Spawning usually takes place from February through July, with peak activity occurring in April and May. It is preceded by a diffuse migration to the upper Delta and the lower reach of the Sacramento River, where spawning takes place in side channels and sloughs. Spawning usually occurs at night during periods of low tidal activity. Although never observed in the field, it has now been extensively observed in laboratory rearing tanks, especially at an experimental Delta Smelt hatchery near Tracy, California. Females lay an average of 2,000 eggs, which they broadcast over the bottom substrate while attended by one or more males. Each egg attaches to the substrate by a minute adhesive stalk. My impression when I viewed these at the hatchery was that they resemble a tiny golf ball on a tee. Eggs hatch within two weeks, and the larval smelts, which at this stage look nothing like the adult, begin feeding near the bottom on very small zooplankton, especially rotifers. As they grow, the larvae move down through the estuary to the mixing zone, where most remain to continue feeding on zooplankton in the water column. Some larvae do occasionally get washed out of this area, however. During a recent three-year survey of several sloughs in Suisun and San Pablo Bays, I collected several larvae from San Pablo Bay in late spring. The young smelts grow rapidly and are sexually mature by the next spawning season, after which most die.

This type of one-year life plan can be advantageous for a species because reproduction occurs before extended adult mortality has a chance to significantly diminish the breeding stock. Coupled with high egg numbers and group-spawning behavior, this reproductive strategy ensures high, annual recruitment for a population if environmental conditions remain favorable for that species. Such conditions were apparently very favorable for the Delta Smelt during its evolution in the prehistoric Delta area of California. During the past half-century, however, there have been numerous, large-scale changes in this system, and organisms that operate on the one-year plan are usually the most susceptible. Indeed, after just one season of greatly reduced reproduction due to new, human-wrought impacts, next year's breeding population (which is also the entire present population) could be reduced to near nothing. If additional impacts further reduce the existing population, a likely outcome is extinction.

Unfortunately, the Delta Smelt may be traveling this disastrous course. As a full-time resident of the Delta, it cannot temporarily escape to the sea, bay, or tributary creeks as other native pelagic species can. Instead, it must meet all impacts to its Delta waters head on, including greatly reduced outflow from excess water diversion, exceptionally high outflow resulting from poor upstream watershed and reservoir management, large losses of larvae and even shoaling adults to canal and power plant pumps, continued dumping of agricultural pesticides and other toxic materials in the Delta sink, and the ongoing introduction of non-native species that compete directly with the Delta Smelt for food or that pursue it for their food.

Of these threats, the nonnative species may be the most serious because unlike the other factors, they are essentially uncontrollable once introduced. The natural piscivores of the pristine Delta were presumably in balance with their Delta Smelt prey and served as a necessary safeguard against its overpopulation. Even the early introductions of the American Shad *(Alosa sapidissima)* and Striped Bass *(Morone saxatilis)* apparently have had little effect on either the adult or larval smelt populations. However, the stocking of the Threadfin Shad *(Dorosoma petenense)* in the Delta in 1958, followed by introductions of the Inland Silverside *(Menidia beryllina)* and Wakasagi *(H. nipponensis)* in the mid-1970s, all of which are highly fecund smelt-size zooplankton feeders throughout life, have most likely overwhelmed the Delta Smelt. The Wakasagi also adds a second dimension to this competition by hybridizing with the Delta Smelt on its own spawning grounds. Add to all of this the massive spread of the foreign Overbite Clam *(Potamocorbula amurensis)* in the mid-1980s, which filter feeds on very small zooplankton, including the rotifer food of larval smelt, and we have an ecological disaster in progress.

The devoted efforts of members of the experimental Delta Smelt hatchery group at the state Fish Rescue Facility near Tracy is the one bright light in this currently dark picture. But even if we develop a successful hatchery program for this challenging species, what do we do with the product? Do we return these fish to the Delta to compete with foreign challengers for the small-fish pelagic zooplankton feeding niche? Once again, with endangered species it is really the habitat that is endangered, and the species are simply along for the downhill ride.

WAKASAGI FOLLOWS ➤

WAKASAGI

Hypomesus nipponensis

INTRODUCED

The Wakasagi was originally considered a subspecies of the Delta Smelt *(H. transpacificus)* but is now recognized as a separate species and more closely related to the Asian Marine Smelt *(H. japonicus)*. It was brought to California from Japan by the California Department of Fish and Game in 1959 for the same reason that the Threadfin Shad *(Dorosoma petenense)* introduction had recently been sanctioned: to provide more prey for hatchery Rainbow Trout *(Oncorhynchus mykiss)* cultigens stocked in reservoirs. Once again, this focus on the hatchery trout stocking program appears to have overlooked the reality of how easily fishes can move through the interconnected waterways of this state and the ecological upheavals that can occur when alien species do so. In this case, the alien is a mirror image of our only native freshwater smelt (pl. 35). Wakasagis were first introduced into six isolated reservoirs, and for a time this seemed a benign field experiment in predator-prey interaction.

In the early 1970s, however, the CDFG made additional stockings in Sierra foothill reservoirs with connections to the Delta system, and in 1990 the Wakasagi was found in the San Francisco Estuary, the ancestral home of the Delta Smelt. It occupies the same zooplankton filter-feeding niche as our native smelt and has also been known to eat Delta Smelt eggs and larvae. The spawning grounds of the Delta Smelt apparently closely resemble those used by the Wakasagi back home, and as a result, hybrids between the two species are appearing in Delta waters.

Plate 35. Delta Smelt (above), Wakasagi (below).

The effect of Wakasagi introduction on other Delta fishes is not yet known, but because nearly all species spend their first year or two eating zooplankton, the competition from this species, especially when coupled with that from the introduced Threadfin Shad and Inland Silverside, could be far reaching. Indeed, in several reservoirs stocked with Kokanee Salmon *(O. nerka)* where Wakasagi have been released, the planktivorous Kokanee has disappeared. I don't believe that this was the sort of change in hatchery salmonid reservoir fishing that the decision makers at CDFG envisioned.

The life history of the Wakasagi is presumed to be similar to that of the Delta Smelt, but no detailed field studies of this fish have yet been conducted in California. It appears to be doing well in the reservoirs where it has been stocked. Furthermore, the Wakasagi is presently conducting a stocking program of its own by way of the California Aqueduct. It has already arrived at the giant San Luis Reservoir in northern Merced County and from there is just a pleasant cruise down the aqueduct to the reservoirs of southern California.

A glance around any large parking lot or at passing traffic attests to the American love affair with Japanese imports. However, after reading the above saga of the Wakasagi, you may agree that this is one import we could have done without.

LONGFIN SMELT ***Spirinchus thaleichthys***

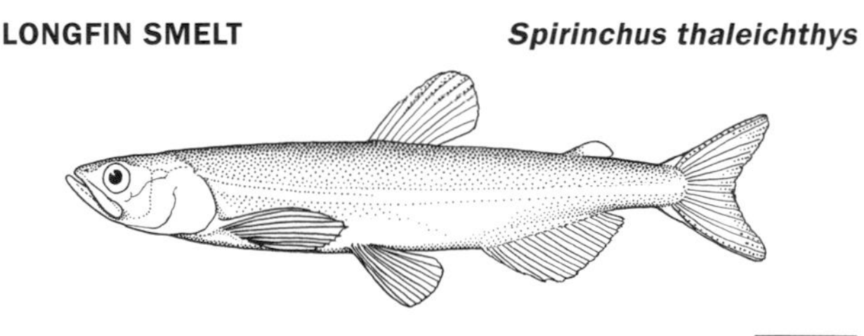

NATIVE

PROTECTIVE STATUS: CSSC.

The Longfin Smelt is another small species, rarely exceeding 12.7 cm (5 in.) in length. In contrast to the Delta Smelt *(Hypomesus transpacificus),* it prefers moderately saline water and thus may be classified as a euryhaline species. In California it is found in all major bays and estuaries from San Francisco Bay northward. It lives in the bay waters throughout summer, and in fall moves into the lower reaches of the rivers that flow into these bays. Spawning occurs throughout winter and spring. The female is far more fecund than the Delta Smelt female and may lay up to 24,000 eggs. The adhesive eggs are scattered over a variety of substrates, including aquatic vegetation. The embryos hatch in about six weeks and move with the current to brackish water in the lower reaches of the estuary, where they begin feeding on zooplankton, the Longfin Smelt's food throughout life. Like the Delta Smelt, most Longfins die after spawning, but a few remain to spawn a second year.

Interestingly, the zooplankton feeding niche of small pelagic fishes was originally divided between the Delta Smelt and the Longfin Smelt. The euryhaline nature of the Longfin allows it to vacate the lower river reaches just as the Delta Smelt returns to this area from its spawning sloughs. It is only when the latter species moves back to its spawning grounds in fall that the Longfin again invades freshwater. Unfortunately, the introduction of Threadfin Shad *(Dorosoma petenense)*, Inland Silverside *(Menidia beryllina)*, and Wakasagi *(H. nipponensis)* into Delta waters has made this natural niche segregation nearly meaningless.

EULACHON *Thaleichthys pacificus*

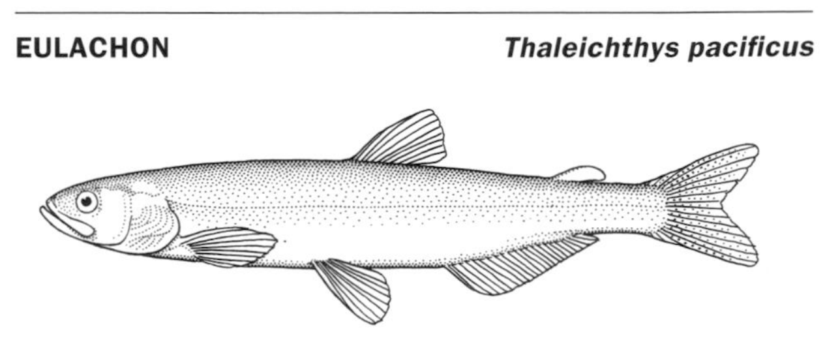

NATIVE

PROTECTIVE STATUS: CSSC.

The Eulachon is the only truly anadromous species of California freshwater smelt. Unlike the Delta Smelt *(Hypomesus transpacificus)* and Longfin Smelt *(Spirinchus thaleichthys)*, it does not operate on the "one-year plan" but instead spends nearly three years at sea before returning to the lower reaches of coastal creeks and rivers to spawn and then die. It is also the largest of the freshwater smelts, with some individuals attaining lengths up to 30 cm (nearly 1 ft), but with most individuals falling within the range of 15 to 20 cm (6 to 8 in.). It is primarily a northern California species, with the major spawning run occurring in the Klamath River. Spawning behavior is much like that of most salmonids. In early spring, Eulachons migrate upstream, where each female lays up to 25,000 eggs. The eggs adhere to the bottom and hatch within three weeks. The larvae are carried rapidly to the sea because the spawning grounds are not very far inland. Once at sea, they begin feeding on small zooplankton but eventually switch to larger krill forms, which they eat for the all of their adult life. At sea the Eulachon is an important food for marine piscivores, including several northeastern Pacific marine mammal species.

ANGLING NOTES: The Eulachon is the only member of the California freshwater smelt group that supports a small sport fishery. Fishing is accomplished by means of large, four-cornered dip nets during massive spawning runs, mainly in the Klamath and Mad Rivers in Humboldt County. Surprisingly few California anglers have taken advantage of this fishery in the past, and not many realize that freshly caught smelts, breaded and fried in butter, are among the tastiest of fish treats. This fact was well known to the Native American groups of the northern California coast, however. They were well aware that the sexes school separately during the run and were able to net males independently of females. The male Eulachon were smoked, and the females, which contain a much higher fat content because of their eggs, were rendered to obtain the much-favored oil. Food was then dipped in this oil in much the same way some Italian foods are brushed or dipped in olive oil. Female Eulachon were also dried by coastal tribes and used for candles because of their high oil content and elongate shape. This ingenious use of this fish has given rise to the popular name of "candlefish" for this species.

Unfortunately, Eulachon runs in California have become scarce since the 1970s, with the exception of good runs in 1989 and 1999. Annual runs in the Columbia River, however, have consistently remained high and continue to support a commercial fishery. This sudden scarcity in the southern end of its range may be due to two severe El Niño events in the latter part of the twentieth century; such events would likely affect the Eulachon because it spends so much of its life at sea. Hopefully, the northern Pacific Ocean will settle into a normal, long-term thermal regime, and this fine fish will again be plentiful in northern California.

MINNOWS (Cyprinidae)

The minnow family is the largest family of fishes in the world, with over 250 species in North America alone. The common name of "minnow" leads you to believe that only small fishes make up this family. However, some members grow to total lengths of over 1 m (over 3 ft) and weight in excess of 23 kg (50 lb). Even so, the majority of species in this family are indeed small fishes, and herein lies a point of much confusion with the name "minnow." It is widely used by many people in the United States to refer to any small fish, no matter what its family. The correct term for small fishes is "fry" or "fingerling." The name "minnow" should be reserved for members of this family only. Unfortunately, this is often not the case, and most dictionary definitions suggest that "minnow" refers to any small fish, and some give examples, such as stickleback (Gasterosteidae) and killifish (Fundulidae). This potential confusion has been reduced by use of the term "cyprinids" when referring to members of Cyprinidae. Like the term "salmonid" used to indicate members of the large and variable family Salmonidae, "cyprinid" is a long-established term and is used throughout the book to avoid any confusion.

In addition to the general confusion about the family name, the cyprinids themselves are a seemingly endless source of frustration for professional and amateur ichthyologists as well as for authors of field guides to freshwater fishes. The professionals are continuously embroiled in heated debates as to exactly how many species, subspecies, and hybrids exist within the many groups of minnows. For the amateur, the task of identifying a fish to genus, let alone to species, is often frustrating. However, California is still a good state to begin a study of minnows because we have only 13 living native minnow species and seven introduced ones. In contrast, an average Midwestern state such as Minnesota has 41 natives and a handful of introductions, and Missouri has over 100 native and introduced species.

Cyprinids are an extremely well-adapted and successful group. Although they tend to be nondescript, minnows possess a number of important adaptations that greatly contribute to the success of the group. This introduction presents the basic biology of minnows and groups the various California species according to broad ecological niche. The individual species accounts are confined to the general distribution and natural history highlights.

One of the most interesting anatomical features of minnows is the absence of teeth on the jaws or in the mouth. Instead, they have only pharyngeal teeth, which are highly modified gill rakers borne on the last or most posterior gill arch. Through muscular action of the throat region, some minnows are able to grind these teeth against a hard, bony plate on the roof of the mouth and thereby break up food items for swallowing. Like the varied dentition of mammals, the pharyngeal teeth of each type of minnow are modified to conform to particular feeding habits. Thus minnow species that feed primarily on plant material, such as the Grass Carp *(Ctenopharyngodon idella)*, have long, serrated pharyngeal teeth similar to those of a hay-cutting machine. Other species such as the Common Carp *(Cyprinus carpio)*, which feeds on small-shelled invertebrates, have pharyngeal teeth with broad, flattened ends for crushing their food (pl. 16). These teeth function much like human molars. Piscivores, on the other hand, have long, pointed pharyngeal teeth that function like the canines and premolars of carnivorous mammals such as the coyote, to hold and slice their prey.

Like suckers (Catostomidae) and catfishes and bullheads (Ictaluridae), minnows have a well-developed sense of hearing, at least compared to other fishes. They have Weberian ossicles, which are small bones derived from vertebrae that are present near the junction of the skull and vertebral column. The function of these in the hearing process is described in the "Anatomy and Physiology of Freshwater Fishes" section. Cyprinids use hearing both to facilitate feeding and to avoid predation. In addition, the ability to hear makes intraspecific sound communication possible, and males of some minnow species have developed the ability to produce sounds during prespawning behavior.

The majority of minnow species remain in schools most of their lives, and the behavior of the school is a key survival factor. One important defensive trait is the school's reaction to a "fright substance" that is released from the skin when a fish is attacked by a predator. Other members of the minnow school detect this substance with their keen sense of smell and react by rapidly fleeing the area.

Spawning occurs in spring at lower elevations and in summer at high elevations. The shallow inshore waters of lakes or the gravel beds of streams are the most common spawning sites.

Although some species are territorial and construct nests, the vast majority of minnows in California are broadcast spawners. Usually a single female is escorted by several males, which fertilize the eggs as they are laid. Some species lay adhesive eggs over vegetation, but the majority lay their eggs over gravel into which they sink and lodge until hatching. Such scattering of eggs and milt over large spawning areas sometimes leads to hybridization when two closely related species live together in the same habitat and have the same preferred spawning time.

One convenient way to present the California species of cyprinids is to divide them into three groups: small native minnows, large native minnows, and introduced minnows. Most small species occupy the broad ecological niche of small invertebrate feeders. The niche of large species, however, ranges from that of filter feeder on phyto- and zooplankton to that of top piscivore, a role not normally performed by North American minnows. Special emphasis is placed on the problems that habitat loss and the presence of introduced species have created for native minnows and other families.

Small Native Minnows

CALIFORNIA ROACH ***Lavinia symmetricus***

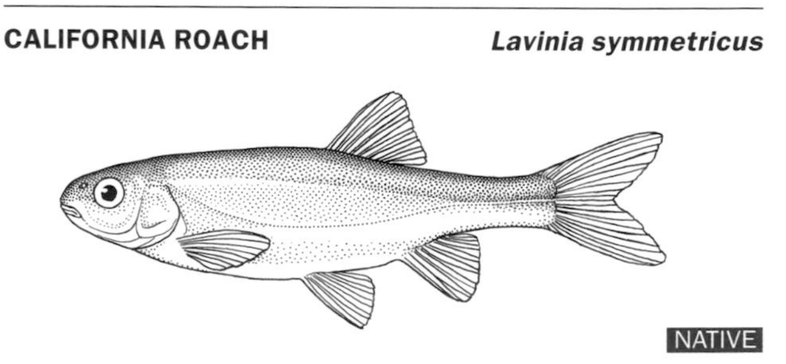

NATIVE

PROTECTIVE STATUS: CSSC.

This is a small cyprinid found in the Sacramento Province, the Russian River, and coastal streams. Next to the Speckled Dace *(Rhinichthys osculus),* it is the most widely distributed native minnow in California. Specimens rarely exceed 10 cm (about 4 in.), and many populations average about 5 cm (about 2 in.) adult size. As of 2005 six subspecies were described, and there will probably be more to come (pl. 36).

Plate 36. California Roach.

The California Roach is primarily a minnow of intermittent, tributary streams. It is able to withstand the rigorous conditions found in shallow, late-summer pools, such as high temperature (up to 35 degrees C [95 degrees F]) and low oxygen levels (over 1 ppm compared to 9 to 10 ppm in mountain streams). As with the pupfishes (Cyprinodontidae), such physiological attributes, coupled with small size, permit this species to occupy a habitat that during part of the year is too restrictive for most other fishes. The California Roach is an omnivore whose diet ranges from filamentous algae to aquatic insect larvae and adults. Zooplankton is also an important part of the diet of the young. Its spawning behavior is much the same as that of other cyprinids, with the female laying up to 2,000 eggs. Sexual maturity is reached after two to three years, and it appears to live to a maximum of six years.

The best place to observe the California Roach is in clear creeks and streams, especially in areas where there are few or no piscivores. In such habitats they usually move and forage in large schools, providing the fish watcher a fine example of such behavior. A pair of binoculars with a short focal length can significantly enhance the viewing experience.

The California Roach is a California Species of Special Concern because many of the small creek systems in which it occurs are degraded by the same impacts cited for the Coho Salmon *(O. kisutch)* and Steelhead *(O. mykiss irideus)*. These two species are listed as Federal Threatened under the Endangered Species Act, and this in turn protects their spawning creek habitats, which they often share with the California Roach. The introduction of piscivorous sunfish (Centrarchidae) species such as the Green Sunfish *(Lepomis cyanellus)* in warm-water creeks has also been highly detrimental to the California Roach. I have seen this minnow completely eliminated from several of my former ichthyology class sampling sites in Napa County as a result of Green Sunfish introduction. Like the California Roach, this sunfish is very tolerant of the higher water temperatures and lower oxygen content of isolated creek pools in inner areas of the Coast Ranges during late summer. When predator and prey are confined in such sites, eventually only the piscivore remains.

HITCH *Lavinia exilicauda*

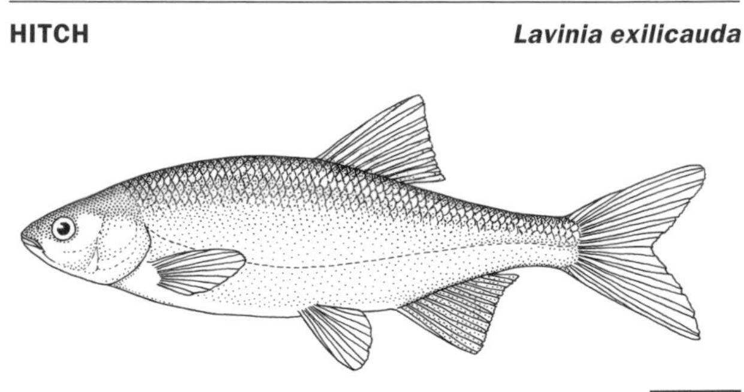

NATIVE

The Hitch is found in sloughs, streams, and reservoirs connected to San Francisco and Monterey Bays (pl. 37). It also occurs in the Russian River and in Clear Lake where it uses the lake's feeder creeks for spawning. Although it has the potential for large growth (up to 35 cm [13.8 in.]) in lakes, its normal length in streams usually falls within the range of 15 to 25 cm (6 to 10 in.). The Hitch follows the normal native minnow life plan, achieving sexual maturity at two to three years and surviving no more than six years. Spawning takes place in creeks and streams, including those that feed lakes and reservoirs. This medium-sized minnow may produce over 60,000 eggs, but more commonly about half that many. The female broadcasts the tiny, nonadhesive eggs over stream gravel beds where they then swell and become lodged within the gravel lattice. Eggs hatch within a week and the larvae soon move to either deep stream pools or an adjacent lake or reservoir. This early evacuation of the spawning site is important because by midsummer many such areas have dried.

Plate 37. Hitch.

The stream-dwelling Hitch feeds on aquatic insect larvae and adults and, like the California Roach *(L. symmetricus)*, also eats filamentous algae. A large Hitch in my home native fish pond regularly takes insects from the water surface much the way a trout or Bluegill *(Lepomis macrochirus)* does. In lakes, it filter feeds on zooplankton, and the introductions of Threadfin Shad *(Dorosoma petenense)* and Inland Silverside *(Menidia beryllina)* are likely one of the main reasons for this minnow's slow decline throughout its range. This type of feeding niche competition has been compounded in Clear Lake by the degrading of Hitch spawning creeks through gravel mining, artificial barrier construction, and a general lowering of the surrounding water table that in turn causes early creek drying. For these reasons, the Clear Lake subspecies *(L. e. chi)* has been designated a California Species of Special Concern.

LAHONTAN REDSIDE FOLLOWS ➤

LAHONTAN REDSIDE *Richardsonius egregius*

NATIVE

This species is without a doubt the most beautiful of the California cyprinids (pl. 38). Both sexes develop a lateral rose (female) to scarlet (male) stripe during the breeding season, which is offset above by a brassy olive back and black lateral stripe, and below by a silver belly. When the breeding season is over, the scarlet reverts to a shade of dark brown. This is also one of the smallest California cyprinids, with the mature adult rarely exceeding 8 cm (3 in.). The Lahontan Redside is abundant throughout the Lahontan Province and has also been introduced into a handful of other

Plate 38. Lahontan Redside: gravid female above, male below.

areas, presumably by bait anglers who possibly felt that this flashy morsel would increase their probability of success.

In Lake Tahoe this species remains in large shoals throughout most of the year. It is a minnow of the inshore area, where it feeds on a variety of insect larvae and other aquatic invertebrates. Spawning takes place in early summer throughout most of its range and entails large aggregations in shallow, gravel-bottom areas of the lake or in tributary creeks. About 1,000 adhesive eggs are shed by each female as large groups of both sexes continuously circle a spawning site. Upon hatching, the young seek out calm backwater areas and begin feeding on zooplankton. The Lahontan Redside becomes sexually mature at three to four years and lives to be five or six years. Two other native minnow species, the Tui Chub *(Siphateles bicolor)* and the Speckled Dace *(Rhinichthys osculus)*, also use such spawning sites in Lake Tahoe and hybridize occasionally with the Lahontan Redside.

SPECKLED DACE *Rhinichthys osculus*

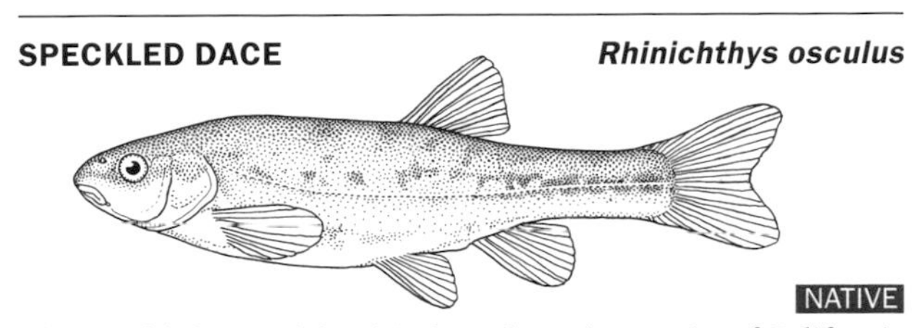

NATIVE

The Speckled Dace (pl. 39) is the only native species of California freshwater fishes that occurs in all six fish provinces (fig. 1). It is also the most widely distributed fish west of the Rocky Mountains. Although numerous subspecies exist, all exhibit the same basic body plan and habits. This is a small cyprinid, rarely exceeding 8 cm (3 in.) in length, with a slightly inferior mouth adapted for feeding on a large variety of small, bottom-dwelling invertebrates. The success of this species lies partly in its predator-avoidance behaviors such as nocturnal feeding coupled with the ability to hide among the bottom rocks during daylight hours. It also avoids forming large aggregations, except at spawning time, and instead forages in small groups that can easily blend into the bottom rocks at the approach of a potential predator. This universal method of defense apparently works very well against all types of predators, including several piscivorous avian species.

Spawning occurs throughout the summer months. As the fishes work their way among the stream bottom rocks and gravel, they lay and fertilize the eggs. Like most other minnow fry, the young seek out calm inshore areas where zooplankton is available.

As a species, the Speckled Dace continues to maintain good populations throughout most of its range. As with several other groups whose distribution includes desert creeks and southern California coastal rivers, however, two isolated populations are in trouble and listed as a California Species of Special Concern. One

Plate 39. Speckled Dace.

is the Owens Speckled Dace, which is being decimated by introduced Brook Trout *(Salvelinus fontinalis),* and the other is the Santa Ana Speckled Dace whose habitat in Los Angeles area streams continues to be converted to flood-control channels. Neither of these proposed subspecies has been formally described yet, but if and when they are, Federal Threatened or Federal Endangered status may be assigned.

TUI CHUB *Siphateles bicolor*

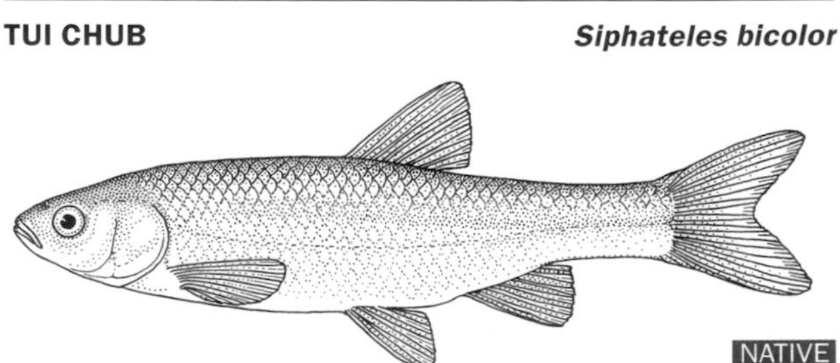

NATIVE

Cyprinids of the genera *Gila* and *Siphateles* are, in general, of small to medium size (20 to 30 cm [8 to 12 in.]) with a heavy build. The name "chub" is a potentially confusing term because it is used for a variety of fish types, including several species of whitefish salmonids in the Great Lakes region where smoked chub is a delicacy. These are about a foot long, and the whole, gutted fish is smoked to a golden brown. In California the only "chubs" are minnows of the genera *Gila* and *Siphateles,* which contain more species than any other native minnow genus.

Barnes & Noble Booksellers #1944
3600 Stevens Creek Blvd
San Jose, CA 95117
408-984-3495

STR:1944 REG:002 TRN:3214 CSHR:Yvette C

BOOKFAIR #10076487

2010 Trout of North Amer
9780761153269
(1 @ 12.99) 12.99
Field Guide to Freshwate
9780520237278
(1 @ 24.95) 24.95
2010 California Wild & S
9781421654249
(1 @ 14.99) 14.99

Subtotal 5
Sales Tax (9.250%) 4.
TOTAL 57.8
VISA 57.8
Card#: XXXXXXXXXXXX5874
Expdate: XX/XX
Auth: 03567B
Entry Method: Swiped

A MEMBER WOULD HAVE SAVED 5.30

Thanks for shopping at
Barnes & Noble

19 11/13/2009 01:35PM

CUSTOMER COPY

prior to the date of return, (ii) when a gift receipt is presented within 60 days of purchase, (iii) textbooks returned with a receipt within 14 days of purchase, or (iv) original purchase was made through Barnes & Noble.com via PayPal. Opened music/DVDs/audio may not be returned, but can be exchanged only for the same title if defective.

After 14 days or without a sales receipt, returns or exchanges will not be permitted.

Magazines, newspapers, and used books are not returnable. ***Product not carried by Barnes & Noble or Barnes & Noble.com will not be accepted for return.***

Policy on receipt may appear in two sections.

Return Policy

With a sales receipt, a full refund in the original form of payment will be issued from any Barnes & Noble store for returns of new and unread books (except textbooks) and unopened music/DVDs/audio made within (i) 14 days of purchase from a Barnes & Noble retail store (except for purchases made by check less than 7 days prior to the date of return) or (ii) 14 days of delivery date for Barnes & Noble.com purchases (except for purchases made via PayPal). A store credit for the purchase price will be issued for (i) purchases made by check less than 7 days prior to the date of return, (ii) when a gift receipt is presented within 60 days of purchase, (iii) textbooks returned with a receipt within 14 days of purchase, or (iv) original purchase was made through Barnes & Noble.com via PayPal. Opened music/DVDs/audio may not be returned, but can be exchanged only for the same title if defective.

After 14 days or without a sales receipt, returns or exchanges will not be permitted.

Magazines, newspapers, and used books are not returnable. ***Product not carried by Barnes & Noble or Barnes & Noble.com will not be accepted for return.***

Policy on receipt may appear in two sections.

Return Policy

With a sales receipt, a full refund in the original form of payment will be issued from any Barnes & Noble store for returns of new and unread books (except textbooks) and unopened music/DVDs/audio made within (i) 14 days of purchase from a Barnes & Noble retail store (except for purchases made by check

Plate 40. Tui Chub.

The Tui Chub is the most abundant and widespread species of this genus and occurs as a number of subspecies throughout the state (pl. 40). It is a minnow of slower-moving streams or of lakes and is usually associated with abundant aquatic vegetation. In such areas, this species travels in small groups, feeding on a variety of plant-dwelling invertebrates. In more oligotrophic lakes it forms large, pelagic schools and feeds on both zooplankton and bottom invertebrates. It also has a high tolerance for alkaline conditions (up to pH 11), and both Eagle Lake and Goose Lake contain subspecies that bear their names.

Because of its large body size, the Tui Chub female is very fecund, with egg numbers approaching 50,000 in the ***Mojave Tui Chub*** subspecies ***(S. b. mohavensis).*** Spawning takes place anywhere from midspring through summer, depending on the habitat, and follows the basic minnow format. Two desert subspecies, the ***Owens Tui Chub (S. b. snyderi)*** and the Mojave Tui Chub, have been designated as both State Endangered and Federal Endangered. Each spring for the past two decades, I have journeyed to California State University's Desert Studies Center at the former Zzyzx Springs health resort on the "shore" of extinct Soda Lake to teach students about desert fishes and wildlife, including the Mojave Chub, which colonized several small natural ponds here during a historic Mojave River flood. The Arroyo Chub *(Gila orcutti)* was later introduced into this river and completely replaced the Mojave Chub. This small population at Zzyzx Springs represents the last survivors. Except for a few pupfish (Cyprinodontidae) species and subspecies, no other California native freshwater fish has been reduced to such a small, isolated habitat (pl. 12).

BLUE CHUB *Gila coerulea*

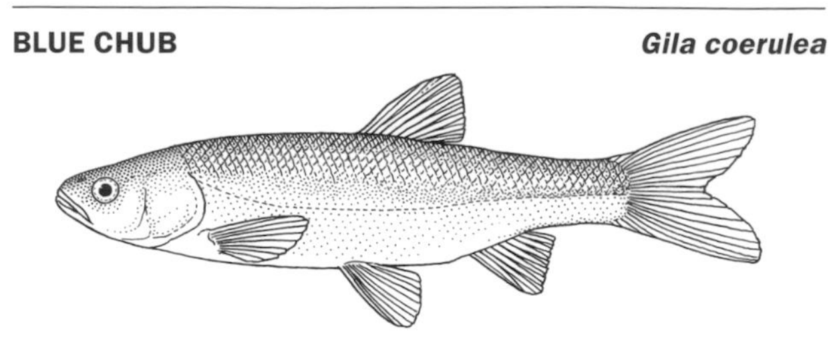

NATIVE

The Blue Chub is abundant throughout the upper Klamath drainage and closely resembles Tui Chub *(Siphateles bicolor)* subspecies in that area, although it grows larger (to near 40 cm [15.7 in.]) and lives longer (17 years or more). The Blue Chub is normally found in large schools in deep water or in rocky inshore areas, apparently avoiding the very shallow shoreline habitat. By this means, it apparently avoids direct competition for food with the Tui Chub, which often occurs in the same habitat. Competition for feeding niches is also reduced because of the Blue Chub's omnivorous diet, which often includes a high percentage of filamentous algae. This species has adapted well to the reservoirs that have been constructed in the Klamath system, despite the presence of a number of introduced piscivorous species. Spawning is similar to that of the other minnows.

ARROYO CHUB *Gila orcutti*

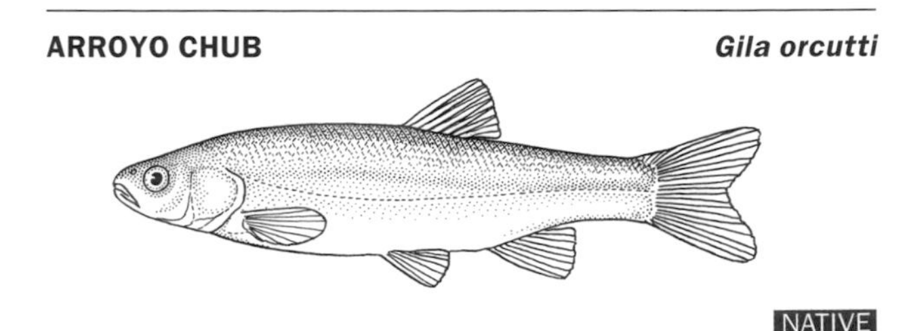

NATIVE

PROTECTIVE STATUS: CSSC.

This is the chub species of southwestern California and is found in the coastal streams of Los Angeles, Orange, and San Diego Counties. Like the California Roach *(Lavinia symmetricus)*, it is adapted for the rigors of the seasonal changes in coastal streams, particularly the intermittent summer stage. This ability most likely gave it a competitive edge when it was introduced into the

Mojave River during a time when water diversions created intermittent creek conditions; it eventually replaced the Mojave Chub *(Siphateles bicolor mohavensis)* there. The Arroyo Chub is small as chub-type cyprinids go, with a maximum size of only 9 cm (3.5 in.). It is also short lived, reaching sexual maturity at one year and living no more than four. It feeds extensively on aquatic vegetation and on invertebrates associated with such plants. Breeding follows the classic minnow pattern and takes place in spring when water conditions are optimal for fry production.

THICKTAIL CHUB and BONYTAIL — *Gila crassicauda and Gila elegans*

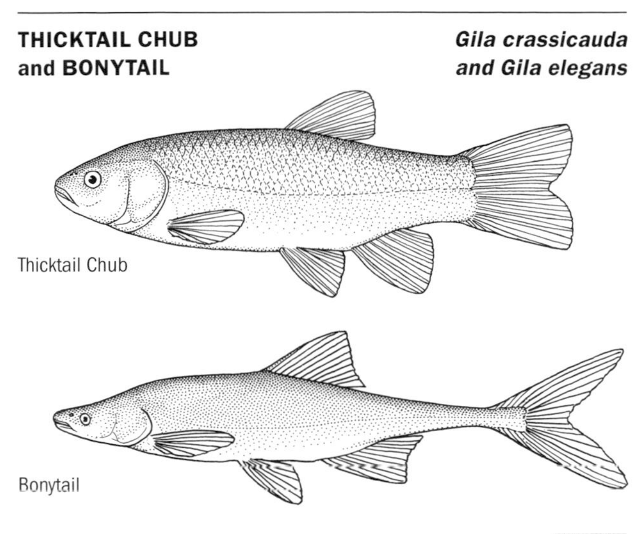
Thicktail Chub

Bonytail

NATIVE

PROTECTIVE STATUS: SE, FE (for Bonytail).

Obituary: These two species, beloved natives of the California fish fauna, are here no more. The Thicktail Chub (top) appears to be extinct. A few wild Bonytails (below) may still exist in portions of the upper Colorado River system, but today the only known populations are the result of hatchery fishes planted in Havasu Reservoir in Nevada and the Upper Colorado and Green Rivers. The Thicktail Chub was a cyprinid of the slow, backwater areas of the Delta and lived in close association with tule beds. The extensive channelization of the Delta system has virtually destroyed this habitat and, with the introduction of numerous predatory fishes, apparently has led to the disappearance of this native minnow. The loss is especially ironic in light of recoveries of fish bones from the Patwin Native American village of Tsaki in southern Colusa County. Of the 11 species identified from the middens, by far the most numerous was the Thicktail Chub.

The Bonytail, like other native species of the lower Colorado River, has been unable to adjust to the drastic, human-caused changes in this habitat that have taken place over the past half-century. It prefers the flowing, silty water that impoundments have eliminated from the lower Colorado River. The Bonytail's unique body design with long, caudal peduncle and large, deeply forked tail suggests that it was adept at handling the periodic high flows of the pristine, unregulated Colorado River. A recent study of a population in Utah shows the Bonytail to be a surface feeder on invertebrates and a variety of plant material. This is the biggest of the western chubs, with the larger sex, the female, reaching nearly 40 cm (about 16 in.) total length. Spawning in the upper Colorado River takes place in late spring over gravel beds in the normal cyprinid fashion. Given the present state of the lower Colorado River, it is unlikely that this species will ever again become established in California waters.

The Thicktail Chub and the Bonytail have been granted State Endangered and Federal Endangered status, and in 1980 the California Fish and Game Commission declared the Thicktail Chub "extinct" (the last known specimen was caught in the Sacramento River in 1950).The philosophy of this author, however, is that without continued full treatment in a field guide, the possibility of an "eleventh hour" discovery is extremely unlikely.

SACRAMENTO SPLITTAIL *Pogonichthys macrolepidotus*

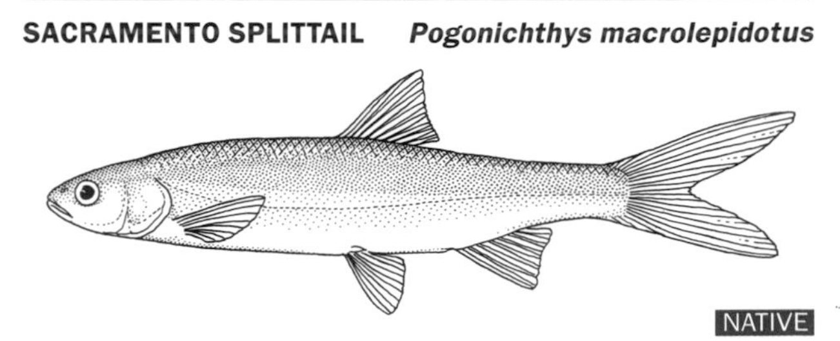

NATIVE

PROTECTIVE STATUS: FT, CSSC.

The Sacramento Splittail is a medium to large (up to 45 cm [about 18 in.] total length) minnow of the lower Delta (pl. 41). It frequented the backwaters, sloughs, and spring floodplains of the pristine Delta area, but with the loss of nearly all such habitat due to the diking and draining of the great Central Valley marshland, it has become scarce enough to merit Federal Threatened designation. Fortunately, the Sacramento Splittail appears to have the ability to produce large year classes of young during very wet years when former backwater areas temporarily flood and levees

Plate 41. Sacramento Splittail.

occasionally break. It can also tolerate salinities into the mid-20-ppt range and is often found in brackish water in the lower reaches of Suisun Bay. For instance, in 2001, I observed adult splittails in a brackish-water slough just east of the Benicia Bridge in Contra Costa County, and the year before, I inadvertently captured and immediately released two adults in a small slough on the San Joaquin River near Modesto. This ability to utilize its preferred backwater habitat throughout the wide geographic and salinity ranges of the greater Delta system may be the saving grace for this species.

The Sacramento Splittail feeds on a wide range of invertebrates, including some larger forms such as small clams and earthworms. Spawning takes place during spring in backwater areas where shore vegetation has been temporarily flooded. Because of her large size, the female is very fecund, producing up to 100,000 eggs. This high egg number is also good insurance that a large year class will be produced during occasional high-rainfall years, which seems to allow this species to "bounce back." Adhesive eggs are scattered over the flooded vegetation and assorted organic debris. The larvae forage in these flooded areas where populations of small zooplankton erupt in spring and then move to the main slough channels. With its life span of eight years or more, the best hope for this native species is that years of heavy rainfall will continue to occur with enough frequency to keep the it afloat until we restore some of its vital inshore habitat.

CLEAR LAKE SPLITTAIL *Pogonichthys ciscoides*

NATIVE

The Clear Lake Splittail was a lake-adapted form of the Sacramento Splittail *(P. macrolepidontus)* found only in Clear Lake, Lake County, and its tributary streams. It has been declared extinct, a near-certain fact given its restricted habitat and that much recent collecting in the Clear Lake system has produced no specimens.

It was primarily a surface feeder in the littoral zone and relied heavily on gnats during summer. It became scarce as the Clear Lake feeder creeks where it spawned were diverted and their banks channelized. Its biggest population reduction coincided with the introduction of the Bluegill *(Lepomis macrochirus)* and the Bluegill's subsequent dominance of the inshore area. The final blow to the Clear Lake Splittail may have been dealt in 1967 with the introduction of the Inland Silverside *(Menidia beryllina)* into Clear Lake. Within a few years the Inland Silverside replaced the Bluegill as the most abundant inshore species and in the process may well have completely eliminated the Clear Lake Splittail.

Large Native Minnows

HARDHEAD ***Mylopharodon conocephalus***

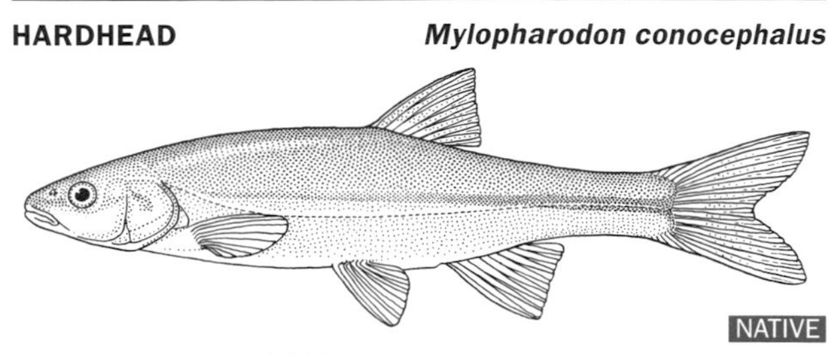

PROTECTIVE STATUS: CSSC.

The Hardhead (pl. 42) is still fairly numerous in the Sacramento–San Joaquin and Russian River drainages (fig. 8). It may attain nearly 60 cm (about 2 ft) total length and is currently the second-largest cyprinid in California. This large size coupled with omnivorous feeding behavior allows it to feed on a wide variety of foods, ranging from zooplankton and filamentous algae to snails and crayfishes. In its native stream habitat the Hardhead occupies deep pools and seldom strays into adjacent shallows. Its spawning behavior has not been studied but is presumed to be similar to that of the Sacramento Pikeminnow *(Ptychocheilus grandis)*. Although still abundant, the Hardhead is presently en-

Plate 42. Hardhead.

countering stiff competition in its stream-pool habitat, much of which has been eliminated through water diversions and upstream blocking. The introduction of large-stream piscivores such as the Smallmouth Bass *(Micropterus dolomieu)* has also eliminated or greatly reduced Hardhead populations in larger rivers. Given these impacts, the Hardhead deserves its designation as a California Species of Special Concern and the possibility that this status will result in the management of some streams specifically for this species.

SACRAMENTO BLACKFISH *Orthodon microlepidotus*

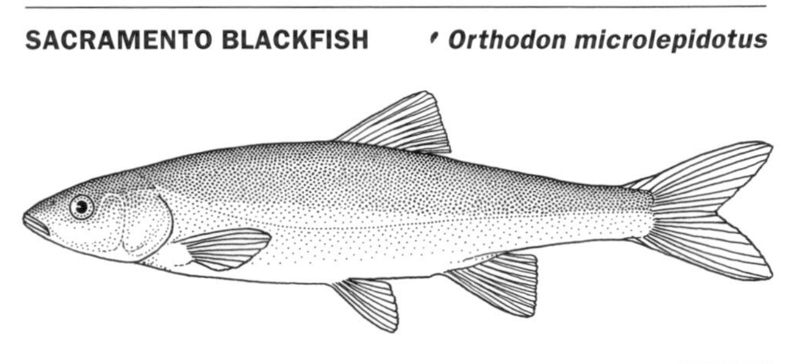

NATIVE

The biology of this large native cyprinid vividly reflects the great shallow lake and backwater habitats that once occupied much of California's Central Valley. It feeds heavily on phytoplankton, which must have been very abundant in these vanished habitats. It also eats a variety of small zooplankton species and detritus suspended in the water column, which it accomplishes by pumping large volumes of water in its mouth and out its opercular opening. This is a very pronounced, continuous activity in which the fishes appear to be chewing or gasping for air. This action moves far more water than normally needed for respiration, but it is necessary for feeding because of the small ratio of suspended food to water volume. Like other filter feeders, the Sacramento Blackfish has many long, thin gill rakers. In this case, however, the gill rakers do not strain out food but instead act as deflectors to channel the water flow to the roof of the mouth. There a continuous film of mucus is secreted to which the minute food particles adhere, and eventually bands of mucus and food are swallowed. Using this method, the relatively large Sacramento Blackfish exploits a feeding niche normally occupied only by very small fry and larvae.

Complementing this unique feeding behavior is its equally impressive physiological adaptation to the summer physical environment of shallow Central Valley lakes and sloughs. The Sacramento Blackfish has a higher preferred temperature range (22 to 28 degrees C [72 to 82 degrees F]) than any other native species except those of the pupfish family (Cyprinodontidae), and the young (pl. 43) can survive at temperatures near that of the human body (about 37 degrees C [98.5 degrees F]). This tolerance allows Sacramento Blackfish fry to feed in very warm, shallow inshore areas where the fry of other natives cannot. Perhaps even more astounding is this species' ability to survive and thrive at extremely low water oxygen levels. In the discipline of

Plate 43. Juvenile Sacramento Blackfish.

fish physiology, a standard graphic representation, the oxygen dissociation curve, compares the oxygen loading or "grabbing" ability of the blood of different vertebrate species. Sacramento Blackfish blood (hemoglobin) is featured on such representations as the epitome of this ability as it can glean oxygen from water in which other species would suffocate. Added to all this is the ability to tolerate high alkaline levels (up to pH 10) that is also found in several other native fishes of the Central Valley floor.

The Sacramento Blackfish is currently the third-largest native minnow in California. It can grow to about 50 cm (20 in.) and weigh up to 1.5 kg (3.3 lb). Females five years or older approaching this upper size can produce up to 300,000 eggs. Spawning occurs in late spring and summer when inshore waters have warmed. Eggs are broadcast over a variety of bottom objects, including vegetation, to which they adhere. Newly hatched fry remain within dense aquatic vegetation where they are able to subsist solely on phytoplankton if small zooplankton is temporarily scarce. Growth is fairly rapid, and after two years most fry have grown to about half the maximum length (about 25 cm [10 in.]). This puts them beyond the prey size range of several of the smaller introduced predators about the time they reach spawning age. Maximum age appears to be 10 years.

The Sacramento Blackfish is far less abundant today than in pristine California, but good numbers still occur in the remaining backwater areas of the Delta, Clear Lake, and several reservoirs. A small commercial fishery for the Sacramento Blackfish, mainly in Clear Lake, has existed for many decades. It is centered around the Chinese-American food trade, and the large display tanks in Asian fish markets often contain this fish. These tanks are an excellent place for a close-up view of the pronounced water-pumping action of the mouth. If you like exceptionally fresh fish, bring a five-gallon bucket and take one home. One suggested recipe is provided in the culinary section.

SACRAMENTO PIKEMINNOW *Ptychocheilus grandis*

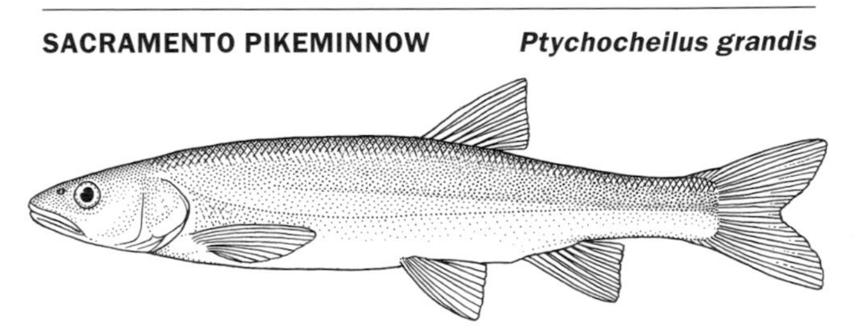

NATIVE

Pikeminnows attract more attention than any other cyprinids in California or, for that matter, North America, largely because of their new name. With their minnowlike fin structure and pharyngeal teeth, they certainly qualify as a member of Cyprinidae, but their elongate mouth and body coupled with their aggressive piscivorous behavior are reminiscent of the pike family (Esocidae)(pl. 95). What we really have here is a beautiful example of how major feeding niches are filled through the natural selection process regardless of the basic stock from which the final fishes are derived. To understand this process, first recall that several

Plate 44. Sacramento Pikeminnow.

million years ago two families noted for large piscivores (Ictaluridae and Centrarchidae) became all but extinct in California. Only a few medium-size freshwater piscivores (Sacramento Perch [*Archoplites interruptus*], Rainbow Trout [*Oncorhynchus mykiss*], and Cutthroat Trout [*O. clarki*]) remained, leaving room "at the top." Through natural selection the Sacramento Pikeminnow gradually evolved to fill this role. With their large size, elongated mouth, and modified pharyngeal teeth that look and act like sharp cutting knives, these minnows do indeed resemble true pike in both form and behavior. I occasionally conduct piscivore-feeding demonstrations in my ichthyology class using small "feeder" Goldfishes (*Carassius auratus*) and an 8- to 10-in. Largemouth Bass (*Micropterus salmoides*), an introduced species known for its aggressiveness and gluttony. During one such demonstration, I happened to have on hand a Sacramento Pikeminnow of equal size, and so a dual demo was held in separate aquariums. After the demonstration, the vote for Piscivore of the Day was unanimous in favor of the Sacramento Pikeminnow, whose only failing was that it didn't swallow its prey as fast as the bass because it sliced the goldfish up as it swallowed each one.

This species is still well distributed in creeks throughout the Central Valley, Sierra foothills, and inner portions of the Coast Ranges. In streams, it prefers large pools from which it makes foraging trips, and to which it always returns. Its tolerance to high water temperatures (up to 36 degrees C [97 degrees F]) allows it to survive in intermittent creek pools through late summer and fall until the rains begin. It can also tolerate slightly brackish water. I have collected pikeminnows in a slough in lower Suisun Bay where high tide salinities were in the 5 to 6 ppt range.

The Sacramento Pikeminnow becomes sexually mature by age four when it averages 25 cm (nearly 10 in.) (pl. 44). At this time the male acquires a faint red orange tint to its fins. A large female may lay up to 40,000 adhesive eggs, which she broadcasts over bottom gravels and rocks. The fry grow fast and feed on the usual variety of zooplankton and other aquatic invertebrates. When between 10 and 20 cm (4 and 8 in.) the young turn piscivorous and remain so throughout life. With a maximum age upward of 12 years, and because fishes are indeterminate growers, old Sacramento Pikeminnows can become very large. The average length of an old fish for most habitats is about 60 cm (2 ft). The record for this species, however, is 115 cm (45 in.) and 14.5 kg (32 lb). And remember, this is a "minnow"!

ANGLING NOTES: No angling tips have been offered in the minnow section so far for obvious reasons. Here, however, is a cyprinid with all three qualifications of a top game fish: it is large, it is a piscivore, and it feeds aggressively. Anglers do occasionally hook a large Sacramento Pikeminnow and thoroughly enjoy the experience until they reel in the exhausted fish and find that it is not the bass or trout they had so ardently sought. My one angling tip for this situation is, do not throw it back! Take it home to the kitchen and try it—you may very well like it. I and a few friends once held a panfish cook-off that featured Largemouth Bass, Bluegill *(Lepomis macrochirus)*, and 10- to 12-in. Sacramento Pikeminnows. The bass, of course, was dead last in the voting, but the Bluegill and Pikeminnow tied for first place. Of course, in many parts of the world, large cyprinids are a favorite food, and some carp species play a prominent role in holiday feasts.

It would be nice to see the Sacramento Pikeminnow promoted as a game fish in this state and a number of deep-pool stream habitats managed for this fish. So far though, a small handful of introduced and hatchery-stocked piscivorous gamefish (Largemouth Bass, Striped Bass *[Morone saxatilis]*, Channel Catfish *[Ictalurus punctatus]*, and large Rainbow Trout cultigens) have held the California Department of Fish and Game's attention and with it an ensured future in California's freshwaters. Maybe it is time that a similar insurance plan is written for our native top piscivore, the Sacramento Pikeminnow.

COLORADO PIKEMINNOW *Ptychocheilus lucius*

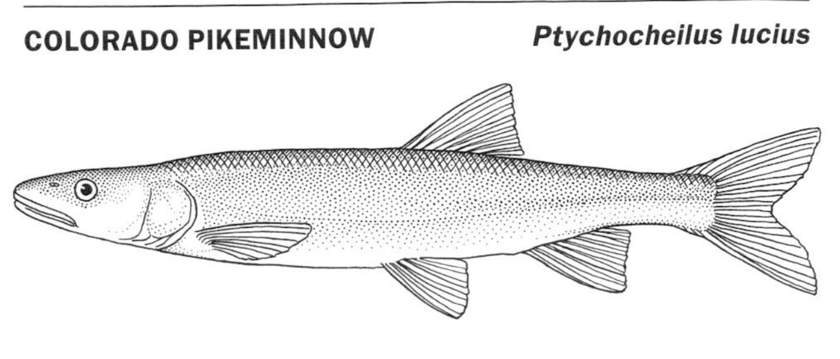

NATIVE

PROTECTIVE STATUS: SE, FE, FPS.

This species, which resembles the Sacramento Pikeminnow *(P. grandis)* in both appearance and behavior, has the distinction of being the largest native cyprinid in North America, and in the pristine lower Colorado River it has reportedly attained lengths up to 1.5 m (5 ft). It also has the dubious distinction of being pre-

Plate 45. Colorado Pikeminnow.

sumed extinct in California. Fortunately, populations of Colorado Pikeminnow still exist in upper portions of the Colorado River, and some hatchery fishes have been released in the lower Colorado (pl. 45).

Juvenile Colorado Pikeminnows feed mainly on aquatic insect larvae, but at between 10 and 20 cm (4 to 8 in.) they become piscivorous and feed on suckers (Catostomidae) and other minnow species the rest of their long lives (about 12 years). Like the Bonytail *(Gila elegans)* and the Razorback Sucker *(Xyrauchen texanus)*, this species has not been able to adapt to the extreme human-wrought changes in the lower Colorado. A recovery program now exists for this species, centered on rearing fishes in a New Mexico fish hatchery and releasing them back into areas where they have disappeared. The good part of this plan is that it protects the gene pool through captive rearing, even though a hatchery-adapted form will eventually emerge. The bad part of the plan is the lack of habitat restoration. Until the basic principle that endangered habitats produce endangered species is fully realized, we will not see the Colorado Pikeminnow on the California side of the state line.

Introduced Minnows

Of the 20 species of cyprinids currently living in California freshwater habitats, seven have been introduced, and all but one of these introductions has been promoted at one time or another by state agencies. That is an alarming ratio given that most sanctioned introductions have been of either game-fish species or small, noncyprinid prey species for the former to feed on. This begs the basic question of why foreign minnows should be introduced into a state that already has its own unique native complement. I believe this can be answered by reviewing cyprinid introductions under three categories: (1) the Great Carp Hard Sell, (2) the "free Goldie" syndrome, and (3) the CDFG bait-minnow boondoggle.

COMMON CARP *Cyprinus carpio*

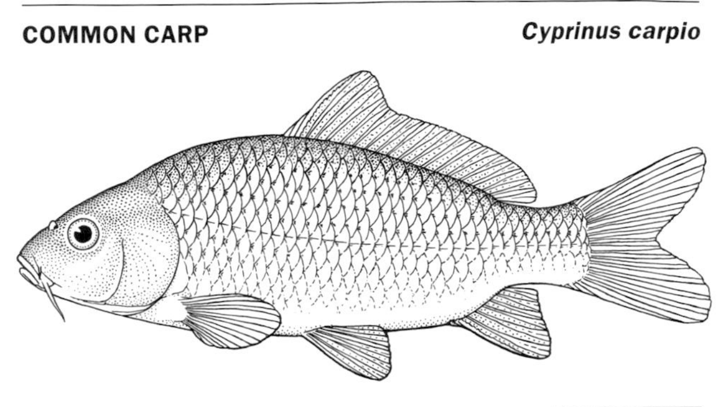

INTRODUCED

In 1872, Mr. Julius Poppé imported five Common Carp from Germany and stocked them in a pond in Sonoma County. This seemingly trivial event has historic significance in that it was one of the first freshwater fish introductions in California. Later in that decade, the California Fish Commission began a carp-rearing program with stock imported from Japan. That same year, a similar program using German carp was begun in Washington, D.C., and from these sources an intensive stocking program was launched across the United States.

It may be hard for people today to understand this nineteenth-century love affair with the Common Carp without briefly considering the history of this species and man. The Common Carp is a native of Asia and the waters of the Black Sea basin, including the Danube River of Europe. It was probably the first fish ever cultured or farmed by humans. European monasteries had Common Carp ponds and distributed breeding stock to English monasteries in the thirteenth century. The moats surrounding castles were among the many habitats utilized for carp propagation. On both continents it was menu fare for the royal and the rich. The propagation efforts in this country in the 1880s were accompanied by an advertising campaign that hailed the Common Carp as better tasting than any North American species. This claim was further supported by European and Asian immigrants who, though they may not ever have tasted carp, were acquainted with its royal banquet reputation in their homelands. All this culminated in what might be called "carp fever," with people actually standing in line to obtain live carp for stocking in their favorite pond, lake, or stream. Unfortunately, this particular American dream was short-lived.

First was the problem with the taste and texture of carp flesh. Americans were already accustomed to the generally excellent taste and consistency of most North American game fishes. They had very few small bones in the flesh, and simple pan frying brought out their fine taste. Carp flesh, on the other hand, is riddled with small bones, and if taken from warm, eutrophic waters with large blue-green algae populations it is rather muddy tasting. In contrast, Common Carp held in clean, cool water for a week or two prior to harvest rarely has this poor taste. A number of Oriental and European recipes are designed to counteract these bad points (see the cooking section), but relatively few immigrant groups were aware of these recipes. The end result was a lot of disappointing fish dinners and a general disenchantment with the "Great Carp Hard-Sell."

An even ruder awakening to the disadvantages of this introduction came as the Common Carp displayed its feeding behavior in countless North American freshwater habitats. Like most fry, young carp (pl. 46) feed on zoo- and phytoplankton. As they grow larger they concentrate more on bottom-dwelling invertebrates. The Common Carp is also very skillful at "grubbing," a feeding method whereby the bottom substrate is taken into the mouth, invertebrate food is extracted, and silt and sand are rejected. Carp extend this type of feeding to aquatic vascular plants, which also provide a substrate for invertebrate species, and in the process of mouthing, they take in and eat parts of the plants.

Because the inshore or littoral zone of a lake or pond produces the vast majority of carp foods, it is this area that can be literally torn up by a large population of these "freshwater hogs." This is exactly what happened in countless eastern and Midwestern lakes, and as a result, the fry nurseries for most other species in these habitats were severely damaged and water turbidity increased.

In California freshwater habitats, the presence of Common Carp often goes unnoticed because of the lack of a real littoral zone in which to display these feeding habits. The lower Stanislaus River, for example, is home to a Common Carp population that thrives on Opossum Shrimp *(Neomysis mercedi)* and other stream-bottom invertebrates. The combination of stream flow and essentially no backwater areas prevents detection by the nor-

Plate 46. Juvenile Common Carp.

mal "muddy water" method, and only in summer can you observe carp as they cruise through the clear, shallow inshore waters. In many reservoirs, the near-complete lack of a littoral zone seems to sufficiently control the Common Carp population. If an extensive inshore zone does exist, however, this fish's presence is felt in much the same way as in other states. Lake Chabot, Alameda County, has one extensive, shallow, vegetated bay area that is potentially an excellent hatchery ground and adult fish habitat for numerous introduced sport fish species. It is continuously held well below the threshold of this potential, however, by a thriving Common Carp population.

Size and fecundity are two additional factors that aid the Common Carp in its shallow-water dominance. This species is the third-largest minnow now inhabiting western waters (the Colorado Pikeminnow *[Ptychocheilus lucius]* and Grass Carp *[Ctenopharyngodon idella]* are numbers one and two, respectively). The world record Common Carp, taken in South Africa, was 75 cm (29.5 in.) total length and weighed almost 40 kg (88 lb). The California record Common Carp, from Lake Nacimiento, San Luis Obispo County, weighed 26.3 kg (about 58 lb). And as with the Sacramento Pikeminnow, we need to occasionally remind ourselves that this is a minnow!

Common Carp reproductive potential is also enormous. A large female may lay up to two million eggs in one season. The eggs normally are deposited over vegetation to which they adhere.

This spawning behavior permits the Common Carp to successfully compete with substrate nest builders such as the centrarchids (sunfish and blackbass), whose bottom-spawning sites are severely damaged by continuous carp foraging. Common Carp eggs hatch in three to six days, and the fry feed on zoo- and phytoplankton for the first several months of life, after which they switch to the adult feeding mode.

The Common Carp in North America is a most controversial fish. Some strains, such as the beautiful, domesticated Koi Carp, are highly prized by fish collectors, with individual specimens commanding prices as high as several thousand dollars. This fish remains highly accepted in many areas of the world, but not, at the present, in North America. Like it or not, one thing seems certain: the Common Carp is here to stay, and in biomass it is the most abundant freshwater fish in North America. As with other successful introductions, complete eradication by netting or rotenone poisoning is impractical, if not impossible, in any but the smallest pond habitats. Small commercial fisheries for Common Carp exist in Clear Lake and San Luis Reservoir, which help to moderately control it. But even if carp fisheries were to increase significantly, they would tend to concentrate on specific, high-yield sites, leaving the bulk of California carp habitats untouched. A better solution may be a constant and intensive sportfishing pressure, as discussed below.

ANGLING NOTES: Izaak Walton, the father of American sportfishing, called the Common Carp the "Queen of the Rivers," and a widely scattered group of carp anglers in the United States agree with this royal title. Recent evidence supports the view that a revival of this fishing philosophy is now under way, and for the past two decades carp derbies have been held in several Central Valley towns.

Carp derbies were modeled on the better-known bass derbies and resulted in good catches for good prizes such as money and fishing equipment. Coinciding with this small but steady revival is the sporadic appearance of pamphlets and newspaper articles offering tips and recipes for cooking and smoking carp. In addition, the ever-growing interest in Asian cuisine has brought into the American home the only class of cookbooks regularly calling for carp. Over a century after its introduction in the United States, the Common Carp may finally realize some of the lofty goals it inspired. If so, an even more important result—the partial reduction of the carp's destructive nature in countless California and other North American habitats—may be achieved.

Carp fishing in Europe is taken seriously, and some devotees adhere to a catch-and-release philosophy. England even has a carp angling society. In spring just before spawning, the Common Carp becomes rambunctious and aggressive, often leaping out of the water for no apparent reason. At this time it will often attack artificial lures. In ponds where tuffed cottonwood seeds fall on the water surface, a carp will suck in these bits of fluff. Such behavior has stimulated interest in fly fishing for the Common Carp, using large white flies that simulate cottonwood blossoms.

As for catching the Common Carp, bottom fishing with bait is the only truly productive method. Do not use bobbers, because a carp likes to carry bait before swallowing it, and such line floats tend to deter this behavior. The range of effective baits is one of the widest found in fishing: soft corn kernels, marshmallows, cheese, and a variety of insects seem to be among the most popular. The most consistently successful bait is a doughball prepared from cornmeal, sugar, and water, with added tempting items such as chicken livers worked in for good measure. A tight, sticky consistency is important, because the bait must stay on the hook throughout the initial mouthing that usually accompanies this species' food intake.

Once hooked, the Common Carp gives the angler a good tug, although it is not in the trout-bass category of surface leapers and splashers. Instead, it usually takes a series of long, strong runs, with the apparent intention of heading for the nearest cover where your line may become hopelessly entangled. This behavior can be as challenging to the angler as the surface antics of other species as they attempt to shake the bait loose. The Common Carp has great stamina and puts up a long, admirable fight.

Sportfishing is not limited to the rod-and-reel approach. As of 2002, California fishing regulations allow bow-and-arrow fishing for "carp" and six other species. The Common Carp is the one to pursue, because five of the other six are native species, including the Hardhead *(Mylopharodon conocephalus)*, a California Species of Special Concern. Two others, the Pacific Lamprey and Goldfish *(Carassius auratus)*, might be too challenging for even an expert archer. As for the sport itself, conventional bows as well as crossbows may be used, and a line must be attached to the arrow point or shaft. A reel is often used to hold the excess line, which may be needed to "play" the fish after it is hit. Hunting is best in relatively clear water or at sites where carp are breaking the surface, such as shallow bays during the late spring spawning season. The real skill in bow hunting lies in accurate correction of the false visual input regarding the fish's exact location. Because the refractory index of water is not the same as that of air, light rays bend at the water surface, with the result that an underwater object appears farther away than it actually is. You must account for the 22 degrees of bend at the air-water interface and lower the

Plate 47. Spearfishing for Common Carp.

aim accordingly. In other words, you aim at a point a little closer than where you think you see the fish.

I cheated a bit on my first bow-and-arrow fish hunt, which was for the Duck-billed Spotted Ray *(Stoasodon narinari)* in clear Florida Keys water. With a circular target about a yard in diameter, even a novice can hit it once in a while. The big Common Carp in the Lower Stanislaus River at the end of my country road are a far more challenging quarry, however, and need to be carefully stalked as well as properly aimed at. I like the stability of a small, flat-bottom pram for this sport, with the archer standing in the bow and another person at the oars. For the beginner, it pays to practice before actually going after the big ones. A Styrofoam carp anchored to the bottom by a weight and chord so it is suspended a few feet beneath the surface of a pond or concrete swimming pool makes a good target.

Spearfishing is still another way to catch a carp and is permitted by current fishing regulations for the same group of seven fishes (pl. 47). This method is defined by the California sportfishing regulations as "the taking of fish by spear or hand by persons who are in the water and may be using underwater goggles, face plates, breathing tubes, scuba, or other underwater breathing devices." At first, the idea of spearfishing for a "big, old sluggish carp" may not seem exciting. True, many Common Carp are indeed big and old, but they are far from sluggish and can rival many of the more sought after marine species in alertness and speed. Visibility in many Central Valley foothill streams and lakes in late summer and fall is at least equal to inshore Pacific waters. For the many spear-fishing men and women who do not live near the Pacific Ocean and never seem to find the time to make all the trips to the coast they would like, some exciting sport may be within a short drive of home.

GOLDFISH *Carassius auratus*

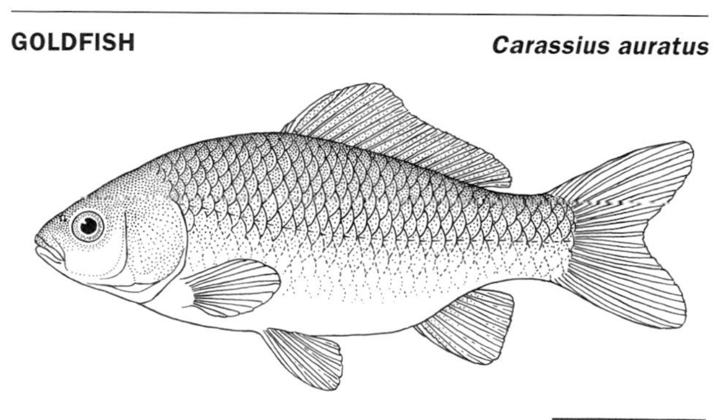

INTRODUCED

The Goldfish needs no introduction, as it is a fair assumption that the majority of Americans have shared their home with at least one member of this species. It is perhaps the most studied fish in the world, for throughout decades of laboratory research it has usually been easier for the lab biologist to make a quick trip to the variety or pet store than to wade nearby streams and lakes in search of research fishes (pl. 48).

Plate 48. Goldfish captured in a stream.

Plate 49. Silver Carp (above), Common Carp (below).

The Goldfish is found in most California freshwater habitats almost entirely because of the release of pet fishes ("free Goldie"). It is easy to understand the psychological motive behind releases of unwanted pet fishes into natural habitat as opposed to alternatives such as flushing them down the toilet. However, these private stocking operations are not only harmful to natural habitats but are also illegal. Luckily for native species, the Goldfish has not caused significant destruction of the inshore habitat, as has its cousin, the Common Carp *(Cyprinus carpio)*. The Goldfish is primarily a filter feeder on zoo- and phytoplankton and bottom organic debris and thus competes most directly with the fry of other species. Some small lakes, such as Lake Temescal in Alameda County, are completely overrun by Goldfishes, presumably through continuous dumping by residents of the surrounding city of Oakland and through natural reproduction. Fortunately, most larger lakes and streams with a good piscivore complement have not been affected in this way.

In the wild, Goldfish may grow to over 40 cm (16 in.). Such specimens often begin as pet-store runts restricted in their growth by the confines of the fishbowl. They can also hybridize with the Common Carp, both in captivity and the wild. During extensive sampling in a restored marsh in Contra Costa County, I found more hybrids, or "Silver Carp" as they are called, than either Goldfish or Common Carp (pl. 49). A small commercial harvest for Silver Carp in California has developed, primarily for Asian fish markets. In the wild or large backyard ponds, Goldfish progeny usually revert to a dull, cryptic coloration in only a generation or two. Spawning is much like that of the Common Carp, and a large Goldfish lays an average of 70,000 eggs in several spawning sessions, distributing the adhesive eggs over vegetation, preferring plants with many fine leaves.

GOLDEN SHINER *Notemigonus crysoleucas*

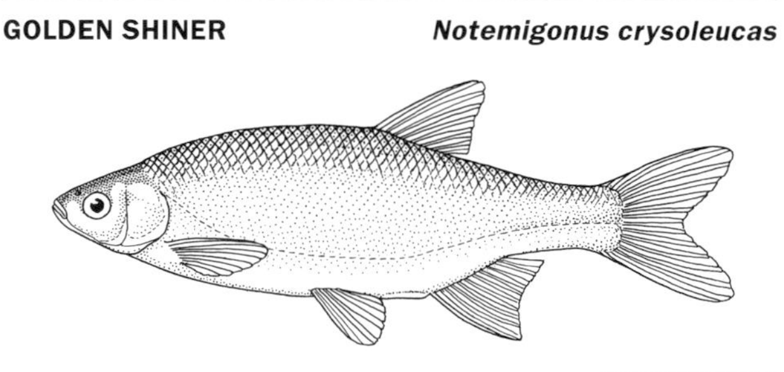

INTRODUCED

The Golden Shiner competes strongly with the Common Carp *(Cyprinus carpio)* and the Goldfish *(Carassius auratus)* for the dubious title of "most widely introduced minnow in California." This time, however, the introductions have been largely courtesy of the angler's bait bucket. The Golden Shiner is a small, widespread minnow of the eastern United States (pl. 50). It is a long-time favorite of bait dealers east of the Rockies, and with the advent of sportfishing in the West, eastern bait minnows came to California with the eastern bait dealers. This situation is ironic because many native California species could easily have been cultured for bait. The Speckled Dace *(Rhinichthys osculus),* for instance, would have been at home in nearly any habitat into which excess fishes were released from the minnow bucket after a not-too-successful day of fishing. With carefully written and enforced bait-sale laws, various subspecies of native bait minnows could be held within their respective geographic regions.

The state of California, however, has chosen to do the opposite and in 1955 declared it legal to use, raise, and sell Golden Shiners for bait in this state. Unfortunately, no such honor was accorded to small native minnows. With very few exceptions, the only legal bait fishes west of the Sierra have been nonnative species, with the Golden Shiner by far the most numerous. Indeed, the bait-fish propagation business is a lucrative small industry in California, and it is based primarily on this species. Over the past half century this has created a situation in which thousands of California citizens take an active role in stocking hundreds of freshwater habitats each year, and they even pay for the fishes they stock!

This situation has resulted in yet another competitive force on native cyprinids and on the fry of some game-fish species. Trout production in certain lake habitats has been lowered considerably by competition with Golden Shiners. A few years ago my ichthyology class conducted a two-day survey of a small private lake in Shasta County. Only a few decades earlier the lake and its feeder creek had supported a thriving Rainbow Trout *(Oncorhynchus mykiss)* population. Despite our full arsenal of fyke nets, gill nets, large and small seines, and several ardent class anglers, we collected only one Rainbow but hundreds of Golden Shiners ranging up to 20 cm (8 in.) total length. Apparently, shiner introduction by past anglers had caused the demise of the Rainbow Trout. With no large piscivore in the lake and the wide-ranging food preferences of the Golden Shiner, which include zooplankton and most other small invertebrate food plus filamentous algae, the Rainbow Trout were literally eaten out of house and home.

The Golden Shiner is a schooling minnow that seldom exceeds 25 cm (10 in.) total length. It is primarily a plankton feeder that feeds by grabbing large, individual zooplankters rather than by filter feeding. In addition, it consumes a variety of small surface invertebrates and filamentous algae.

In California the Golden Shiner breeds from March through August. The exact time for any one site is determined by the inshore water temperature, which must reach a threshold of between 15 and 20 degrees C (59 and 68 degrees F). The prominent schooling behavior of this species persists throughout spawning. Adhesive eggs and milt are shed over vegetation and bottom debris. Fry hatch in about five days and form large schools inshore, where they feed on zooplankton. They grow approximately 2.5 cm (1 in.) a year and live up to 8 eight years.

Plate 50. Golden Shiner.

A recent effort by the California Department of Fish and Game to curb foreign minnow introductions by anglers was summed up in a list of requests on the back outer page of its 2002–04 fishing regulations. These ask anglers to avoid using nonnative bait and to empty unwanted bait into garbage cans. They have been replaced by ads in the 2005 edition. Although this is a small step in the right direction, it fails to address the reality that a certain percentage of the angling public will always free unused bait minnows. The only sure way to stop this problem is to reverse the 1955 decision and make the sale of all bait fishes illegal. In fact, given the phenomenal assortment of lures today, many of which look like a minnow, swim like a minnow, and when coated with the right substance even smell like a minnow, the time may be right to ban all live-fish-bait fishing in California.

RED SHINER *Cyprinella lutrensis*

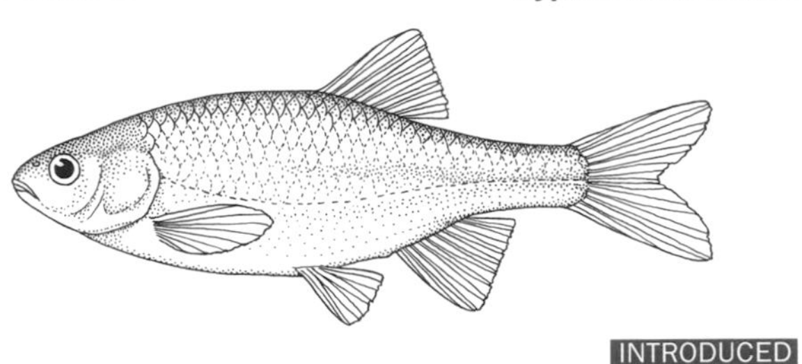

INTRODUCED

Like the Golden Shiner *(Notemigonus crysoleucas)*, the Red Shiner has been used extensively as a bait fish east of the Rockies. It was brought to California in the early 1950s as one of the state's three legal bait fishes. It established rapidly in the Colorado River and the irrigation ditches of the Imperial Valley, but not in other areas of the state.

The Red Shiner is a minnow of quiet backwater areas and intermittent streams. It remains in large schools throughout most of the year and feeds on a broad spectrum of small invertebrates and occasionally algae. Spawning occurs either in small schools or within the territories of individual males. Adhesive eggs are shed over submerged vegetation and other objects on the substrate.

Several aspects of the Red Shiner's physiology make it well adapted to intermittent creek pools and shallow, backwater areas. It can tolerate water temperatures up to nearly 40 degrees C (104 degrees F), salinity to 10 ppt, a pH as high as 11, and very low water-oxygen levels. The combination of these factors has allowed it to readily establish in shallow, well-vegetated sites and outcompete resident native minnows. This is now happening in a number of habitats in the Sacramento Valley and coastal creeks. Unfortunately, current fishing regulations permit the use of the Red Shiner as live bait in all or parts of the state's seven sportfishing districts. It should be pointed out that all such rules and regulations are approved and finalized not by biologists of the California Department of Fish and Game but by the appointed members of the California Fish and Game Commission. The commission also vetoed an attempt by the CDFG to ban the use of the Red Shiner everywhere in the state except the Colorado River. We may only surmise that the live-bait lobby is alive and well in Sacramento, and unfortunately, so is the Red Shiner in the Sacramento Valley.

FATHEAD MINNOW *Pimephales promelas*

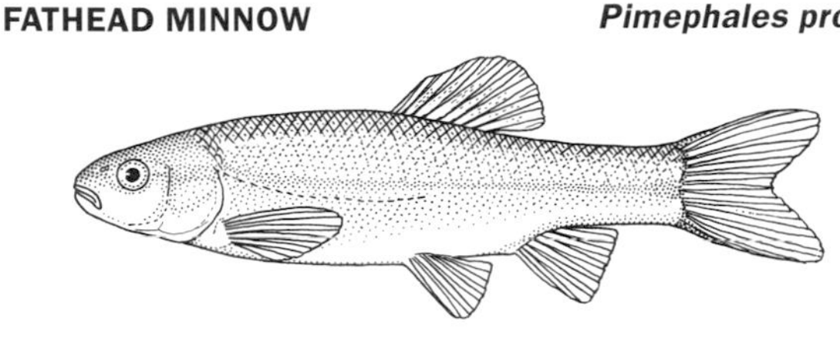

INTRODUCED

This is another small, eastern cyprinid, rarely exceeding 10 cm (4 in.) total length, granted the questionable status of legal bait minnow in California. In its native range, the Fathead Minnow lives in the muddy, backwater areas of streams and ponds. Its adaptation to such habitats includes tolerance for high water temperatures, low oxygen levels, and organic pollution. Thus, it came to California well suited for the minnow bucket, as well as for any backwaters where it might happen to be released.

The Fathead Minnow is a bottom browser, relying heavily on filamentous algae in addition to a variety of invertebrates and organic debris. Breeding differs from that of other introduced minnows in that the male is highly territorial and vigorously defends a nest site, which consists of some flat, smooth object (pl. 51). In the species' natural range, the undersides of lily pads are used extensively. As the female moves into a male's territory, she lays sticky eggs on the undersurface of the lily pad. The male guards the eggs as well as the newly hatched young. This type of "reproductive insurance" is not seen in native minnows in California, which instead scatter their eggs over the substrate where bottom

Plate 51. Fathead Minnow, male in breeding color.

browsers, such as the Fathead Minnow, can eat them. In contrast, Fathead Minnow spawning behavior minimizes egg and larval loss, allowing this species to establish itself in a new habitat in a very short time. This has occurred extensively in pools of intermittent streams and backwater areas and poses a serious threat to several native minnows adapted to such habitats.

Like the Red Shiner *(Cyprinella lutrensis)* and Golden Shiner *(Notemigonus crysoleucas)*, the Fathead Minnow has been introduced throughout the state via the bait dealer–bait angler connection. Another source of Fathead Minnow introduction centers around its ability to withstand relatively high levels of organic pollution. This physiological trait has made it the preferred species for pollution bioassay work, a fact that in itself seems rather questionable. Some years ago I visited the water-quality laboratory of a former Bay Area oil refinery that was using the Fathead Minnow as its "safe water indicator." The water in question had passed through the refinery cooling system and looked and smelled pretty foul; however, if the Fathead Minnow survived in it for several hours, the water was deemed safe and released back into the lower Delta. Unfortunately, so was each group of experimental Fatheads. Recently I concluded a three-year fish-sampling program in a tidal pond area within a mile of the refinery, which is now under new ownership. The Fathead Minnow was one of the dominant species.

TENCH *Tinca tinca*

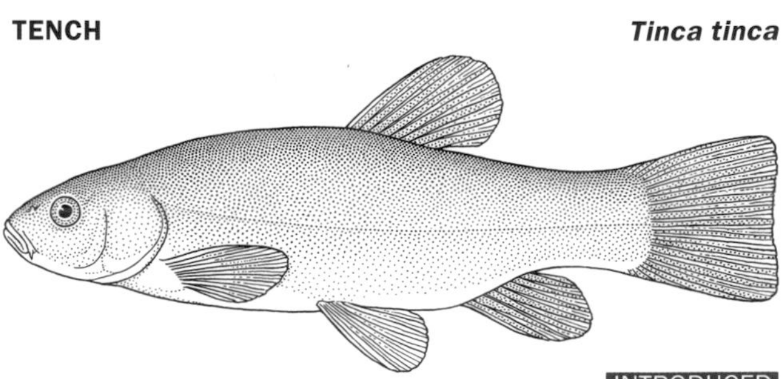

INTRODUCED

The Tench is a carplike fish from Europe that was brought into California illegally in 1922 and planted in a private reservoir in San Mateo County. In England the Tench is a member of the "coarse fish" group and is ardently pursued by well-dressed

Plate 52. Tench.

pond-shore anglers sitting on folding lawn chairs or stools. It is a fish of the eutrophic pond and backwater areas of rivers. Tench are very solitary and sluggish. When kept in an aquarium with other minnows, they simply "lie around," never participating in the usual schooling and chasing behavior common in most other species (pl. 52).

The Tench is a bottom feeder on invertebrates and aquatic plants. It grows to over 60 cm (about 2 ft) and is quite prolific. It spawns in large groups, and a female may produce up to half a million eggs in a season. Spawning is the only schooling activity of the Tench. The eggs are adhesive and are deposited over submerged vegetation in much the same manner as described for the Common Carp *(Cyprinus carpio)* and the Goldfish *(Carassius auratus)*. My first encounter with the Tench was in a large private pond in San Mateo County. I was actually conducting a trapping survey for the endangered San Francisco Garter Snake *(Thamnophis sirtalis tetrataenia)* when I noticed a lot of splashing and thrashing in the pond inshore area by large fishes that did not appear to be Common Carp. On the next trap-check day, I brought along several helpers and my trusty 50-foot seine, and within a few minutes several year classes of Tench were flopping on the shore.

The appearance of the Tench in California is a classic example of an "ethnic introduction"; in this case, Italian immigrants apparently felt that no fish was better than the one back home, where it is used for making a very rich soup. Tolerance to low oxygen levels and high temperatures makes this species ideal for long-range transport in small containers. Fortunately, the Tench has not been distributed widely like other introduced minnows in California. The original stocking sites are not connected to a canal or river system, and apparently the Tench is only in privately owned ponds and reservoirs in San Mateo and Santa Cruz Counties. It did appear in a Humboldt County pond in the 1950s, however, but was eventually eliminated. Incidences such as this support the theory that seemingly isolated introductions are really ichthyological time bombs waiting to explode. Partly as a result of my discovery of the San Francisco Garter Snake along the shores of this large Tench pond, the area has been acquired by the California Department of Parks and Recreation. If the pond is opened to the public and especially to fishing, a conduit for further introductions via bait and pet capture and release will be in place, and we could then witness the second coming of this carp throughout California. The total eradication method used in Humboldt County is really the only way to successfully treat this pending problem, and hopefully such action will be taken before it is too late.

GRASS CARP *Ctenopharyngodon idella*

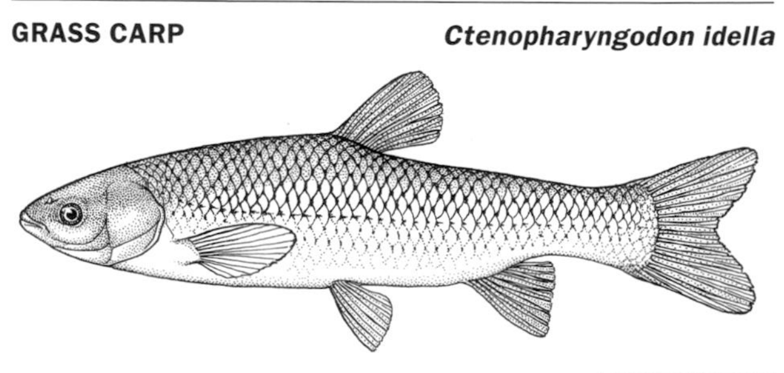

INTRODUCED

The Grass Carp is a native of the large temperate rivers of China and Southeast Asia (pl. 53). It can grow to 1.5 m (about 5 ft) and weigh up to 30 kg (66 lb). It differs from other North American cyprinids in that it grazes on aquatic vascular plants and can denude vegetated inshore zones in a relatively short time when fish numbers and water temperature are high. Long, serrated pharyngeal teeth that function like canine premolars to slice and crush vascular plant leaves and stems make this type of feeding possible (pl. 16) since no minnows have teeth in the jaw. If the Grass Carp runs out of its preferred plant food, it can switch to invertebrate food until it locates new, productive grazing areas, which it does by making long movements of 100 miles or more in rivers. This behavior is reminiscent of that displayed by plains-dwelling East African antelope species and the former wild North American Bison *(Bison bison)*. The Grass Carp possesses a true vegetarian feeding niche, which is rare among fishes, both freshwater and marine.

Plate 53. Grass Carp.

The Grass Carp is very fecund, with a large female producing well over a million eggs. Spawning is also different in that the eggs remain suspended in the water column rather than attached to the substrate, as in most minnows. The young feed initially on small aquatic invertebrates, but by 4 cm (1.5 in.) they switch to the herbivorous feeding mode that they maintain throughout their lives, which can be as long as 15 years. Grass Carp can also tolerate a wide range of water temperatures (2 to 38 degrees C [35.5 to 100 degrees F]), very low oxygen levels, and moderately brackish water, all of which provide this species with the basic ingredients for success when introduced into new habitats, including large portions of the United States and Mexico.

The Grass Carp was imported to the state of Arkansas in 1963, where an aquaculture program is now well established. It has been sold throughout much of the United States for both food and aquatic vegetation control and as a result has spread throughout the Mississippi River basin during the past three decades. Unfortunately, Grass Carp salespersons have been knocking on California's door, and this species in now in the Golden State. Sanctioned introductions of sterile triploid Grass Carp have been made in the irrigation canals of southeastern California for vegetation control. Unfortunately, normal Grass Carp have turned up in a number of California ponds, particularly those on golf courses. The California Department of Fish and Game has been eradicating these illegal introductions as they are discovered, but it may be just a matter of time before this exotic, which has the potential of destroying vast areas of the inshore nursery habitat for nearly all species of fishes, becomes established in the Central Valley. Even with full discovery and complete elimination of Grass Carp stocked in private ponds, the continued sanction of triploid imports is still risky. Are chromosomal counts being made on each shipment? What if a dealer runs a little short on sterile Grass Carp? If this occurs, will the order be rounded out with some normal fertile specimens? These are relatively expensive fishes. Realistically, the only reliable way to prevent a Grass Carp catastrophe in California is to ban any further importation, including through Internet trade, which is now a major source. Even then, agriculture check stations at state borders may have to modify the standard question "Do you have any fresh fruit and vegetables?" to include "and how about Grass Carp?"

SUCKERS (Catostomidae)

California has 11 native species of suckers. Their body form and fin arrangement is similar to that of the cyprinids except for one key difference: the mouth. It is in an inferior (ventral) position, and the soft, protruding lips coupled with powerful mouth suction have allowed suckers to successfully exploit the bottom-substrate feeding niche. A single row of comblike pharyngeal teeth allows suckers to break up the shells and exoskeletons of the many invertebrates that they glean from the bottom muds and gravels. The pharyngeal teeth also act as fencelike guards that prevent large nonfood items entering the mouth during the suction-type feeding from passing into the narrow digestive tract.

Suckers are almost exclusively North American, and their basic feeding niche is exploited by relatively few other species. In Europe and most of Asia, which completely lack suckers, cyprinids such as the Common Carp occupy a similar bottom-feeding niche, and the widespread introduction of carp throughout the United States has resulted in severe competition for some sucker species. On the other hand, the sucker family is one of the few large groups of California native fishes that has not been subjected to introductions of nonnative species, perhaps because most people do not give much thought to suckers, even forgetting that they exist.

Contrary to a popular belief that categorizes these species as sluggish "trash" fishes, suckers are active, strong-swimming animals. Many inhabit mountain streams and lakes and closely resemble trout in their ecological requirements. Despite the range of habitat preferences within the family, our native suckers are a uniform group and exhibit little diversity. For this reason, the most abundant species in California, the Sacramento Sucker *(Catostomus occidentalis),* is discussed in detail and serves as a "type species" to which the others are briefly compared.

ANGLING NOTES: Although suckers as a group possess a number of features in common with recognized game fishes, such as large size, strong, vigorous swimming ability, and good abundance of some species in many habitats, they do not command much angling interest. This situation results from several factors. First, they are considered "trash" fishes by many anglers and are therefore ignored. Just what the title of "trash" fish specifically denotes has never been fully explained, but descriptions of this category usually do not apply to suckers. Most American anglers, however, tend to draw a very sharp distinction between fish species that are worthy of their angling attention and those that are not. In contrast, in Europe and the British Isles, large freshwater fishes other than salmonids are generally grouped together as "coarse fish," and despite this somewhat dubious title, a distinctive subsport of "coarse fishing" has developed around them.

A second consideration is the general difficulty involved in hooking suckers. For most species, the bait and hook must be small and fished exclusively on the bottom. Large spawning concentrations provide the best opportunities for continuous fast action. Worms work as well as any bait, but remember that suckers have very sensitive lips and have a habit of "mouthing" a larger food item during which time they may detect the hook. It is therefore best to react after the first couple of light tugs on your line.

Finally, there are the matters of taste, flesh texture, and bones. The first two items vary with the seasons. Cold spring and fall waters produce much better "kitchen fish" than do the warm, summer months. When taken from cold mountain streams or lakes, however, suckers taste very much like trout. Many years ago, I camped my ichthyology class on the shore of Fallen Leaf Lake for a weekend of fish collecting. A small gill net soon produced four hatchery Rainbow Trout *(Oncorhynchus mykiss)* and six Tahoe Suckers *(Catostomus tahoensis)*, all within the 10-inch size range. After all specimens were measured and weighed and their stomach contents sampled, we heated the griddle and held a taste test. Despite the ever-present small bones in the Tahoe Sucker (which are characteristic of all suckers and minnows), we agreed it was a draw for best-taste honors.

SACRAMENTO SUCKER *Catostomus occidentalis*

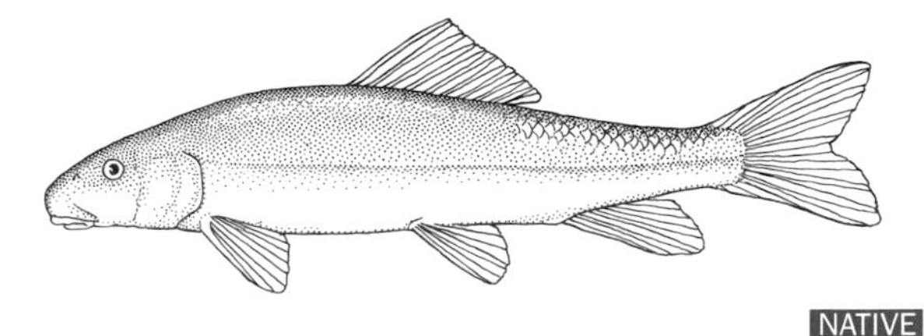

NATIVE

This species is found throughout the varied habitats of the Sacramento–San Joaquin drainage, where four subspecies are recognized. One of these, the **GOOSE LAKE SUCKER** ***(C. o. lacusanserinus)*** is listed as a California Species of Special Concern. Adults feed on a variety of bottom foods ranging from organic debris and algae to aquatic insect larvae. In the now-common stream-reservoir complexes of central California, adults inhabit the lakes, whereas juveniles occupy the shallow, upstream waters. In large streams and rivers, the adults show a strong preference for deep pools, whereas the young occupy the shallow areas. When

Plate 54. Sacramento Sucker.

alarmed, young suckers in shallow streams with loose stone and gravel bottoms react in an interesting manner. Using their strong swimming muscles, they burrow their way between large pebbles until they are completely hidden from view. This behavior is no doubt an adaptation that enables them both to withstand high, seasonal flows and to avoid surface predation during daylight hours. Small suckers have also been observed using this gravel-burrowing behavior to successfully survive a rotenone eradication effort to eliminate "trash" fishes in the Trinity River.

The Sacramento Sucker is one of the largest suckers in California (pl. 54) and is long-lived. Five-year-old fish may exceed 30 cm (about 1 ft) in length, and 10-year-old specimens may exceed 40 cm (16 in.). The oldest-recorded Sacramento Sucker was 30 years old and 56 cm (22 in.) long. It is a slow-growing fish that does not spawn until it is at least four years old. Spawning begins in late February and somewhat resembles trout spawning. The

Plate 55. Sacramento Sucker, juvenile.

Sacramento Sucker has distinct spring spawning runs to upstream gravel-bed sites. A mature female lays an average of 20,000 eggs over gravel substrate, where they are fertilized by one to several attending males. The eggs sink into the gravel where they adhere and remain for about a month before hatching. The newly hatched yolk-sac fry are moved slowly downstream by the current. The young fry have a terminal instead of an inferior mouth and feed on a variety of drift and free-swimming invertebrates for a number of weeks until their mouths develop the fleshy lips and inferior position of the adult. The beginning fish watcher may have difficulty distinguishing small Sacramento Suckers (1 to 3 in.) from small cyprinids because their fin arrangement is identical, and the small sucker's inferior mouth is often hard to detect when viewed from the creek bank. These very small fishes have several diffuse dark blotches on their sides, however, that apparently help them blend with the bottom substrate, whereas small minnows are almost always uniformly colored (pl. 55).

The Sacramento Sucker remains widely distributed in California and, given the profound changes to the state's freshwater habitats, has fared better than most natives. In many small tributary creeks, young of this species are one of the most abundant fishes. Healthy adult populations also persist in foothill streams and reservoirs with permanent tributary creeks. The large size of the Sacramento Sucker qualifies it as a potential sport fish.

TAHOE SUCKER *Catostomus tahoensis*

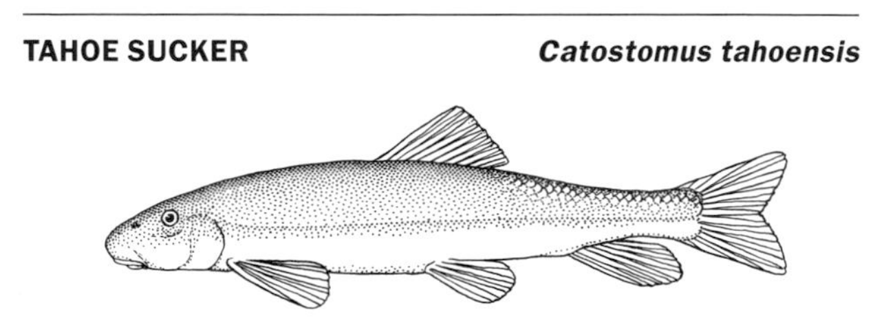

NATIVE

This native California sucker species is perhaps second only to the Sacramento Sucker *(C. occidentalis)* in abundance. It is also the largest sucker, with some individuals growing to 60 cm (about 2 ft). It may be viewed as the Sacramento Sucker's mountain counterpart, for the habitat of the Tahoe Sucker is the lakes and streams of the Tahoe Basin. Feeding and reproductive behavior are similar in both species, except that much larger spawning aggregations have been noted for the Tahoe Sucker, and spawning begins later in April. This species occupies a key position in the food chain of its habitats and constitutes a large portion of the diet of both Lake Trout *(Salvelinus namaycush)* and Rainbow Trout *(Oncorhynchus mykiss)*. Because of its size and the coolness of its habitat, this species receives special attention in the cooking section.

OWENS SUCKER FOLLOWS ➤

OWENS SUCKER *Catostomus fumeiventris*

NATIVE

PROTECTIVE STATUS: CSSC.

This species is similar to the Tahoe Sucker and inhabits the Owens River system to the south. Like the Tahoe Sucker *(C. tahoensis)* and the Sacramento Sucker *(C. occidentalis)*, it is a very successful species that has withstood the water manipulations of the Owens River far better than other native fishes of that area. The main population of this species is located in Crowley Lake in Mono County. Reservoirs have replaced streams as the dominant habitat for the Owens Sucker, but the spectrum of introduced fishes that inhabit these lakes often compete with the Owens Sucker for bottom invertebrate food. For this reason, the Owens Sucker has been listed as a California Species of Special Concern.

LOST RIVER SUCKER *Catostomus luxatus*

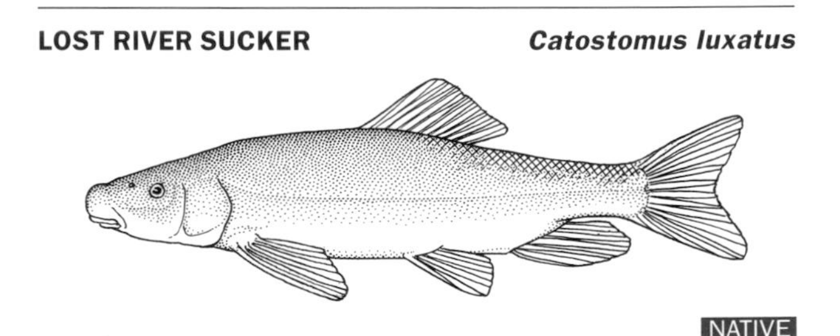

NATIVE

PROTECTIVE STATUS: SE, FE, FPS.

This is a large sucker of the upper Klamath and Lost River systems in northern California and southern Oregon. It grows to about 1 m (39 in.) and weighs up to 4.5 kg (10 lb). Its large size and formerly large numbers made it an important food source of the Klamath and Modoc tribes. This big fish was also utilized by immigrants from the eastern United States who found it easy to catch during spring spawning runs, and a small sport fishery for this species existed up to 1974 when it was designated a State Endangered and Federal Endangered species. In addition to feeding on a spectrum of small, bottom-dwelling invertebrates like other suckers, the Lost River Sucker may also take small invertebrates from aquatic plant leaves and stems. Its unusual V-shaped lip configuration and a slightly flexible snout apparently aid in this type of feeding.

The Lost River Sucker is a clear-water species, and the extensive drainage of natural lakes within its range in the 1920s resulted in extensive eutrophication of many, thereby greatly reducing its numbers. This, along with the former harvest of large, breeding individuals and other impacts which have been leveled upon fishes of the Klamath drainage, including nonnative introductions, has more than justified its dual listing as a State Endangered and a Federal Endangered species. To insure this species' preservation, we need protective measures such as finalization of the critical habitat designation proposed in 1994 for the Lost River Sucker's home waters. If and when this is made official, special management considerations or protection for specific habitat areas may be required, even if some are not currently occupied by the Lost River Sucker.

KLAMATH LARGESCALE SUCKER *Catostomus snyderi*

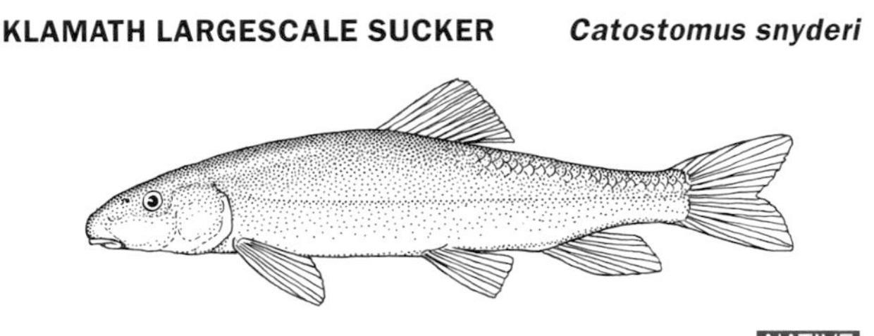

NATIVE

PROTECTIVE STATUS: CSSC.

This species is similar to the Sacramento Sucker *(C. occidentalis)* in size and appearance, but little is known about the details of its life history. It occupies a range similar to that of the Lost River Sucker *(C. luxtus)* and Shortnose Sucker *(Chasmistes brevirostris)* and, like these species, is not abundant in California today. Its geographic range extends into the Upper Klamath drainage along the Oregon border; fortunately, this species is far more numerous in Oregon. Unlike that of the Lost River Sucker, the Klamath Largescale Sucker's mouth is in the full inferior position, which suggests the feeding habits for these two species are different, perhaps permitting the coexistence of these two large sucker species throughout the same range. Its designation as a California Species of Special Concern is well deserved and will perhaps spark interest in further study of this little-known species.

KLAMATH SMALLSCALE SUCKER FOLLOWS ➤

KLAMATH SMALLSCALE SUCKER *Catostomus rimiculus*

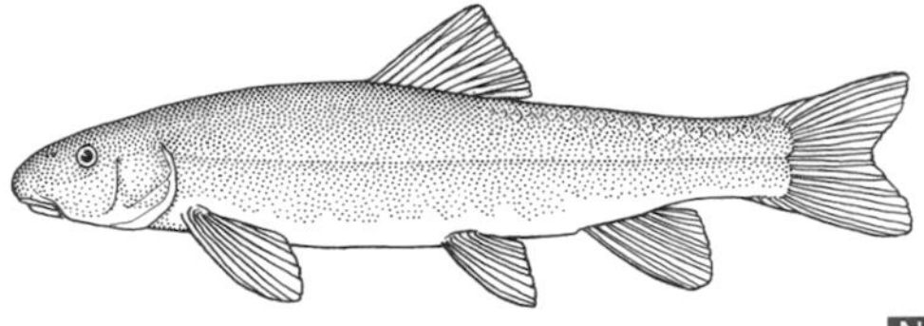

NATIVE

This is the smallest sucker species of the Klamath drainage. Adults average about 35 cm (14 in.) and occupy slow water and pool areas in the lower Klamath and Trinity Rivers, where they are quite common. Small tributaries of these rivers are used for spawning, and the reproductive biology of this species is presumably similar to that of other native suckers. The young often serve as a food source for large trout. Like several other sucker species, the Klamath Smallscale Sucker in is need of in-depth study of its natural history.

MODOC SUCKER *Catostomus microps*

NATIVE

PROTECTIVE STATUS: SE, FE, FPS.

The Modoc Sucker is another small, native sucker with a highly restricted distribution. It occurs only in creeks, primarily in Modoc County. It is normally found in still-water pools with mud rather than gravel substrate and may share this habitat with several other species, including two large piscivores, the Sacramento Pikeminnow *(Ptychocheilus grandis)* and the introduced Brown Trout *(Salmo trutta)*. In this habitat, the Modoc Sucker, along with the Sacramento Sucker *(C. occidentalis)*, occupies the bottom "vacuum" feeding niche, thereby avoiding competition with several other invertebrate feeders that glean the bottom substrate surface.

The Modoc Sucker is another example of a species isolated within a relatively small aquatic complex that has suffered a number of adverse impacts (Brown Trout introduction, watershed degradation from logging and grazing, and stream diversion and channelization). The Modoc Sucker's declining population deserves its status as a State Endangered and Federal Endangered species, and its home waters deserve their official critical habitat designation.

MOUNTAIN SUCKER *Catostomus platyrhynchus*

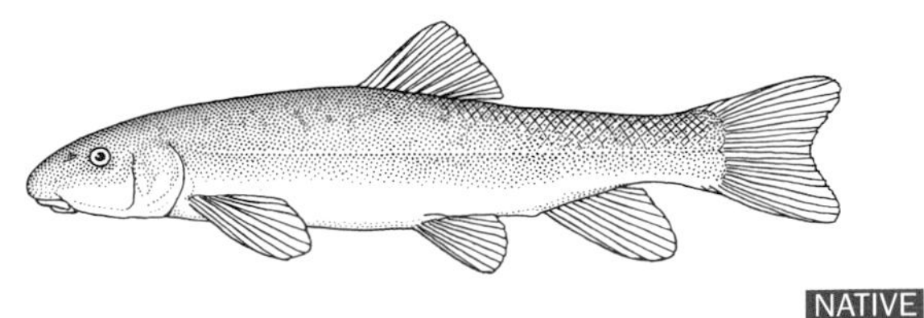

NATIVE

PROTECTIVE STATUS: CSSC.

Unlike other native California sucker species, the Mountain Sucker (pl. 56) is widely distributed through the western United States and Canada. In California it occurs in the Lahontan Fish Province on the eastern slopes of the Sierra. This small sucker rarely exceeds 25 cm (10 in.) total length. It is more herbivorous in its feeding than other suckers and is specialized for algae scraping by the presence of a hard plate on the inner edge of its lower lip. Its reproductive habits are similar to those of other native suckers.

Although the Mountain Sucker is relatively abundant throughout its wide range, populations are declining in California. One reason lies in this species' inability to adapt well to reservoirs, which are becoming more numerous throughout its range and are all created by damming mountain streams. The Mountain Sucker's preferred habitat loss coupled with its poor adaptability to lakes is proving disastrous for this fine mountain fish. Only serious creek preservation and enhancement programs can effectively reverse this trend.

Plate 56. Mountain Sucker (above), Tahoe Sucker (below).

SANTA ANA SUCKER *Catostomus santaanae*

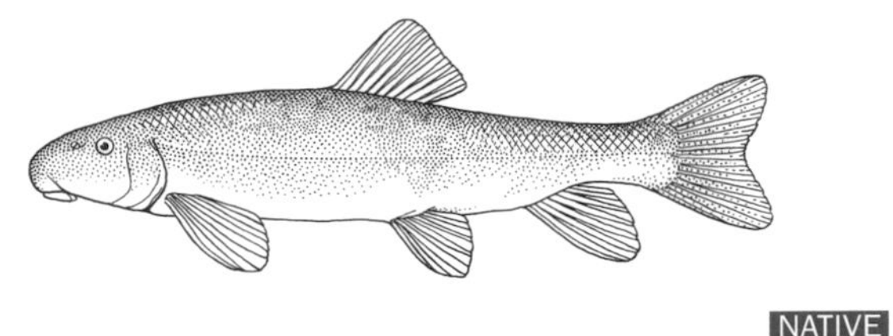

NATIVE

PROTECTIVE STATUS: FT.

This is the native sucker of southwestern California, where it is found only in the Santa Ana, Santa Clara, San Gabriel, and Los Angeles Rivers. Like the Mountain Sucker *(C. platyrhynchus),* it is primarily an algae feeder and also has hard inner-lip edges that are used to scrape algal growths off the substrate. Habitat manipulations and pollution such as that which has ruined much of the Los Angeles River also threatens its existence in the other three river systems where this species still exists.

It is the smallest sucker species in California, rarely exceeding 15 cm (6 in.) standard length. This small size has probably been selected for by the great annual water fluctuations of the species' native coastal streams. In late summer, many portions of these habitats are reduced to little pools, where small fishes that can tolerate warm water and low oxygen levels have the best chance for survival. Given the very limited distribution of this species and the ongoing introductions of competitive nonnatives such as the Red Shiner *(Cyprinella lutrensis)* and piscivores such as the Brown Trout *(Salmo trutta)* and the Green Sunfish *(Lepomis cyanellus),* the Santa Ana Sucker deserves additional protective status, including the all important critical habitat designation for at least some portions of its home streams.

RAZORBACK SUCKER *Xyrauchen texanus*

NATIVE

PROTECTIVE STATUS: SE, FE, FPS.

The Razorback Sucker, formally called the Humpback Sucker, is a large sucker of the Colorado River drainage. It differs conspicuously from members of the genus *Catostomus* because of the high keel on its back, anterior to the dorsal fin. This "dorsal keel" apparently serves to stabilize the body during periods of high-velocity flow. Most of the time it is relatively useless because this sucker prefers the shallow, vegetated, backwater areas of the Colorado River where it feeds on the usual assortment of bottom-dwelling invertebrates and algae. It may occasionally filter feed when zooplankton is abundant, using its long, closely spaced gill rakers for this purpose.

Because of its large size (length up to about 1 m [3.3 ft], or more; weight to 7.3 kg [16 lb]), the Razorback Sucker was an important food for a number of Native American tribes along the Colorado River. At one time, a commercial fishery existed for this species, but with the continued manipulations of the lower Colorado River water flow during the past half century, its numbers greatly declined and the fishery was discontinued. This species in now quite rare, but its longevity (up to 40 years) has masked its steady decline. Spawning is rarely successful except in hatcheries. In California it now may exist in only the form of non-breeding wanderers from upstream areas. The Razorback Sucker is listed as a State Endangered and Federal Endangered species, and its preferred river backwater habitat was granted critical habitat status in 1994. The restoration or re-creation of shallow, backwater habitats on large rivers is very feasible, and if successfully accomplished, we may see a comeback for the Razorback Sucker in California. Addressing egg and fry predation by non-native fishes must be a major part of any restoration plan. The need for this is supported by the observations of Ken Aasen, CDFG biologist at Senator Wash Reservoir, Imperial County, from 1978 to 1980. He observed Bluegill *(Lepomis macrochirus)*, Smallmouth Bass *(Micropterus dolomieu)*, Common Carp *(Cyprinus carpio)*, and a catfish species all feeding on eggs at the only Razorback Sucker spawning site in the reservoir. He reported that of the razorbacks he observed, all were mature adults, which suggests similarly severe predation on fry and juveniles.

SHORTNOSE SUCKER *Chasmistes brevirostris*

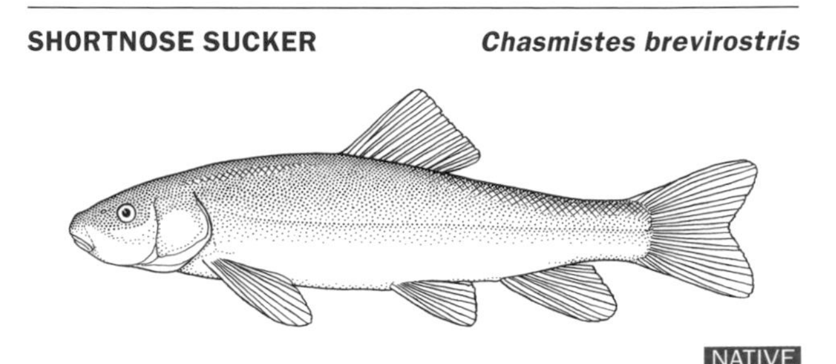

NATIVE

PROTECTIVE STATUS: SE, FE, FPS.

This is a medium-size sucker (45 cm [about 18 in.]) of the Upper Klamath and Lost River drainages. The name of this species is misleading because its actual nose area is no shorter than the nose or nostril area of most other fishes of similar size. What is shorter is the snout area ahead of the nostrils. This gives the species an oblique or slightly terminal mouth much like that of the Lost River Sucker *(Catostomus luxatus)*, to which it is probably closely related. This mouth position coupled with numerous, fine gill rakers adapts it well for feeding on zooplankton and surface drift. This major deviation from normal sucker feeding behavior makes it well suited to the shallow lake habitat where the Shortnose Sucker is normally found. Its range in California is also similar to that of the Lost River Sucker, but its numbers are far smaller.

Unfortunately the array of human-wrought impacts that have led to a marked decease of other sucker species in the Upper Klamath region has similarly affected this species. Its relatively long life span (over 30 years) and high fecundity of large females (about 38,000 eggs per spawning) has so far allowed the Shortnose Sucker to survive short-term impacts, but long-term protection of its shallow lake habitat is greatly needed. Hopefully, a pending 1994 critical habitat proposal for this species will soon be finalized.

CATFISHES AND BULLHEADS (Ictaluridae)

The large catfish and bullhead family, which comprises over 40 species, is native to the eastern United States, particularly the Mississippi drainage and parts of Mexico. All seven species in California have been introduced, although several million years ago now-extinct catfish species did exist west of the Rocky Mountains. Ictalurids (catfish and bullheads) possess some unusual characteristics not found in other groups, the most apparent of which is the complete lack of scales. The skin, however, is amazingly tough and proof that scales are not essential to the overall fish body plan. To compensate for any lack of predator protection from the lack of scales, ictalurids have evolved a sharp, serrated spine at the leading edge of the dorsal and pectoral fins. When poked or grabbed, the spines are erected and locked in place, making consumption of catfish and bullheads both difficult and painful for small piscivores. Pain is also inflicted in those anglers who do not know how to correctly handle a catfish or bullhead. The handling method described for the Brown Bullhead *(Ameirurus nebulosus)* works well on all except the very large catfishes. With these it is better to simply grab the front of the lower jaw with your thumb inside the mouth. Ictalurids have bands of small teeth on the roof of the mouth but none on the jaw edge. The spines also serve a an important function in catfish and bullhead research. Normally, fishes are aged by counting the various bands on concentric rings in the scales, a technique not possible with in this family. Fortunately, the growth of the spines also has an annual cycle, and by cross-sectioning one at the base and counting the faint rings, the age of a specimen can be determined.

Catfishes and their smaller relatives, the bullheads, have eight "whiskers," or barbels, which contain numerous taste buds on their surfaces. Six of these barbels are positioned so that they drag along the bottom substrate, and additional taste receptors are located on the skin and fins. This feature allows ictalurids to locate and taste the smallest food items in the bottom substrate, even at night or under turbid water conditions.

This family has a great tolerance for low oxygen and high

carbon dioxide levels. This not only promoted their occupancy of bottom-feeding niches but also has adapted them to backwater areas unsuited to most other species. Besides providing pleasurable fishing for thousands of California anglers, catfishes are one of the few warm-water fishes raised commercially in this state. Their ability to tolerate warm, shallow, farm-pond water conditions has made this fishery possible.

BROWN BULLHEAD *Ameiurus nebulosus*

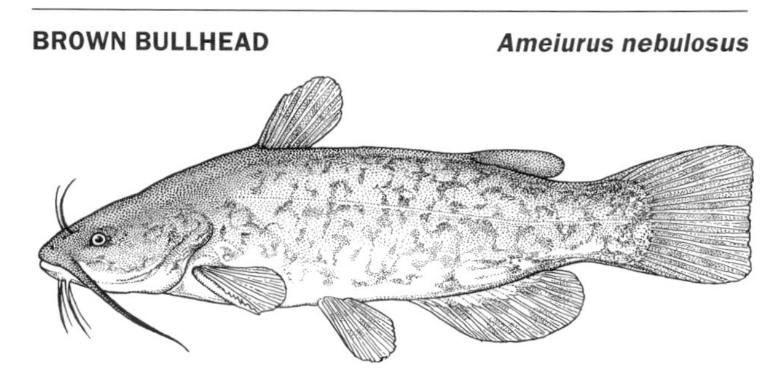

INTRODUCED

This is the most abundant of the three bullhead species in California (pl. 57). It was also the first ictalurid to be introduced into the state. It took only one introduction into the San Joaquin River near Stockton in 1874 by the California Fish Commission to eventually establish this species throughout the Central Valley, Klamath basin, southern coastal streams, and numerous other locations. The Brown Bullhead exhibits the same broad spectrum of adaptations to a variety of habitat conditions in California that it does in its native range east of the Rockies. It is particularly well adapted to large, eutrophic lakes at lower elevations and to the delta regions of rivers, and it is also present in stunted form in mountain streams and lakes such as Lake Van Norden in the Donner Summit area. Like most members of the family, it is a bottom forager. It is interesting to observe a captive aquarium specimen use its food location system. The detection of a food item by the barbel taste buds triggers an immediate grubbing action by the mouth, which can secure even the smallest food morsel. A by-product of this sort of feeding is ingestion of a fair amount of detritus and sand, which is usually passed through the intestine along with the food.

This species and other ictalurids occasionally are found to have small fishes in their stomachs. We do not normally associate the usual sort of bullhead-catfish feeding with active predation, but apparently these fishes are capable of swimming rapidly through large schools of small forage fishes such as Threadfin Shad *(Dorosoma petenense)* or Inland Silverside *(Menidia beryllina)* and taking an individual here and there. The preference for night feeding in these species may also account for their success as predators, because this habit would give them an advantage over diurnal prey.

Plate 57. Brown Bullhead.

Breeding behavior in all ictalurids is very similar and extremely interesting. The Brown Bullhead's behavior is described here in detail and serves as an example for all other catfish and bullhead species presently in California. Breeding begins when a female selects a suitable nest site, usually among aquatic vegetation, under submerged roots, or sheltered by undercut banks. She builds the breeding nest by mouth. Captive Brown Bullheads have been observed moving up to a gallon of gravel in a single nest-building endeavor. When the nest is complete, the male and female lie head to tail and move slowly in a circle. The female lays up to 14,000 eggs, and after spawning, the male, physically exhausted, lists over on his side. The eggs are adhesive and clump together in a large mass in the nest. They are fanned by the tail movement of one parent while the other parent patrols the periphery of the nest, and the parents take turns. In experimental nests, eggs that were not aerated died within 24 hours. Even under trout hatchery conditions, egg losses are heavy. One theory suggests that the heavy gelatinous coat surrounding the eggs causes the need for continuous aeration, an idea enhanced by the observation that the parents "yawn" at about 15-minute intervals with their heads near the nest. Such mouth movements increase water flow over the eggs, and any egg that drifts from the nest is quickly retrieved in a parent's mouth and returned to the nest.

Hatching occurs in about 17 days, usually in May or June, when surface water temperatures reach about 20 degrees C (68 degrees F). After hatching, the yolk-sac fry are still protected by the parents, even to the extent of being carried orally. The sac fry huddle together in a tight school and do not leave the nest area until the yolk sac is absorbed. Once they leave the nest, the fry remain in a tight school as they move through the shallow portions of the habitat. An adult remains with the fry school for up to two weeks after hatching. This high degree of parental care, exceptional for a North American fish group, is rivaled in California only by the recently introduced cichlids from Africa. Parental care is certainly a major factor in the success of the catfishes and bullheads. The fry grow rapidly, feeding mainly on zooplankton during their first summer. By the third year they average 17 cm (about 6 in.), at which time they are capable of spawning.

ANGLING NOTES: The catfishes and bullheads are second only to the salmonids in the numbers caught by sport anglers in California, and over half the annual catch is taken in the Sacramento–San Joaquin Delta. Here, as in other areas, the take of Brown Bullhead is comparatively light compared with its high reproductive success. In Clear Lake, Lake County, for instance, the exploitation or catch rate is a mere 7.5 percent of the estimated total population. In Folsom Lake, Placer and Eldorado Counties, the catch rate is twice that but still relatively low, an indication that there is plenty of opportunity for the Brown Bullhead angler.

In fishing for any member of this family, the basic biology and behavior of a species must be seriously considered. All adult bullheads and catfishes are highly nocturnal in their feeding habits. During the day they normally retreat to deeper portions of the habitat and remain relatively inactive. As dusk approaches they move into shallow, nearshore areas where the bulk of their food is found. They have also been observed to initiate feeding runs through shallow areas during dark, cloudy days, which gives the impression that they feed nocturnally to avoid high light levels rather than to take advantage of increased availability of food at night.

Bear in mind also how these fishes locate their food. Their eyes appear to be little used, and instead they rely on their sense of smell and the profusion of taste buds on the barbels and other external body areas. Because feeding is accomplished during continuous exploration of the bottom, the angling technique of seeking out a potential lair or territorial retreat, as in trout fishing, is not applicable. A far better alternative is to "read" the shoreline carefully and determine the most likely areas for ictalurid foraging. In lakes, these are usually the gradually deepening shores with a moderately clear bottom. These fishes avoid extensive rock rubble or dense stands of aquatic vegetation, probably because of the difficulty of actually capturing bottom food once it has been located in such areas. In rivers and streams, the preferred feeding sites normally are the slow-current and backwater areas, where food items are not scoured continuously from the substrate but instead may build up in good numbers. Once you've decided on a fishing site, the best method is to sit and wait for the catfishes and bullheads to come to you. Herein lies one of the joys of bullhead and catfish angling. There is no need for nonstop casting of lures or constant adjustment of the trolling motor, just sit back in the shoreline lawn chair or seat in your anchored boat and relax, read a book, and wait for that wiggle at the tip of your pole.

The question of bait for bullheads and catfishes has always been an interesting one. Some of the most unlikely morsels, such as marshmallows, cheese, and popcorn, have produced good catches. Many anglers swear by the theory of "the smellier the bait, the better" and use chicken liver, guts, or rotting meat. Considering the taste and olfactory capabilities of these fishes, such extremes are probably not necessary in productive waters. The secure attachment of the bait to the hook should be carefully considered, and the most effective method of presentation is placing it directly on the bottom, either with an unweighted line in still water or with an appropriate-sized sinker a short distance above the hook in currents. In both cases, it is difficult to detect the initial strike because of the slack line, and if the bait is too lightly attached, the first bite may be the last. For this reason, freshwater clams are a popular bait in California, especially in the Delta area. The soft body or stomach area of the clam produces ample aroma and also serves as a loose morsel to be "mouthed" during the bait-inspection stage, but the hook is solidly entrenched within the tough foot area of the clam and usually stays there should the fish decide to take the entire offering.

After a capture has been made, take special precautions to avoid the notorious dorsal and pectoral spines. These spines, once extended, actually lock in place and evolved as a deterrent against being swallowed by large piscivores. Instead of grabbing randomly at the fish, as most people who get "spiked" do, spread your thumb and index finger and move the palm of your hand from tail to head along the top of the back. Adjust your hand so that the notch between your thumb and forefinger comes to rest immediately behind the dorsal spine. At the same time, let the right or left pectoral spine, depending upon your hand preference, move between the index and middle finger. This spine should now be erect, which makes the task easy and generally safe. Finally, squeeze down, using the dorsal and pectoral spine for leverage and support. By using this hold, you cannot be spiked by a live bullhead or catfish, as long as you apply adequate counterpressure when needed.

A further bit of ictalurid postcapture biology that never ceases to amaze the angler is the ability of these fishes to remain alive for hours out of water. Their hemoglobin has one of the highest affinities for oxygen of all families of California freshwater fishes, supplying enough oxygen from a poorly functioning (out-of-water) gill system to sustain life for hours on land.

BLACK BULLHEAD *Ameiurus melas*

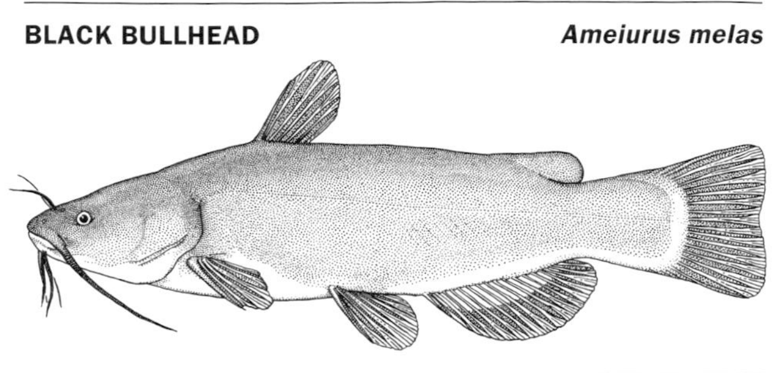

INTRODUCED

The Black Bullhead is more highly adapted to warm backwater areas of lakes and to small, shallow ponds than other ictalurids. An exceptional tolerance of high water temperatures (up to 35 degrees C [95 degrees F]) and low oxygen and high carbon dioxide levels make possible its year-round occupancy of such habitats. The Black Bullhead is also tolerant of moderately brackish water (10 to 12 ppt), which allows it to live and move easily through drainage canals and other interconnecting ditches, resulting in its spread through the Central Valley and adjacent Sierra foothills, coastal streams from the Bay Area southward, and much of southwestern California. This has all occurred since it was first "discovered" in California in 1942, the result of an earlier, unknown introduction. Feeding and reproductive behavior is similar to that of the Brown Bullhead. In clear ponds, the parent Black Bullheads are excellent subjects for the fish watcher because they are especially attentive to their young. In fact, the best way to determine the species of a particular school of bullhead fry is by observing the attending adult, because all bullhead fry and some catfish fry are mostly black.

Angling procedures for the Black Bullhead are the same as for the Brown Bullhead *(A. nebulosus)* and most other members of this family. Although this species may reach a length of about 60 cm (2 ft), it produces stunted populations when confined to pond or small-lake habitats. Though egg numbers per female are relatively low (2,000 to 3,000), the spine erection defense against predators combined with good parental care can produce a large population in a short period of time. For this reason the Black Bullhead is not a good choice to stock farm ponds or small reservoirs. Most anglers do not like to catch small fishes, and when the word gets out that they are small in a particular location, population control by fishing effectively ceases there.

Attributes of a particular fish that are unfavorable to anglers are sometimes quite useful to other fish enthusiasts. The Black Bullhead, for example, makes an excellent freshwater aquarium specimen. It is nicely preadapted to many of the normal physical aspects of the small aquarium (low oxygen, high carbon dioxide, etc.), and its habit of stunting is a distinct advantage.

YELLOW BULLHEAD *Ameiurus natalis*

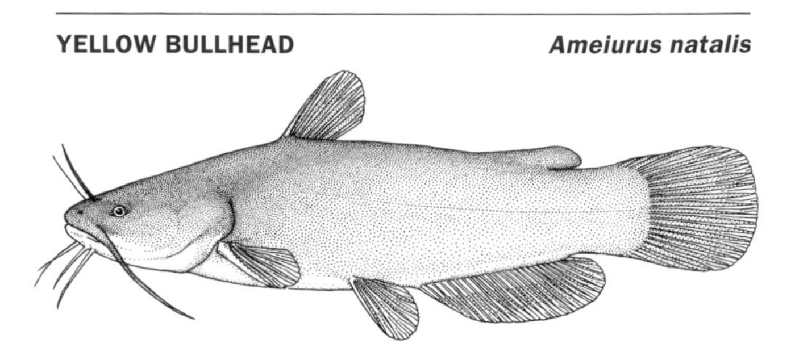

INTRODUCED

The Yellow Bullhead is a stream dweller that prefers clear water and rock substrate. Like the Black Bullhead *(A. melas)* it can withstand warm, summer water temperatures and very low water-oxygen content. In its native eastern U.S. habitats, it feeds on fishes and crayfishes more frequently than the other two bullhead species. In such habitats it can grow to about 48 cm (nearly 19 in.) and attain a weight of 1 kg (2.2 lb). In most other respects, its biology is the same as that of the Black Bullhead and Brown Bullhead *(A. nebulosus)*. The Yellow Bullhead is still rare in California, although all three species of bullhead may have been introduced as early as 1874. Its preferred native habitats are warm, permanent streams with clear, well-vegetated inshore areas, and this may have limited its spread here because these habitats have become relatively scarce in California. Its present range in this state is limited to streams and a few reservoirs in Orange and Riverside Counties and the Colorado River.

CHANNEL CATFISH *Ictalurus punctatus*

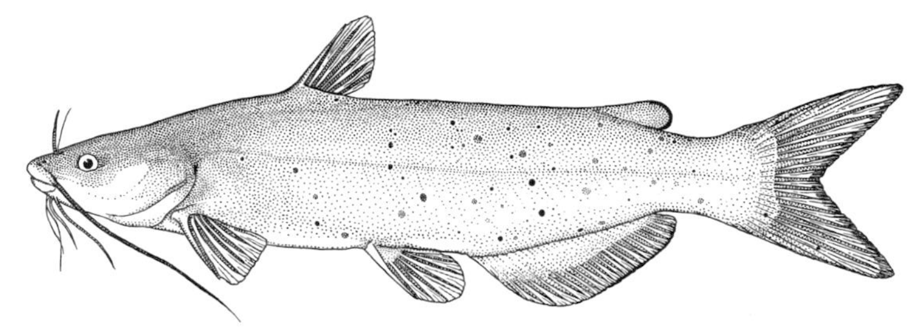

INTRODUCED

The Channel Catfish may well be called the trout of the ictalurid world (pl. 58). It is a strong swimmer, and in stream habitats it feeds in midcurrent areas. Like the Brown Trout *(Salmo trutta)* and Brook Trout *(Salvelinus fontinalis),* it prefers to "hole up" under banks or beneath root tangles in deep pools when not feeding. Sequestered areas seem necessary for the successful establishment of a spawning site. In many California habitats where such lairs are not present, the species often does not spawn and must be maintained by hatchery stocking. Its excellent hatchery record, however, is the key to its success in this state. Given suitable hatchery ponds with nesting sites, it is easy to raise and quite prolific, with females laying up to 70,000 eggs. In captivity, sunken barrels or other large containers provide nest sites, but in nature this fish uses undercut banks or shoreline floating debris jams. Absence of such sites from most California reservoirs is perhaps the major reason for its lack of reproduction in these lakes.

Hatchery fry grow fast and may be stocked as fingerlings (up to 10 cm, or 4 in.) after their first year of life or as small catchables (20 cm or 8 in.) after two years. Unlike that of put-and-take trout, Channel Catfish survival appears excellent. In addition, these fishes live a long time (record: 39 years) and grow up to 1 m (39 in.) and 26 kg (57 lb). Larger fishes are increasingly piscivorous as they grow older but still retain their wide preference of food items.

The first introduction of the Channel Catfish was in 1891 when a group that was responsible for many early stockings of nonnatives in California, the U.S. Fish Commission, brought 500 young and adults from Illinois to San Diego County and the

Plate 58. Channel Catfish.

Feather River in northern California. This and subsequent southern California introductions failed, but in the late 1920s the Channel Catfish was planted in the American River, which likely started its spread through the Central Valley. Today, questions about the sources and stockings of this fish are moot because at this writing California has 106 fish farms where you can purchase this and other live species. Twelve farms are devoted exclusively to the Channel Catfish, making it the most commercially cultured fish (in terms of biomass) in this state. Given the massive catfish farms in Mississippi, Louisiana, and Arkansas, it is also the only nonsalmonid cultured on a large scale in the United States.

The physical attributes of the Channel Catfish, along with its tolerance for brackish water and ready acceptance of commercial food, make it ideal for fish farming. Such farms have become especially popular in the saline-alkaline lands of the Central Valley where, normally, several years of soil reclamation are required before crops such as alfalfa can be successfully raised. If well or canal water is available, however, you can stock shallow pond basins with large fingerlings and raise a salable crop in only one year. The average production is around 1,360 kg (3,000 lb) of catfishes per acre of pond, and with the market price of all fish steadily edging upward, raising Channel Catfish can be a rewarding business.

This kind of success story demonstrates the great potential of freshwater fishes as everyday food. With demand for protein growing in this country and the rest of the world, fishes such as the Channel Catfish may well become the steer of tomorrow. Such a switch would relieve pressure not only on overgrazed lands, but also on carotid arteries.

Of course, the alternate way to obtain your catfish meal is to catch it yourself. Fishing for Channel Catfishes follows the same rules as for the Brown Bullhead *(Ameirurus nebulosus)*, with a caution that the tackle should be adequate in case you hook into a big "lunker." Channel Catfish now occurs in every drainage system and in most lower-elevation reservoirs in the state, and given its large size and the excellent fight it puts up when hooked, it is rapidly emerging as a top game fish in California.

BLUE CATFISH *Ictalurus furcatus*

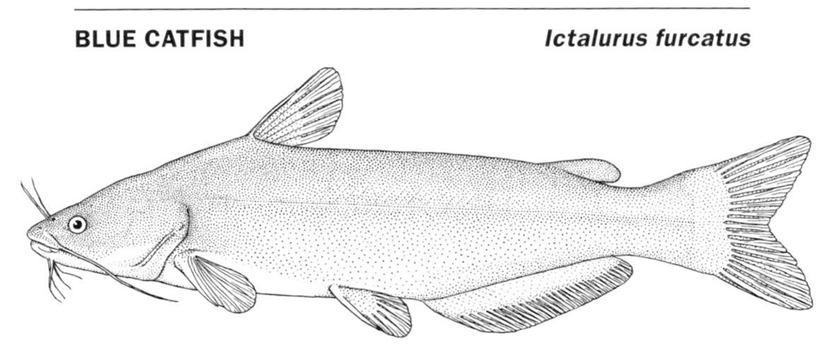

INTRODUCED

The Blue Catfish is a giant among North American catfishes. It is a native of the Mississippi River, where it occupies the deep main channels. It attains lengths of up to 1.5 m (5 ft) and weighs up to 45 kg (99 lb). Anyone who has grown up or traveled extensively in states bordering the Mississippi River has undoubtedly heard stories of giant Blue Catfishes, Channel Catfishes *(I. punctatus)*, and Flathead Catfishes *(Pylodictis olivaris)*, and the long battles that they give the local anglers. Like other ictalurids, they eat a wide variety of food but tend to be increasingly piscivorous as they get larger and grow older. Their reproductive behavior is known only from observations in their native range but appears to be similar to that of the Channel Catfish. As with the Channel Catfish, the lack of sequestered spawning sites is a major limiting factor.

The ecological separation of the Blue Catfish and the Channel Catfish in their native habitat is distinct, with the Blue Catfish inhabiting the deep main channels and the Channel Catfish occupying the deep, shoreline pools and shallower, faster water. The Blue Catfish was introduced into a number of southern California reservoirs in 1969. A decade later it was found in the Delta region, where it is now reproducing. Because this species is not related to the famous walking catfish, you can presume that it hitched a ride up Interstate 5 from one or more helpful citizens sometime during that decade. Note also that you can buy live Blue Catfish from a half-dozen fish farms in the state.

As more Blue Catfish are introduced into more California habitats, we can expect to see direct competition between this species and the well-established Channel Catfish. Relatively few sites exist in this state where both species can simultaneously find ideal environments. The Channel Catfish has done moderately well in large reservoirs, for example, but will probably give way to the Blue Catfish, which is more at home in deep, turbid conditions. Both species will also play an important role in future California fish farms, but the Channel Catfish will likely emerge as the major domestic species, primarily because of its hatchery adaptability and fast growth.

WHITE CATFISH *Ameiurus catus*

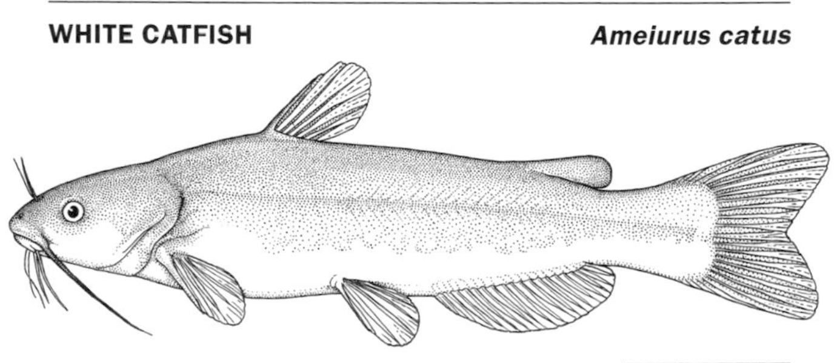

INTRODUCED

The White Catfish (pls. 59, 60) is the smallest of the fork-tailed catfishes in California, seldom attaining a length of more than 60 cm (23.6 in.) and a weight of 3 kg (6.6 lb). Nevertheless, the species is larger than any of the bullheads. It evolved in the eastern seaboard states as a delta species, inhabiting that portion of large coastal rivers where the flow is slow and the depth moderate to shallow, and where tidal action occasionally creates a slightly brackish condition. This description also fits a major habitat in California, the Sacramento–San Joaquin Delta, and it was here

Plate 59. White Catfish, juvenile.

Plate 60. White Catfish, adult.

that the White Catfish was first planted by the California Fish Commission in 1874. Today this species is the most important sportfishing catfish in this system.

The White Catfish is highly piscivorous as an adult and feeds on small schooling fishes such as the Threadfin Shad *(Dorosoma petenense)* and the Inland Silverside *(Menidia beryllina)*. In mature individuals, the head becomes extremely large and appears out of proportion to the body. With this head enlargement comes a massive gape, which may aid this species when it rushes into a school of prey fish and grabs for food.

Spawning is similar to that of the bullheads. Unlike the two larger catfishes previously discussed, the White Catfish needs no special spawning sites and reproduces successfully in California lakes and rivers. Fishing for this species is much like fishing for other ictalurids, and despite its taste for small fishes, it readily takes a wide variety of bait offerings.

FLATHEAD CATFISH ***Pylodictis olivaris***

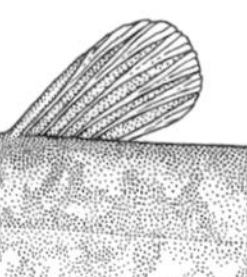

INTRODUCED

This is another large catfish native to the Mississippi River drainage. Its world record sportfishing size of 56 kg (123 lb 9 oz) is second only to that of the Blue Catfish *(Ictalurus furcatus)*. Unlike our other imported catfish and bullhead species, this one came to California courtesy of the Arizona Game and Fish Department, which stocked it in the lower Colorado River in 1962. This river is the gateway to the Imperial Valley via the extensive irrigation canal system in that area, and the Flathead Catfish is now established in waterways throughout this region. Its ecological niche in California seems to be very similar to that of the Blue Catfish. The Flathead is more solitary, however, with mature individuals taking up permanent residence in deep pools. It is unfortunate that yet another large foreign piscivore was set loose in the lower Colorado River system as the first glimmers of hope for native fish habitat restoration began to appear. The basic scheme of restoring an endangered species habitat and restocking it with the protected fishes is theoretically sound. However, if during the process one or more foreign piscivorous or otherwise competitive species are thrown into the mix, all such efforts can be for naught.

This species is reproducing successfully in its present California range but will probably never provide the sportfishing that the other two large catfishes provide because of its strong preference for night feeding, its slowness, and its occasional reluctance to take bait. Long night hours of sitting and waiting, coupled with extreme patience when the first light bites finally come, are the only proven keys to successful flathead fishing.

PUPFISHES (Cyprinodontidae)

The pupfish family comprises a small group (about 100 living species) of little, stocky fishes adapted to an amazing variety of habitats, many of which have thermal and salinity conditions well beyond the limits of most other fishes. Geographically, their habitats range from high-elevation lakes in the Andes to below-sea-level pools and creeks in Death Valley. Many of these species closely resemble members of the livebearer family (Poeciliidae), but all pupfishes are egg layers and therefore the males do not possess a gonopodium. Because of their small size, they are sometimes mistaken for minnows (Cyprinidae). Unlike the minnows, however, pupfishes' jaws are lined with small, sharp teeth. The mouth itself is usually set at an oblique upward angle that permits the large lower lip to protrude straight ahead when extended, aiding the pupfish in scrapping algae, a large component of its diet.

The four species of pupfishes that inhabit isolated pools and creeks in the desert regions of California are remnants of a past fish fauna once abundant throughout southeastern California. As the extensive springs, creeks, and lakes of this lush, Pleistocene landscape gradually dried up and desert conditions predominated, the pupfishes competed for survival with other native species. They emerged from this competition as the sole victors because of two primary factors: their small size and their physiologic tolerance to a range of extreme habitat conditions. The adult size range of the California species is 60 to 75 mm (about 2 to 3 in.) total length. The advantages of small size in very shallow, overpopulated habitats with periodic food scarcity is evident.

Among the physiological adaptations of pupfishes, perhaps the most outstanding is the ability to withstand water temperatures up to 47 degrees C (116.6 degrees F) for short periods, well above the lethal body temperature for nearly all vertebrates. Only a few desert lizards, such as the Desert Iguana *(Dipsosaurus dorsalis)*, can attain this level and still live. Of even greater importance is that most pupfish species spend several summer months in water that varies between 35 and 40 degrees C (95 to 104 degrees F), the lethal range for nearly all other California freshwater fishes. Well-developed thermoregulatory behavior is crucial for pupfish survival in such habitats. In summer, they undergo a daily vertical migration from shallow inshore water to deeper areas as the former sites warm to and beyond their thermal lethal limit. At dusk, when inshore waters cool, they return to feed in these productive sites. When the cold winds of winter sweep across the California deserts, water temperatures in some pupfish habitats may drop to 7 degrees C (45 degrees F), whereupon some species become inactive on the pool bottom, often burrowing into organic debris piles. This may be a defense against predation by migratory wading birds that might otherwise be able to catch pupfishes in a cool, less-active state.

Pupfish tolerance of high-salinity conditions is no less astounding. Pupfishes can withstand salinities a little more than twice that of seawater (68 to 70 ppt). In some segments of habitats such as Cottonball Marsh in Death Valley, salinity may rise to 4.5 times that of seawater, but the resident pupfish subspecies avoids those sites. Also of importance is pupfish tolerance of very low oxygen levels and high degrees of organic pollution, both of which characterize their habitats in summer. In all ways these midgets of the North American fish world are, in reality, physiologic giants.

Reproductive behavior is similar in all species. The adult breeding male is very aggressive and takes on a metallic blue coloration at spawning time, which lasts from April to October, actively defends a spawning territory that range from 1 to 2 m in diameter. Females and nonbreeding males remain in schools, which constantly pass over the territories of the breeding males. A ripe female leaves the school and enters a male's territory, where she is courted and then quickly spawns. Her spawning is unusual among California freshwater fishes in that only a few eggs (one to four) are deposited during each laying session. Egg numbers are low for most pupfish species, rarely exceeding 50 to 100 per spawning, although a female may spawn several times during a season. This type of breeding pattern, in which a male displays and a female comes into the display area, or "lek," is reminiscent of the reproductive behavior of North American grouse species and provides quite a show for the fish watcher (pl. 100). Some years ago I was the temporary custodian of a large aquarium containing Amargosa Pupfishes *(Cyprinodon nevadensis),* and during spring breeding season, it was a nonstop arena for displays, chases, courtship, and spawning.

Under normal high water-temperature conditions of late spring and summer, fry grow quickly. This is especially true in species such as the Amargosa Pupfish, which inhabits hot springs. With these constantly high water temperatures, sexual maturity may be reached within one month after hatching, and hot spring pools may foster as many as 10 generations per year. Among vertebrates, this feat is equaled only by some fast-breeding rodents of the meadow vole and lemming subfamily. The scientific research value of a species such as the Amargosa Pupfish is only now being realized. It is easily kept and bred in captivity and, if commercially reared, could easily become a popular research and home aquarium fish. Please note, however, it is a California Species of Special Concern and can be kept only under provisions of a memorandum of understanding from the California Department of Fish and Game.

Considerable public concern has recently been directed toward these waifs of the desert. This response is somewhat unusual, because small, nonsport fish species are often among the last vertebrates to receive any sort of public recognition. Perhaps the extreme direness of the ecological situation—that is, one small desert pool containing one species of amazingly adapted little fishes—has been the deciding factor. Americans love an underdog, and the pupfish's triumph over the high odds against surviving as desert Pleistocene lakes dry up certainly qualifies them for this role. Public concern took an organized form in 1969 with the formation of the Desert Fishes Council, a private group dedicated to the preservation and study of a number of species, including the Desert Pupfish *(C. macularius)*.

Habitat preservation is, of course, the dominant factor in pupfish conservation. Within its tiny universe, the population of each species thrives and may reach densities greater than those of most other fish species in California. Their existence, however, is totally dependent upon what humans decide to do or, better yet, not to do to these fragile habitats. Currently, the California Department of Fish and Game and the U.S. Department of the Interior's Pupfish Task Force have joined with the Desert Fishes Council to keep a close watch on season-to-season habitat changes. Further protection has been accorded by listing two California species as State Endangered and Federal Endangered, a subspecies of a third (Cottonball Marsh Pupfish *[C. salinus milleri]*) as State Threatened, and a fourth as a California Species of Special Concern.

AMARGOSA PUPFISH *Cyprinodon nevadensis*

NATIVE

PROTECTIVE STATUS: CSSC.

California populations of this species (each population is technically a subspecies) occur at the following sites: Saratoga Springs, Death Valley National Monument; the Amargosa River and ditches associated with Tecopa Hot Springs and Tecopa Bore, San Bernardino County; and Shoshone Springs, Inyo County. A fourth subspecies, the Tecopa Pupfish *(C. n. calidae)*, is now extinct as the result of "developmental improvements" at the Tecopa Hot Springs site. Two additional endangered subspecies occur in Nevada.

This is one of the most herbivorous of the California pupfish species, thriving on warm-water-adapted blue-green algae throughout much of the year. The Amargosa Pupfish also eats small invertebrates, including mosquito larvae, where salinities permit their existence. Some years ago, I found Western Mosquitofish *(Gambusia affinis)* coexisting with Amargosa Pupfish in a ditch that drains Tecopa Hot Springs. Unfortunately, this widely introduced nonnative has thermal and salinity tolerances that permit its survival in some pupfish habitats. Hopefully, the desert-area mosquito abatement folks now understand that "locals" can accomplish mosquito control just fine, without any help from their friends.

Amargosa Pupfish spawning follows the general pattern described in the family description, except that in some habitats a specific lek area is not maintained by the male (pl. 61). This species operates on an accelerated one-year plan as previously described for the Delta Smelt *(Hypomesus transpacificus)*. Here, however, sexual maturity is reached in only a few months, so that a habitat may contain several spawning generations all produced within the past year. Even though the life span of the Amargosa

Plate 61. Amargosa Pupfish male in breeding color.

Pupfish is only about one year, the continuous and rapid production of breeding stock ensures that a successful spawning may take place at least once during a year, even if habitat conditions are unfavorable for reproduction during most months. This system also permits rapid recovery from an impact that may reduce the population to a very few.

Of the four California pupfish species, the Amargosa Pupfish, with its relatively wide distribution in several types of habitat, has the best chance for continued survival. Given that all of its habitats are dependent on springs for their existence, however, any significant lowering of the water table could quickly eliminate a population.

DESERT PUPFISH *Cyprinodon macularius*

NATIVE

PROTECTIVE STATUS: SE, FE.

This is the largest of the California pupfishes. It may grow up to 75 mm (3 in.) and, unlike the hot-spring pupfish species, usually does not reach sexual maturity until its second year. The Desert Pupfish occurs in segments of San Felipe Creek (a tributary of the Salton Sea) and several irrigation drains that empty into the Salton Sea. It formerly occurred in shallow inshore areas of the sea itself, but the presence of the piscivorous Redbelly Tilapia now precludes it from these sites, even though its upper limit of salinity tolerance (68 ppt) is well beyond the current salinity of this habitat (45 ppt and rising). A number of small populations also occur in Mexico. A population has also been established at a spring pond in Anza-Borrego State Park, San Diego County.

This species feeds on a wide variety of small invertebrates and algae. Spawning follows the pupfish pattern described in the family description. Fry possess the ability to tolerate salinities up to nearly three times that of seawater and rapid salinity changes comparable to moving from freshwater to seawater. These characteristics and its tolerance of high water temperatures permits it to forage in the very shallow portions of a habitat that adults cannot normally invade.

The Desert Pupfish adapts well to life in the aquarium and to captive breeding, so in this sense its genetic future is ensured. The establishment of introduced populations at sanctuary facilities also helps in this regard. The ongoing introduction of piscivores into its remaining natural habitats, however, make its future in such areas uncertain. Given that the Salton Sea will be "dead" early in the twenty-second century, the most appropriate site for habitat restoration work is San Felipe Creek or segments thereof. Any restoration program, however, must begin by removing piscivores and reducing introduced food-niche competitors such as the Western Mosquitofish *(Gambusia affinis)* and Sailfin Molly *(Poecilia latipinna)*.

SALT CREEK PUPFISH *Cyprinodon salinus*

NATIVE

The original habitat of this subspecies is Salt Creek in Death Valley National Park, approximately 49 m (160 ft) below sea level. An elaborate boardwalk system allows visitors to view these fishes in the shallow, clear water of their native habitat (pl. 100). A second subspecies, the Cottonball Marsh Pupfish *(C. s. milleri)*, has been designated a State Threatened species because of its limited habitat. Although both subspecies occur within the protected area of the national park, the latter's entire habitat consists of approximately 26 ha (64 ac), the total area of Cottonball Marsh. Although the realm of this subspecies is small, it is by no means the smallest among naturally occurring vertebrate animals. That record is held by the famous Devils Hole Pupfish *(C. diabolis)*, of western Nevada, where the entire population of several hundred individuals exists in a deep pool about 20 m^2 (215 ft^2) and connected to an underground aquifer.

Both subspecies feed on algae but also utilize the few small invertebrate species found in their habitat. Reproduction follows the two-year plan of the Desert Pupfish *(C. macularius)*. Salt Creek is an ideal site to view pupfishes. Initially, the creek appears to be a clear, cool, freshwater stream flowing into the desert from the distant hills, where it indeed originates at McLean Springs.

One touch of the warm water and then a fingertip taste to check the salinity quickly dismiss the illusion of a cool, freshwater stream. The clarity of the water, however, is genuine and allows for a long session of viewing the antics of male pupfishes as they scurry around their little leks in a jerky fashion. You can best view the fishes from the prone position, but if you wear shorts, be sure to bring sunscreen for the backs of those calves and thighs.

OWENS PUPFISH — ***Cyprinodon radiosus***

NATIVE

PROTECTIVE STATUS: SE, FE, FPS.

This species once occurred in the Owens River and in springs within the old Owens Lake bed on the east side of the Sierra. Problems began for this species in the first half of the twentieth century when Los Angeles decided it needed Owens River water more than the pupfish, and accordingly took most of it. The introduction of the Largemouth Bass *(Micropterus salmoides)*, Common Carp *(Cyprinus carpio)*, and Western Mosquitofish *(Gambusia affinis)* into the remaining Owens River pools was thought to have spelled the end for the Owens Pupfish. But three noted ichthyologists, Carl Hubbs, Robert Miller, and Phil Pister, discovered a small population and developed plans for a sanctuary at the slough site where they were found. During the planning process, however, the slough began to dry, and in an eleventh-hour rescue attempt, Pister (a California Department of Fish and Game biologist) and his crew relocated about 800 fishes to holding cages in a deeper slough pool. A day later, however, he discovered them dying from poor water quality. Without his helpers of the previous day, he carried the last remnants of this entire species to safety in two buckets. To have an entire species' fate literally in your hands, even for one brief moment, must be the highlight of any field biologist's career. With this dramatic physical rescue completed, Phil Pister and his colleagues began the challenging task of garnering multiple agency support for the creation of two refuges, one of which still supports the Owens Pupfish today. It was also introduced into several additional sanctuary sites.

The Owens Pupfish is the least herbivorous of all California species, feeding mainly on a variety of aquatic insect larvae, snails, and small crustaceans. Its spawning behavior and other aspects of its life history are similar to those of the Desert Pupfish *(C. macularius)*.

KILLIFISHES (Fundulidae)

The killifish family is small both in species numbers and in body size, with less than 50 species whose total length rarely exceeds 75 mm (3 in.). Killifishes are closely related to the pupfish and until recently were grouped into the pupfish family (Cyprinodontidae). The dorsal anterior portion of their head is flattened and the mouth is in an upward oblique position. Like pupfishes, they are oviparous (egg layers) and can tolerate a wide range of salinities, temperatures, and water oxygen levels. They can also withstand relatively high levels of organic pollution, a trait that adapts them well to home aquariums, where several of the more brightly colored Central American forms are often seen.

Killifishes are found from the Great Lakes region to Central America. Some eastern U.S. species are commonly called Topminnows, but of course they are not cyprinids. California has one native and one major introduced killifish species. Two other species introductions were attempted in southern California, but fortunately appear to have failed.

CALIFORNIA KILLIFISH *Fundulus parvipinnis*

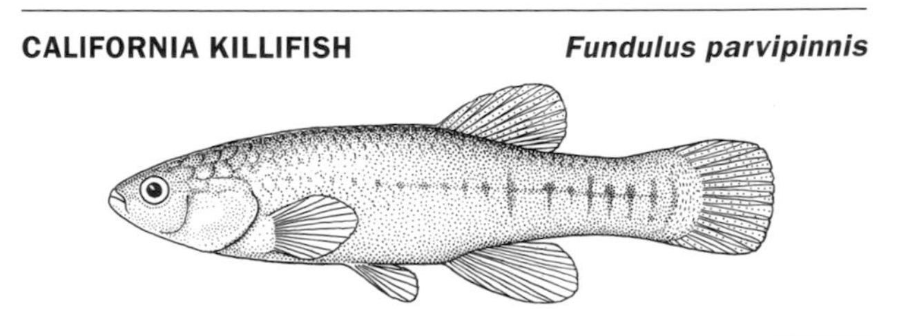

NATIVE

The California Killifish inhabits coastal streams from Monterey Bay southward to southern Baja California and the rest of Mexico. It is strongly euryhaline and can complete its entire life cycle in either brackish water or freshwater. Physiological tests have shown that it can tolerate extremely high salt concentrations and water sulfide levels that would be lethal to most other fishes. These characteristics, along with a tolerance for high temperatures and low oxygen levels, make the California Killifish the model warm, brackish-water marsh species (pl. 62).

Plate 62. California Killifish.

This species spawns continuously for several weeks in late spring. Large adhesive eggs are shed over aquatic vegetation, and the newly hatched fry are also anchored to plants by adhesive bands. These attachment methods have most likely evolved to keep the eggs and young fishes from being carried away from the productive shallow-water spawning habitat by tidal currents. California Killifish live no more than three years and breed after one year of age. The ability of this native species to function well in habitats where most other species, especially foreign invaders, cannot should ensure a future for this freshwater species that bears the state's name.

RAINWATER KILLIFISH *Lucania parva*

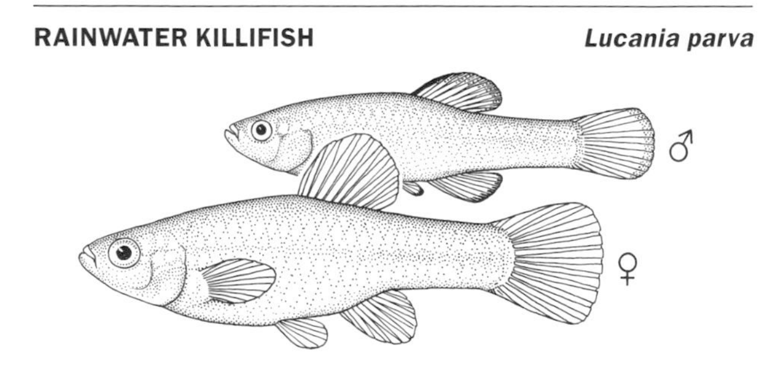

INTRODUCED

This is the only introduced killifish to establish extensive populations in California. The Rainwater Killifish is native to the eastern seaboard and states bordering the Gulf of Mexico, and it thrives in brackish water as well as freshwater. The details of its introduction into San Francisco Bay are uncertain, but it may have arrived either as an adult fish in water ballast from eastern ships or as a hitchhiker in a shipment of game fishes brought by the U.S. Fish and Wildlife Service from a federal hatchery in San Marcos, Texas. Whatever the source, this species first appeared in field collections in and around San Francisco Bay in the late 1950s and has been expanding its range in both marine and freshwater habitats ever since. I first collected this species in 1966 from Lake Merritt, Alameda County, where it feeds primarily on small copepods. It appears to be a small-invertebrate feeder throughout its California range.

Plate 63. Rainwater Killifish, gravid female.

This is one of the smallest killifish species in the United States and skillfully utilizes the inshore waters to escape predation. The Rainwater Killifish spawns over algal mats, which are abundant in its California habitats during spring and summer seasons. Its eggs hatch within a week, and the young grow rapidly, in some habitats attaining sexual maturity within three months.

The introduction of the Rainwater Killifish presents an identification challenge for the ichthyology student because of its close resemblance to the Western Mosquitofish *(Gambusia affinis)*. If you have the fish in question in hand or in a container, see if it has a gonopodium, in which case it is a male livebearing Western Mosquitofish. If it does not, it could be a female of either species (pl. 63). Now look at the position of the dorsal and anal fins. If the anterior edge of both are located at the same point along the body, it's a Rainwater Killifish. If the anterior edge of the dorsal fin is positioned noticeably posterior to that of the anal fin, it's a Western Mosquitofish.

LIVEBEARERS (Poeciliidae)

This is one family that needs little introduction to most people, because it contains well-known tropical aquarium fishes such as the Guppy *(Lebistes reticulatus)* and the Platy *(Xiphophorus maculatus)*. This family is most abundant in Central America, and only a few species, such as the Western Mosquitofish *(Gambusia affinis)* and the Sailfin Molly *(Poecilia latipinna)*, have extended their range well into the United States. The family is noted primarily for its small size and livebearing reproductive habit. Species in this family are ovoviviparous (the retention of yolked eggs and giving birth to young), giving them several advantages over other fish species. They do not have to rely on the presence of a particular substrate for spawning, and the eggs are not subject to a wide spectrum of predators, especially those found in tropical habitats. In addition, the female is able to exercise control over the incubation temperature of the eggs by shifting her position in the habitat with changes in water temperature; this ability is especially important in northern and mountain populations.

A disadvantage of livebearing is the greatly reduced number of young a female can produce at any one time because of the space requirements of the embryos in the body. The problem is circumvented, however, by the ability to bear several litters of young per year. This trait, coupled with a very short maturation time, makes the system a successful alternative to egg laying.

The retention of developing eggs requires internal fertilization. The anal fin of the male is modified at maturity to form a spike-shaped appendage known as a gonopodium (fig. 5). Muscular control of the gonopodium causes it to be swung forward and inserted into the vent of the female, and a groove in what is then its upper surface serves to guide the sperm into the female's reproductive tract. Sperm may be stored by the female for the fertilization of future egg groups, giving one fertilized female the ability to generate an entire population.

Of the many species in the livebearer family, only two are well established in California: the Western Mosquitofish and the Sailfin Molly. Two other species, the Porthole Livebearer *(Poeciliopsis gracilis)* and the Shortfin Molly *(P. mexicana)*, have recently appeared in drainage ditches along the north shore of the Salton

Sea. The former is apparently the result of fishes escaping from a nearby tropical fish farm, but the origin of the Shortfin Molly is unknown. Temporary populations of livebearer species popular with home aquarium enthusiasts are also found in selected warm-water habitats in California, especially near urban areas. These fishes are often released into nearby aquatic habitats in the same manner as Goldfishes *(Carassius auratus)* when people lose interest, or the species in question has reproduced beyond the space limits of the tank. Unlike the Goldfish, however, most of these fishes, fortunately, pass on to that great fishbowl in the sky with the onset of cool winter temperatures.

WESTERN MOSQUITOFISH *Gambusia affinis*

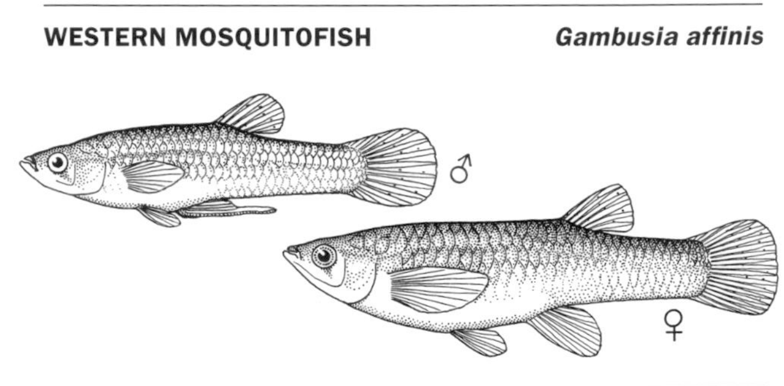

INTRODUCED

The Western Mosquitofish is native to the southern Midwest and was brought to California for mosquito control. In this role it has been introduced into nearly every low- and middle-elevation freshwater and brackish-water habitat in the state, and it may well be the most widely distributed and numerous freshwater fish species in California (pl. 64).

The Mosquitofish is quite small, rarely exceeding 5 cm (2 in.) total length. It does not feed solely on mosquito larvae but is a wide-spectrum omnivore that can do well on a variety of other invertebrate and alga species without ever eating a mosquito. As a small, opportunistic feeder, however, it normally includes the larval stage of the mosquito in its diet. This species is the mosquito-control species of choice for two main reasons: (1) it is a livebearer (300 young per birth sequence), allowing it to reproduce in any habitat and eliminating the need for specialized spawning habitats; and (2) it tolerates a wide variety of unfavorable water conditions, including high pesticide levels.

Plate 64. Western Mosquitofish.

Unfortunately, what mosquito abatement districts throughout the state have overlooked is that many small native species along with the fry of large fish will readily feed on the larval stage of the mosquito. I have surveyed a number of mosquito-larvae-free Threespine Stickleback *(Gasterosteus aculeatus)* ponds and have maintained a breeding population of California Roach *(Lavinia symmetricus)* in a backyard pool that never exhibited larval presence.

The wholesale dumping of the Western Mosquitofish throughout California has led to the very same problem that has resulted from other exotic introductions: the Western Mosquitofish competes with species whose fry also occupy the shallow shore edge and with adults of endangered species such as the pupfishes. Furthermore, its impacts on other species are not limited to the fishes. The Western Mosquitofish nibbles at and eats the tiny external gills of newly hatched frog larvae, including the Federal Threatened California Red-legged Frog *(Rana aurora draytonii)*, especially when both are highly concentrated in a small aquatic habitat.

In temporary habitats such as flooded rice fields that contain no other fishes but plenty of mosquito larvae, the Western Mosquitofish is an ideal answer to the problem—a beautiful form of biological control that avoids the use of sprays and oils. In habitats that already have resident fish populations, however, the addition of this species often compounds the unending problem of the introduction of exotic species into California freshwater habitats. Perhaps human nature is really the culprit here: the act of physically dumping a large number of fishes into a body of water is such a positive-appearing, immediate action that it too easily substitutes for a thorough, thoughtful analysis of the ecological problem at hand.

SAILFIN MOLLY *Poecilia latipinna*

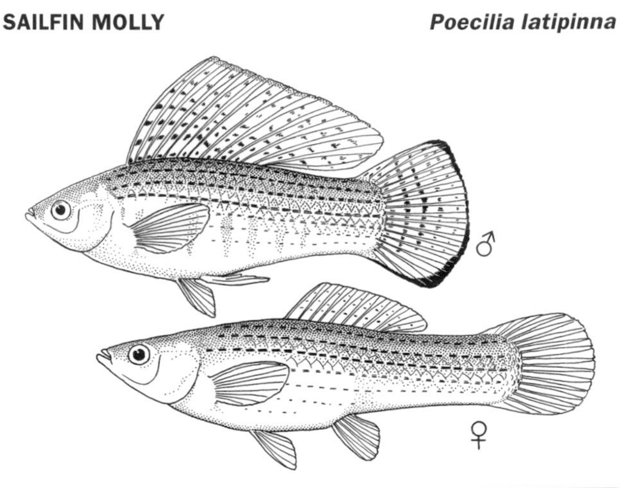

INTRODUCED

This species is native to the southeastern portion of the United States and northeastern Mexico. It is the wild form of the species from which the well-known aquarium fish, the Black Molly, is derived. The Sailfin Molly is a beautiful fish in its own right, with two male color phases, plain and checkered. Mollies are primarily vegetarians and feed on a variety of algae and detritus. The Sailfin Molly attains a total length about three times that of the Western Mosquitofish *(Gambusia affinis)* (up to 15 cm [6 in.]) but has only about half the number of young (up to 140 per birth sequence).

The Sailfin Molly was introduced into the Salton Sea and Death Valley areas, where it now thrives and, unfortunately, competes directly with the Desert Pupfish *(Cyprinodon macularius)*.

Not only are the Sailfin Molly's food habits similar to that of the Desert Pupfish, but it can also withstand salinities approaching three times that of seawater (up to 87 ppt). The adult Sailfin Molly, however, is a strict vegetarian and is not known to prey on fish eggs. In a competition between pupfishes and mollies, the lack of egg predation by the molly could tip the scale in favor of the pupfishes. Even so, Sailfin Molly eradication projects are needed in desert-spring habitats to ensure protection of native pupfish species.

At this writing, a second species of molly, the Shortfin Molly *(P. mexicana)*, has appeared in several Salton Sea drains where the Sailfin Molly is established. It is easily distinguished from the latter by its comparatively short dorsal fin and olive-drab color. We will have to wait and see what role this latest actor will play in the ongoing Salton Sea saga.

PORTHOLE LIVEBEARER *Poeciliopsis gracilis*

INTRODUCED

This is another "tropical fish" from southern Mexico and Central America that possesses many of the attributes necessary for life in the home aquarium. Its only attractive physical feature is a series of bold dark spots ("portholes") along the lateral line on each side. The Porthole Livebearer is small, with a maximum total length of 7.5 cm (3 in.). It also prefers the warm water of the heated aquarium and can tolerate the less-than-pure water often found within. Tropical fish breeders favor this species for another characteristic—the ability to bear live young—and it was apparently from a tropical fish farm in Riverside County that some "escaped" into several Salton Sea drains where they are able to withstand the moderately brackish water.

Like other tropical fish species that have found their way into southern California waters, the Porthole Livebearer requires room-temperature water (20 to 22 degrees C [about 70 degrees F]), which prevents their spread northward or even beyond their immediate warm habitat. Unfortunately, the Porthole Livebearer shares such habitats with small groups of the endangered Desert Pupfish *(Cyprinodon macularius)*, and thus yet another competitor for the food resources of the Desert Pupfish is now present.

SILVERSIDES (Atherinidae)

Silversides are small, elongate, trim fishes that normally inhabit coastal marine waters. Wherever they occur, they are usually abundant and travel in large schools. They look much like smelts but differ in a number of ways, the most important being their small, pincerlike mouth. Unlike smelts, which have large mouths and are essentially filter feeders, silversides grab individual large zooplankters and small, bottom-dwelling invertebrates with their strong, efficient lips and tooth-lined jaws.

There are three native species of California silversides, and all are marine. One of these, the Topsmelt *(Atherinops affinis)*, is found occasionally in the lower reaches of coastal streams, but its limited penetration into freshwater does not merit discussion here. In 1967, however, a freshwater silverside, the Inland Silverside *(Menidia beryllina)*, was authorized for introduction into Blue Lakes, Lake County, by the California Department of Fish and Game.

INLAND SILVERSIDE *Menidia beryllina*

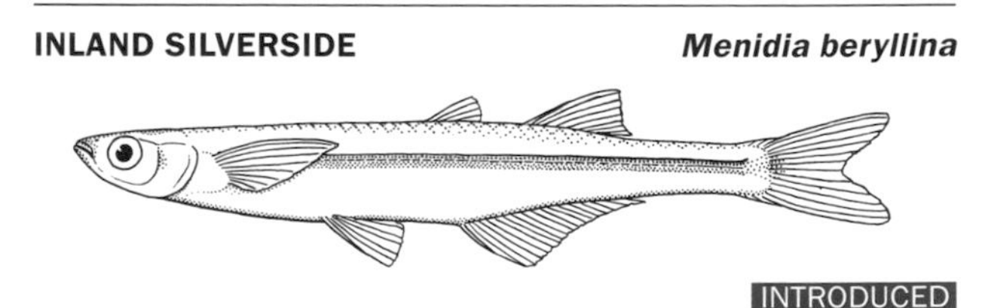

INTRODUCED

The Inland Silverside is native to the lower Mississippi Valley and was originally called the Mississippi Silverside (pl. 65). It feeds on zooplankton and small invertebrates in the inshore zone. It rarely grows beyond 15 cm (6 in.) and lives only two years. However, it makes up for this short life span by reproducing often. The female spawns several times a year and when mature may lay up to 15,000 eggs in one season. Spawning takes place over aquatic vegetation and, like other activities of this species, is conducted in schools. The Inland Silverside is able to withstand a wide range of water conditions, including high temperature, low oxygen, and moderate organic pollution, which are conditions many other species cannot tolerate.

One year after the initial introduction of this species in Lake

Plate 65. Inland Silverside.

County, the California Department of Fish and Game made two further experimental introductions in Alameda and Santa Clara Counties. The stated purpose was to use the Inland Silverside to control gnats and midges, particularly in Lake County; however, the preference of this fish for zooplankton and bottom-dwelling invertebrates, plus its ecological restriction to the inshore zone, makes its gnat and midge control potential less than that of a number of other California species.

Far more serious was the assumption that the Inland Silverside would remain at the initial introduction sites because the sites were not connected to other aquatic habitats. Once a small potential bait-fish species is available, especially a highly visible, inshore-schooling form, some anglers will capture and use it. Indeed, the fishing regulations of that era allowed the netting of nongame species for use as bait. Thus unauthorized introductions were almost a certainty and unfortunately did occur. By far the most serious unauthorized introduction, likely a result of bait dumping, was the planting of the Inland Silverside into Clear Lake in 1967 or 1968. Once established in California's only large, low-elevation natural lake, this species promptly underwent an population explosion. During a collecting trip to Clear Lake in August 1972, my students and I caught several thousand Inland Silversides with each haul of a 31-m (100-ft) beach seine. This species has replaced the Bluegill *(Lepomis macrochirus)* as the dominant inshore zone fish and in the process probably provided the final competitive blow for the extinction of the native Clear Lake Splittail *(Pogonichthys ciscoides)*.

A much farther-reaching consequence of the Clear Lake introduction has been movement of the Inland Silverside via creek and irrigation ditch connections to the Delta system. For nearly two decades, my ichthyology classes have conducted surveys of a small portion of the San Joaquin River near Manteca in San Joaquin County. Prior to 1971 we never collected this species, but that year we collected one. The numbers of Inland Silversides in our seine hauls increased rapidly over the next several years, and by 1976 it was the dominant inshore species in the area, surpassing Striped Bass *(Morone saxatilis)* fry and Threadfin Shad *(Dorosoma petenense)* by a ratio of at least two to one.

A study I conducted on the food habits of Striped Bass fry and Inland Silversides in the middle San Joaquin River revealed that the most preferred food item for both species was the Opossum Shrimp *(Neomysis mercedis)*. Because of the general lack of extensive inshore zone in this river area, these two fish species do not segregate by depth, and thus their competition for this food resource is rigorous. Speculation as to the role of this competition in the recent decline in Striped Bass numbers is presented in the "Angling Notes" section for that species.

Presently, the Inland Silverside is found throughout the Sacramento–San Joaquin Delta system and most connecting reservoirs. The inclusion of this fish in the legal bait-fish section of the California Sport Fishing Regulations for years after its appearance in Clear Lake continued to promote this spread. The fine print of the regulations did caution that the Silverside may be used only in waters where it is captured, but the history of freshwater fish movement in California, and of this species in particular, raises serious doubts that this rule will be followed.

The history of the Inland Silverside in California is a classic example of the speed with which an introduced species can spread. Beginning with a small population in an isolated lake some 160 km (100 mi) north of the San Francisco estuary, it became one of the most numerous species in this giant river-lake system in a little over a decade. Like a substance injected into a remote capillary bed of the body then soon appears in every vessel of the circulatory system, introduced fish species spread through the interconnected aquatic arterioles of central California. The 2005 state fishing regulations still permit the use of live Inland Silverside bait in most areas of the Central Valley, southern California, and north-central sportfishing districts, but only in the latter is such use restricted to the area of capture.

STICKLEBACKS (Gasterosteidae)

This family is represented by one native and one introduced species in California.

THREESPINE STICKLEBACK *Gasterosteus aculeatus*

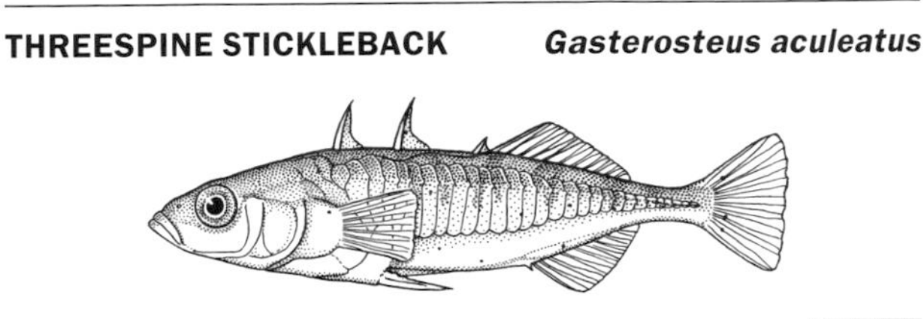

NATIVE

This is a small fish, never more than 8 cm (3.2 in.) in length, which when viewed from above first appears to be one of the smaller minnows (Cyprinidae). Further inspection, however, reveals pronounced differences. The most apparent, when viewing a live fish in water, is the presence of three separate dorsal spines anterior to the soft dorsal fin (pl. 66). In addition, the pelvic fins are modified into long, sharp spines. When held in the hand after capture in a net, this species often lowers these spines, a habit that can cause a moment of confusion for first-time collectors. A little tap of your finger on the lateral body wall, however, usually causes

Plate 66. Threespine Stickleback, nonbreeding.

immediate spine erection. Note that the spines are not just raised but also locked in place and are very hard to force down. Indeed, the pelvic spines often break before the locking mechanism is released.

The selective advantage of this spine complex is apparent when this species' ecological niche is considered. A Threespine Stickleback feeds primarily on small invertebrates it picks off the substrate after a series of jerky advances; between advances it hangs motionless in the water. This stationary feeding habit, coupled with its small size, makes the Threespine Stickleback potential food for a number of piscivorous species that also occupy its habitat. The spines, however, seem to function well in protecting this seemingly easy target. I recently set up a cold-water aquarium containing Rainbow Trout parr *(Oncorhynchus mykiss),* California Roach *(Lavinia symmetricus),* and Threespine Stickleback. I was not careful enough in sizing the fishes for this community tank, and the largest Rainbow parr began eating the smaller roach. The unscheduled experiment was allowed to run its course, and after several weeks all but the largest roach had been eaten; however, only one of the Sticklebacks, all of which were smaller than the roach, was missing. It was perhaps this one that was responsible for a bit of fast reward-and-punishment learning on the part of the Rainbow.

Another highly adaptive feature of the Threespine Stickleback is its wide range of osmotic tolerance. Anadromous populations live most of their lives in saltwater and return to coastal freshwater streams along the California coast for spawning only. These have been recognized as a subspecies and given the appropriate name: Anadromous Threespine Stickleback *(G. a. aculeatus).* Numerous freshwater populations throughout the state never go to sea, and these are now called Resident Threespine Stickleback *(G. a. microcephalus).* Populations also differ morphologically, with the most apparent feature being the number of bony plates on the lateral body wall. One group has no plates at all and is designated the Unarmored Threespine Stickleback *(G. a. williamsoni).* It occurs in small numbers in a few coastal creeks in southern California, and because of its low numbers and the ongoing risks to its habitat, it has been listed as both State Endangered and Federal Endangered and is a Fully Protected Species.

A Threespine Stickleback feeds by picking small invertebrates, including earthworms, off the bottom and from aquatic

Plate 67. Threespine Stickleback, gravid female above, male in breeding color below.

vegetation. It often hangs motionless with its snout pointed at the intended prey, then quickly lunges forward and grasps the prey with its small, pointed mouth.

The Threespine Stickleback is one of the most sought after of all California freshwater fishes by beginning aquarium enthusiasts. It is relatively easy to capture with a small hand net because of its stationary behavior and its preference for pondside vegetation. It also tolerates the low oxygen and fluctuating thermal conditions of the home aquarium, and its small size permits a modest population to exist in a 5- or 10-gallon tank. The most impressive aspect of this species, however, is its intricate breeding behavior, which may be observed from the first courtship displays through the maturation of the young. This behavior was described in detail by Niko Tinbergen in his classic book, *The Study of Instinct.* His theory concerning the mechanisms of instinctive behavior was formulated in part on Threespine Stickleback behavior because of its many intricate aspects. I include the following description of this behavior for readers who may choose to try a pair of these exciting fishes in their aquarium.

The breeding cycle lasts two or three months in spring or summer. During this time marked sexual dimorphism develops, which does not exist during the rest of the year. The male develops a red throat and forebelly, blue irises, and often a greenish back (pl. 67). During courtship, the dark patterns of the female's body become more pronounced. As males take on their breeding color, they start to defend territories, whereas the females remain in schools.

The male builds a nest on the substrate within his territory. This nest is constructed out of vegetation and sand grains glued together with a mucus secretion from the kidney. The male continually defends this nest from all other males and also chases off other species of small fishes that get too close. When he has completed the nest, he signals this to the females by pushing through the center of the loosely packed sphere to form a tunnel.

Gravid females now move out of the school into male territories. When a male approaches, either directly or in a series of zigzags, the female adopts a head-up posture and moves toward the male, which usually returns to nest-directed activities such as fanning. If the female follows the male, he leads her to the nest entrance by placing his snout toward the opening of the nest tunnel and moving back and forth. A female ready to spawn will push her way into the nest, leaving only her caudal peduncle and tail protruding from the entrance. The male quivers rapidly while pressing his head against the vent area and flanks of the female. With this final stimulus, she raises her tail and spawns the eggs. She then wiggles forward and exits the nest tunnel while the male enters at the other end and releases milt over the eggs.

After fertilization, the male is very aggressive and drives the spawned-out female away. (At this time it is wise to remove the female from the tank.) He then pushes the eggs deeply into the nest and presses them firmly to the nest bottom with his head and body. Within an hour the male begins courting another female and the spawning procedure is repeated. Each clutch of eggs is pressed down so that each successive clutch can be piled on top. Often a male fertilizes six or seven clutches before he stops courting. After spawning, females return to the school where they feed voraciously as another batch of eggs matures in their ovaries.

With eggs in his nest, the male assumes a parental role. Fanning becomes the predominant nest-tending activity. In this behavior the male ventilates the eggs by driving a fresh stream of water through the nest with his caudal fin. The amount of fanning increases until just before the eggs hatch. Once the fry emerge, the male tears the nest into little pieces. For a few days he retrieves the young that stray from the nest site, but as the young become more active, fill their swim bladders, and become more agile, he can no longer retrieve them, and they disperse. During the parental phase, the male tends to darken in color, a change that makes him less conspicuous to potential predators. The male is also less territorial during this phase. A few days after the young disperse, the male may build another nest, and if more gravid females are present, the spawning cycle is repeated.

During the course of his study, Tinbergen realized that the behavior he was seeing was innate or genetically inherited, not learned. Each new display or step in the intricate nest-spawning process was triggered by a movement, posture, or color of the other fish. He named such triggers for the release of a specific response "sign stimuli." We now know that most of the behavior of fishes, especially behavior involving breeding, is innate and dependent on sign stimuli.

The Threespine Stickleback is one of the most numerous and adaptable small native fishes in California and at least one of the three subspecies is almost always present in creek, stream, marsh, or estuary habitats. It does not appear much affected by the many introductions of nonnative small fishes, probably because of its unique life history.

Unfortunately, another species, the **BROOK STICKLEBACK *(Culaea inconstans)*,** has recently appeared in Siskiyou County near the California-Oregon border. This species, with five dorsal spines instead of three, paired pelvic spines, and an anal fin spine

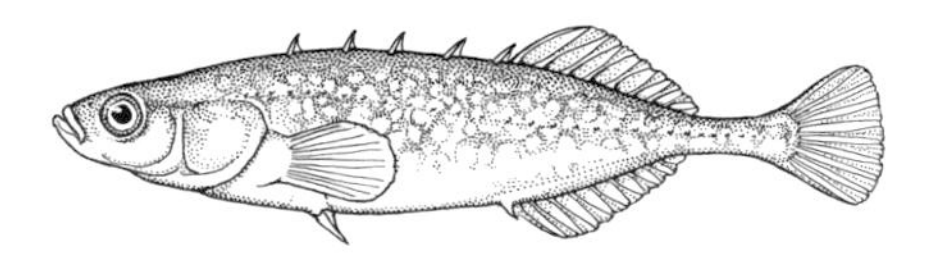

but no body plating, is a native of the northern United States and southern Canada east of the Rocky Mountains. The means of its arrival is still unknown, but the possibility exists that it may spread throughout the Klamath River drainage. As the introduction of the Wakasagi *(Hypomesus nipponensis)* into the habitat of the Delta Smelt *(H. transpacificus)* shows, nothing is more injurious to a native species than confrontation with a "mirror image" of itself. We hope this will not be the case here.

TEMPERATE BASS (Moronidae)

Species in the temperate bass family (Moronidae) were once a part of the larger family Serranidae. Taxonomists felt that Serranidae was becoming too much of a "catch-all" family, and so the family Moronidae, comprising a more uniform group of spiny-rayed fishes, was formed. In California, the family includes several marine basses, plus one anadromous and one freshwater species: the Striped Bass *(Morone saxatilis)* and the White Bass *(M. chrysops)*, respectively.

STRIPED BASS *Morone saxatilis*

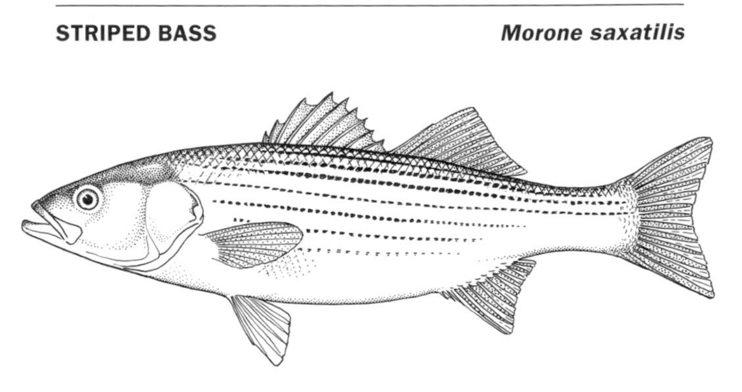

INTRODUCED

Although the vast majority of freshwater fish introductions in California have created more problems than their presence justifies, the Striped Bass has for the most part been a success, at least for the angler. The species is native to the Atlantic coast of the United States and was introduced into the lower Sacramento–San Joaquin Delta in 1879. The population rapidly established itself, and by 1890 a commercial fishery for Striped Bass was well under way and lasted until 1935, when it was discontinued to further develop the Striped Bass sport fishery. Today the Striped Bass is one of the most popular sport fishes in the state. Such popularity usually inspires biological investigation, and accordingly, we know more about this species than about most other fishes in California.

One of the most striking aspects of Striped Bass biology is its ability to move freely between freshwater and saltwater throughout its long life. This means that the osmoregulatory mechanisms (discussed in "Anatomy and Physiology of Freshwater Fishes" section) of an adult Striped Bass must change from seawater adapted to freshwater adapted and then back to seawater adapted each year. This pattern is in sharp contrast to that of most salmon, which make a major osmotic adjustment only twice in their lives. This physiological ability gives the Striped Bass an unusually wide range of feeding opportunities as it moves from river through estuary to bay or sea and back again. The adult Striped Bass is a top piscivore but may also take larger invertebrates such as shrimp. At sea it primarily feeds on small pelagic fishes. During the first several decades after the Striped Bass's introduction in California, the Pacific Sardine *(Sardinops sagax)* was probably its key prey species, but since its decline, the Pacific Herring *(Clupea pallasii)* and Northern Anchovy *(Engraulis mordox)* are its main prey.

During its early decades in the San Francisco Estuary, the wide spectrum of introduced eastern sunfish (Centrarchidae) species and designated forage species such as the Threadfin Shad *(Dorosoma petenense)* that the Striped Bass now dines on either had not yet arrived or were just getting established. Furthermore, prior to the 1870s, the first decade of freshwater fish introduction in California, all potential prey species were natives, and the Delta had no large, pelagic piscivore that fed voraciously during a long annual stay in freshwater. It is therefore possible that the exodus of the native Sacramento Perch *(Archoplites interruptus)* from the Delta system and the elimination of the Thicktail Chub *(Gila crassicauda)* from the planet are partly due to Striped Bass predation.

The Striped Bass is a spring spawner that moves into the lower reaches of the Sacramento and San Joaquin Rivers to spawn as water temperatures rise above 15 degrees C (59 degrees F) in late April or May. The time varies from year to year depending on the amount of cold snowmelt water, which feeds the tributaries of the Delta system. The male spawns for the first time at two years of age or older, whereas the female waits until at least four. Like their cousins the White Bass *(Morone chrysops)*, "stripers," as they

Plate 68. Striped Bass, juvenile.

are called by anglers, are mass spawners. Large numbers of males occupy a nearshore spawning site and wait for individual females to arrive. A large adult female may lay up to two million eggs in a single season. The eggs are only slightly heavier than freshwater, and the moderate currents of the lower river moves them along, suspended above the bottom. They hatch in two days, and the yolk-sac fry continue to drift downstream to regions where tidal inflow counters river outflow. In the Sacramento River, where the current is always stronger than the reverse tidal flow, the fry are carried all the way to the estuary. In parts of the San Joaquin River, however, where upstream tidal current periodically overcomes downstream flow, the fry may remain in the lower reaches of the river. After the yolk sac is absorbed, the fry begin feeding on zooplankton and other small invertebrates and continue to do so throughout their first year of life. By the second year they begin to take the young of small fishes such as juvenile Threadfin Shad and, when available, small adult fishes such as the Delta Smelt *(Hypomesus transpacificus).* By the end of their second year, young Striped Bass are primarily piscivorous and remain so for the rest of their lives (pl. 68).

Growth in Striped Bass follows approximately the following pattern: a one-year-old fish is up to 10 cm (4 in.) long; a two-year-old fish averages 25 cm (10 in.); a three-year-old attains the legal size limit of 46 cm (18 in.); a four-year-old reaches 50 cm (20 in.). From then on, the species grows about 2 cm (almost

Plate 69. Striped Bass, adult.

1 in.) per year. A 20-year-old striper averages 122 cm (48 in.) long and weighs 18 kg (40 lb).

Landlocked populations of Striped Bass occur in Millerton Lake, Fresno County, and in San Luis Reservoir, Merced County. The Millerton Lake population is small but sustains itself through spawning runs into the upper San Joaquin River, which supplies the lake. The San Luis Reservoir population is stocked inadvertently through water delivery from the Delta, which contains young bass, so there is no way of knowing whether the population is spawning successfully in the reservoir. We do know that conditions for adult stripers are pretty good in O'Neill Forebay, the area from which water is pumped to the big reservoir, because in 1992 it produced the California record Striped Bass: 30.5 kg (67.5 lb). Silverwood Reservoir, San Bernardino County, and Pyramid Reservoir, Los Angeles County, along with all other reservoirs that receive water from the California Aqueduct, are also now stocked with Striped Bass.

ANGLING NOTES: Striped Bass fishing continues to be big business in the Delta. Millions of dollars are spent each year on boats, fishing equipment, and all the other accessories that striper anglers require to pursue their sport. The striper provides the California freshwater angler a unique opportunity to catch a large, essentially open-ocean fish without the expense and, if you are a poor sailor like me, the discomfort of going out on the high seas. Although many large Striped Bass are taken from power boats in San Francisco Bay and the lower Delta, many more are caught from the shores of the San Francisco Estuary's waterways, including fishing access areas along two major aqueducts in which Striped Bass, which arrived as larvae through the giant pumps, may grow to a considerable size (pl. 69). The California record Striped Bass traveled to O'Neill Forebay in Merced County via such a concrete stream.

If you are after big fishes, your tackle should be moderately heavy. Striped Bass at or just above the current minimum size of 18 inches are a challenge on light tackle, including fly rods. These are most often young males that have made their first trip back from marine waters and are usually the best "supper stripers." By far the most popular fishing area is the lower Delta region, although large fishes may be taken anywhere from outside the Golden Gate Bridge to the middle reaches of the San Joaquin and Sacramento Rivers. Although trolling with feathered jigs or spoons is quite productive in San Francisco Bay, live bait minnows appear to be the most used bait in the Delta and middle river systems. Pacific Staghorn Sculpins *(Leptocottus armatus)*, confusingly referred to by some anglers as "bullheads," have long been a standard bait in the estuary region. Lately, however, the recently introduced Yellowfin Goby *(Acanthogobius flavimanus)* has produced impressive catches and now commands a high price in Delta bait stores. For striper anglers in the middle areas of the Sacramento and the San Joaquin Rivers, the Golden Shiner *(Notemigonus crysoleucas)* is perhaps the most popular bait. If you do choose to use live bait, please heed the request of the California Department of Fish and Game and do not release your excess bait.

Minnows are usually rigged so that a number 1 or 2 hook is positioned on its own short line about a yard above a large, terminal sinker. This holds the bait low, but off the bottom, which seems to be the preferred cruising depth for Delta stripers. When fishing the middle reaches of the San Joaquin where the river is relatively narrow, usually you obtain the best results by fishing the bait minnow just off the inshore shelf where the water becomes deep and the current picks up. The inshore waters in this area are full of yearling stripers and adult Threadfin Shad, which appear reluctant to enter the main current. Striped Bass are thought to cruise the edge of the inshore shelf and pick off occasional prey fishes that leave this shelter. This idea is consistently supported by good catches using the method described above, especially in mid- and late fall.

Of prime importance in Striped Bass fishing is a knowledge of this fish's basic annual movements. In March or April, legal-size fishes move from San Francisco Bay into the lower Delta. By May they begin spawning in this area and up into the Sacramento and San Joaquin Rivers. By early June, spawners have moved back into the lower Delta, and by late June these fishes are again in the Bay and outside the Golden Gate Bridge. Superimposed upon this classic movement pattern associated with spawning is a variety of secondary movement patterns that as yet are poorly understood. Most of these movements result in large fish in various portions of the freshwater range at other times of the year. The movement of large female Striped Bass into the middle reaches of the San Joaquin River in fall is one example. These fishes do not appear to form schools but instead distribute themselves along the inshore zone of the river at undercut banks or underwater obstacles. Anglers aware of this pattern can do quite well during what is assumed by many to be the off-season for stripers. After overwintering in the San Joaquin River, these mature females apparently join the normal spring spawning groups.

No essay on Striped Bass fishing would be complete without comment on the continuous decline in fish numbers since 1935 when commercial fishing was banned in favor of sportfishing. In those early years, an angler's annual catch was about 20 fishes per person. This catch steadily declined until the 1960s, when 10 stripers per year was the norm. By the early 1970s, the entire population of adult Striped Bass in California was estimated to be less than two million. The all-time low (to date) came in 1994 when the number of stripers 18 in. or larger was estimated to be a little over a half million. Since the beginning of the twenty-first century, the adult striper population appears to have rebounded and is currently estimated at about 1.5 million. As is usually the case when working with a group of animals such as fishes whose habitat and behavior are extremely hard to view, the reasons for this unexpected upswing in numbers are unclear.

Major impacts to California freshwater fishes described in the introduction are most likely involved. For instance, water quality and water flow in California have changed substantially as a result of the diversion of Delta water to the south. During some years, the great pumps at the "headwaters" of the Delta-Mendota

Canal and California Aqueduct pump more water annually than the amount allowed to flow into San Francisco Bay. This in turn affects the brackish-water regime of the lower Delta to which hatchling Striped Bass are very sensitive.

The great pumps that move Delta water southward also ship a large number of hatchlings in that same direction. At the beginning of the 1980s when the Striped Bass population was in full free fall toward the 1994 low, an estimated 80 million fry were sucked out of the lower Delta annually by the pumps of the California Aqueduct near Tracy. Despite the ongoing inventive efforts of the dedicated group of fisheries workers at the federal Tracy Fish Collecting Facility, which screens the water supply for the California Aqueduct, and at the John F. Skinner Fish Protection Facility, which does the same for the Delta-Mendota Canal pumps, very small (larval) fishes less than 35 mm (1.4 in.) in length cannot be fully excluded while retaining the required water flow. As the demand for more water from both canals increases, the accompanying loss of fry increases proportionately. It should also be pointed out that the pumps of the great canals are not the only culprits. At the start of the same major period of Striped Bass decline in the early 1980s, a study estimated that 165 million Striped Bass fry were killed in the intake flows of the Antioch and Pittsburgh power plants in just one year.

A further reason for the reduction of Striped Bass numbers in the Delta has been the accompanying increase of introduced species that compete with striper fry for small invertebrate food resources. The many sunfish species introduced into the San Francisco Estuary since the introduction of the Striped Bass all produce many young annually that for the most part eat the same food as striper fry. So also do the young and adults of dumped bait-minnow species and forage fishes (Threadfin Shad, Wakasagi *[H. nipponensis]*, and Inland Silverside *[Menidia beryllina]*) introduced with the intention of improving sportfishing opportunities. A basic rule of "fish-bowl biology" is that you cannot keep adding more species and expect all to be well in the aquarium without adding more fish food. Unfortunately, the supply of zooplankton and many other small invertebrates is far smaller today than when the Striped Bass was originally introduced. Then there were no massive levee systems that in turned allowed the drying and draining of the great Delta marshes where much of this food was produced.

By looking at just one of these fry competitors, the problem becomes apparent. Both Striped Bass fry and Inland Silversides preferred the opossum shrimp *(Neomysis mercedis)* over all other food items when it was still abundant in the Delta. In a study of these two fishes in the middle San Joaquin River in the early 1980s, I found that at some sampling sites the stomachs of both species often contained just this one item. It is not uncommon to find two species depending heavily on the same food resource, but when this does occur, the two usually divide that resource by feeding in two different areas. The Inland Silverside is an inshore feeder, whereas Striped Bass fry tend to be more pelagic. Unfortunately, most of the real inshore zone disappeared with the channelization of the San Joaquin River, and now the food source and the feeding sites of the two species appear to significantly overlap. The enormous population of Inland Silversides now present in the Delta merits its inclusion in the evolving equation for Striped Bass scarcity. Indeed, it now appears that the key reduction in the annual recruitment of stripers takes place at the end of the first year of juvenile life when heavy feeding on small invertebrates should promote the size increase needed for the conversion to piscivorous feeding. The failure of a significant number of Striped Bass fry to convert each year would certainly mean fewer adults.

Unfortunately, the list of "usual suspects" in this mystery continues to grow, and at present the massive increase of the introduced Overbite Clam *(Potamocorbula amurensis)* in, among other areas, Suisun Bay, where post-yolk-sac striper fry begin feeding, is the latest serious problem. Small clams filter feed on both small zooplankton and phytoplankton, the food of zooplankton, dealing a double blow to striper reproduction.

Because the Striped Bass sport fishery is second only to that for trout and Largemouth Bass *(Micropterus salmoides)* in this state, the California Department of Fish and Game came under pressure to do something. One response was to raise the legal minimal catchable size in 1981 from 16 to 18 in., but of course such a move can have no effect on the real problems affecting fry, discussed above. A second plan initiated that year proposed the catch-all solution for which the state of California is famous: another hatchery program, this one for Striped Bass. Beginning in 1981, a new $3.50 Striped Bass fishing stamp was required of all striper anglers, and revenue from this new stamp supports the propagation of yearling Striped Bass at the Central Valley's hatchery at Elk Grove. Before long, it was decided that far more were needed, and for the next decade about 11 million fingerlings (less than one year old) and yearlings were purchased from private fish farms and released in the Delta. This turned out to be a very expensive, quick-fix attempt to solve the problem, especially because the greatest decline in numbers appears to occur at the end of the fingerling stage. Cost estimates per fish, presented by retired CDFG biologists William Dill and Almo Cordone in their 1997 monograph on introduced California fishes are enlightening. Each hatchery-produced yearling striper that was eventually caught by an angler cost $106. The price per angler-caught fish rose to $237 if stocked as an older fingerling. The really high-priced stripers, however, were those that originated as young hatchery fingerlings: a whopping $1,071 per angler-caught fish. On a recent visit to a fish market, I observed that farm-raised Striped Bass hybrids were selling for $3.99 per pound. At the estimated cost for catchable small fingerlings, an angler catching only one fish from a stream on a given day could have brought home a minimum of 268 pounds of striper if he or she had instead bought them at the store. Of course, most anglers are more interested in catching Striped Bass than eating them, so perhaps the state could provide "pay and fish" ponds for stripers, like those already in operation for Rainbow Trout *(Oncorhynchus mykiss)* and Channel Catfishes *(Ictalurus punctatus)*. At least the cost would be borne solely by the user.

Fortunately, the hatchery stocking program was discontinued in 1992, just two years before the 1994 all-time low Striped Bass population estimate. It has been replaced with a new scheme whereby small stripers are trapped before they are sucked away by the Delta-Mendota Canal pumps, raised to near-catchable size, and released. Hopefully, the cost of these fishes when taken by an angler will be less than those inflated prices of the 1980s.

WHITE BASS *Morone chrysops*

INTRODUCED

The White Bass, a panfish-size version of the Striped Bass *(M. saxatilis)*, lives entirely in freshwater. It is native to the Mississippi River drainage and has been introduced into numerous reservoirs in the South and East. In 1965 the California Department of

Fish and Game made an experimental introduction of this species into Lake Nacimiento, San Luis Obispo County, and today this lake contains an established, reproducing population. The species was also planted in the lower Colorado River, but to date, natural reproduction there seems unlikely. In the early 1980s, an unauthorized introduction was made in Kaweah Reservoir, Tulare County.

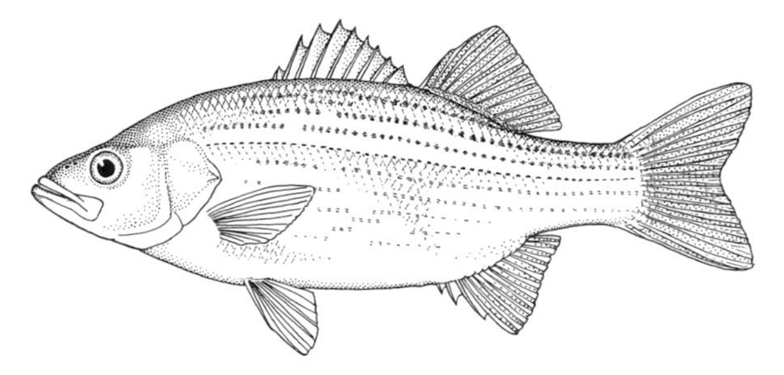

Adult White Bass are pelagic, close-schooling fishes that feed on zooplankton and small fishes. Zooplankton feeding is most prominent during the summer months and takes place mainly at dawn and dusk, when zooplankton aggregate at the surface. A very rich, eutrophic lake condition that produces an abundance of both zooplankton and fish fry is essential for the success of this species. When these conditions are met, the White Bass thrives. Its pelagic, schooling behavior is similar to the crappie species of the sunfish family (Centrarchidae).

White Bass are mass spawners that prefer rocky, shallow waters in lakes with favorable current or wave action. They also utilize similar substrate in rivers at the head of lakes and reservoirs. Spawning behavior begins in midspring when water temperatures rise above 15 degrees C (59 degrees F). Males move into these areas in advance of the females and remain for several weeks. Females come to the area a few at a time, spawn, and leave. Depending on the size of the female, anywhere from 25,000 to 1 million eggs may be deposited. Eggs hatch in about two days, and the yolk-sac fry remain among the rocks, where they begin zooplankton feeding as the yolk sac is absorbed. Growth is extremely rapid under good conditions, and one-year-old fishes may attain lengths from 25 to 31 cm (10 to 12 in.). Maximum length is about 45 cm (17.7 in.) with weights reaching 2.4 kg (5.3 lb).

Plate 70. Striped Bass/White Bass hybrids ("wipers").

Although the White Bass is a valuable sport fish in warm-water lakes with good zooplankton and small-fish production, many California reservoirs do not fit both criteria. Some have high populations of forage species such as Threadfin Shad *(Dorosoma petenense)* and Inland Silverside *(Menidia beryllina)* but are relatively low in zooplankton production compared to lakes in the native range of the White Bass. In addition, both of these prey species, especially the Threadfin Shad, directly compete for the zooplankton resource.

A further fear is that the White Bass might make its way into Delta waters and compete for resources with its close relative, the Striped Bass, or even hybridize with it. Although unknown in nature, a sterile hybrid called a "wiper" (pl. 70) is produced in hatcheries and is a popular game fish in Colorado and several other states. Even more important from the nonangler's viewpoint is that the White Bass would be yet another introduced planktivore-piscivore competing with the struggling native fishes of that region.

This situation is unfortunate, because the White Bass is an ex-

Plate 71. White Bass caught from a surface-filter-feeding school.

cellent large panfish and a special favorite of mine, because during my undergraduate days as a fisheries research assistant at the University of Wisconsin at Madison, it was the primary research species of the fisheries group. Indeed, at first it seemed that I had landed a student's ideal summer job. I arrived at the Lake Lab at dawn, launched a small boat, and motored out on Lake Mendota to look for schools engaged in surface filter feeding. I would then catch about a dozen for stomach content analysis, using the fishing technique described below (pl. 71). The fun abruptly ended upon my return to the lab, however, where, for the rest of my eight-hour day, I counted the zooplankton species consumed that morning by the fishes. To make matters worse, the microscope bench faced a large window that looked out on the campus's beautiful lakeshore where many of my fellow students were spending their day swimming, sailing, and water skiing.

Given my background in White Bass research, I was more than amused to read and hear highly embellished descriptions of the White Bass in 1983 when it was learned that some fishes from the illegal Kaweah Reservoir plant had moved via a drainage channel to a temporarily flooded section of the former Tulare Lake basin. The poisoning of all fishes in the large reservoir might not have occurred without the reactionary "bad press" initiated by both anglers and some so-called environmentalists. The White Bass was loudly proclaimed a voracious predator, which, of course, any good game fish is. The media then had a field day with such statements, which reached a high (or low) point when one major newspaper warned "Beware the White Bass!" and went on to describe this moderate-size feeder of zooplankton and small fishes as "The Great White Bass Threat." You had to read carefully to see that the fish in question was not actually the Great White Shark. When the White Bass removal from Kaweah Reservoir and the drain to Tulare Lake was completed, the bill presented to the taxpayer was $7.5 million. After this costly venture, the CDFG then stocked Kaweah with Florida Largemouth Bass *(Micropterus salmoides)* and Spotted Bass *(M. punctulatus)*, two other nonnative "voracious predators."

ANGLING NOTES: In Lake Nacimiento, White Bass are caught during spring spawning runs to the headwater area. Small live minnows or minnowlike lures or wet flies are often the preferred bait. Throughout summer, trolling in the midepilimnion with small minnows or spinners may also produce fishes.

Another effective and perhaps more exciting way to catch this species is during its dawn and dusk surface-feeding for zooplankton. Schools of filter feeding White Bass are easily located in calm water by the gulping action of their mouths breaking the surface of the water. Schools may number several hundred fishes, making these clustered surface ripples visible for some distance. Surface-schooling fishes are rather wary, and a quiet approach is necessary. Even though the fishes are busy feeding on zooplankton, streamer flies or spinners usually produce a rapid strike. The school normally is not disturbed by the action of a hooked fish, and a number of fishes may be taken before the group moves out of reach or sounds. Midsummer is the best time for this sort of fishing, and calm water is a must for locating the surface-feeding activity. Keep in mind that zooplankton respond negatively to light, and within an hour after sunrise they move down to the lower epilimnion for the rest of the day. The day-active (diurnal) White Bass "know" this and take full advantage of the productive dawn feeding period, and you want to do the same. But please do not take any live White Bass away from this lake. It could cost the California taxpayer another $7.5 million!

SUNFISHES, CRAPPIES, and "BLACK" BASSES (Centrarchidae)

Although this small North American family contains only 30 species, it includes more game fishes than any other completely freshwater family in the United States and Canada. Despite its abundance throughout most of the continent, California has only one native centrarchid, the Sacramento Perch *(Archoplites interruptus)*. This species is a remnant from the Miocene when California had more natural lakes and early sunfish species than today. All other present-day species in this family evolved in the lakes, rivers, and streams of central and eastern North America. The continuous repatterning of major lake systems and drainages in these areas in recent evolutionary time has produced a number of closely related species, particularly in the sunfish family. The centrarchid complement in an average eastern lake today consists of at least two or three sunfish species in addition to the Black Crappie *(Pomoxis nigromaculatus)* or White Crappie *(P. annularis)*, and the Largemouth Bass *(Micropterus salmoides)* or Smallmouth Bass *(M. dolomieu)*. Eleven eastern sunfish species have been introduced into California freshwater habitats during the past century, so instead of a naturally occurring complement of "naturally selected" species in a given lake or stream, we often find an unnatural and unusual mixture of species.

This eastern species complex has produced keen competition for both feeding and breeding habitats. Prior to spawning, males scoop out depressions in the bottom material of the inshore zone by fanning the substrate with the body and caudal fin. The area immediately around the "nest" becomes a breeding territory that males vigorously guard. In group spawning species, you may observe scores of nests in a relatively small area. A male then escorts females that pass through his territory to the depression, where they spawn. A postspawning nest may contain eggs deposited by several females. In this form of territorial polygamy, more aggressive males are better at holding and defending a territory and ultimately produce more offspring. Because successful territorial behavioral traits include color as well as body size and aggressiveness, the sunfish family includes some of the most spectacular coloration found among North American fishes.

Nest defense continues through the first few days after hatching, assuring better survival of sunfish fry than that of many other temperate zone egg layers. The centrarchid's breeding biology is perhaps the prime factor accounting for the great success of its species when introduced in lake-type habitats, both in the western United States and around the world. Sunfish are particularly well adapted to life in a freshwater aquarium, as discussed in "The Wild Freshwater Fish Aquarium and Pond."

The family Centrarchidae is also known for the interesting but often misleading common names of some of its species. For instance, the Bluegill *(Lepomis macrochirus)* has red, not blue, gills like all other live fishes. The "blue" here refers to the color of the opercular apex. By the same token, the Redear Sunfish *(L. microlophus)* does not have red ears. In fact, like all fishes, it does not have the structures that comprise the terrestrial vertebrate ear. Again the "red" in its name refers to its opercular apex color. At least the Pumpkinseed *(L. gibbosus)* has a squash-seed shape to justify its name, and the name "crappie" fits right into some angler's "bottom of the boat" humor when the "a" is not accorded the proper "ah" sound. Finally, our only native sunfish, the Sacramento Perch, is not really a perch or even a member of the perch family (Percidae). "Perch," the most misused of all fish names, will appear several more times in the names of widely diverse species before the end of these species accounts. Such common names are often derived from older folk names that persist in one form or another through the decades, adding a bit of color to the sometimes drab world of fish taxonomy.

SACRAMENTO PERCH *Archoplites interruptus*

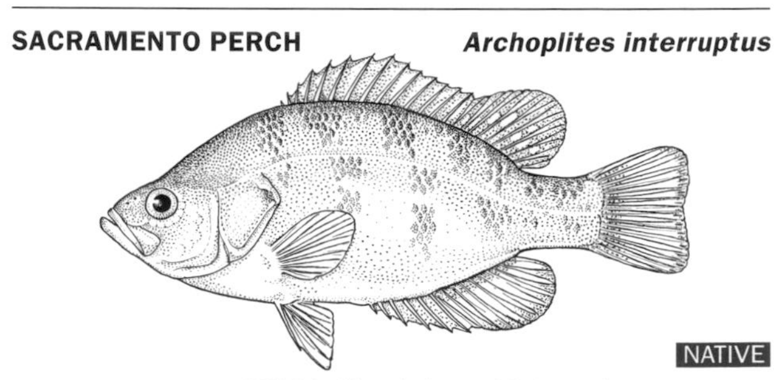

NATIVE

PROTECTIVE STATUS: CSSC in Clear Lake and Delta region.

This was once an abundant food fish for Native Americans inhabiting the Sacramento–San Joaquin drainage. During the con-

Plate 72. Sacramento Perch.

tinuous settlement of this area throughout the 1800s, the Sacramento Perch was commercially fished but later became a top game fish, with reports of specimens up to 61 cm (about 2 ft) and weighing over 1.6 kg (3.5 lb); the current world record weight is 7 kg (15.4 lb). Today, unfortunately, many of the populations in ponds and small lakes comprise stunted fishes (about 15 cm, or 6 in. total length). Although fossils of other sunfishes have been found in the West, the Sacramento Perch is the only member of this family to persist to modern time (pl. 72). Instead of specializing in a relatively restricted feeding niche, as many of its eastern counterparts have done, this species utilizes a broad spectrum of food types throughout life. Fry feed primarily on small, bottom-dwelling crustaceans, yearlings concentrate on aquatic insect larvae, and adults in most habitats are mainly piscivorous. In its pursuit of larger prey, however, this species does not exhibit the vigor and overall aggressiveness of most eastern sunfishes. Unlike those relatives, the Sacramento Perch possesses the physiological ability to cope with the wide range of salinities and alkalinities found throughout the Central Valley pond and backwater habitats. Like all native species in their undisturbed habitats, it was well adapted to and successful in pristine California.

The spawning behavior of the Sacramento Perch is similar to that of other sunfishes. A male fans a depression in the substrate and defends the area against intrusion by other males. Gravid females are allowed to enter and then spawn as the male expels milt over the eggs. A fully mature female is highly fecund and may produce well over 100,000 eggs per season, which are usually laid in the nests of several males. The male continues to guard the nest for several days after hatching until the fry begin to swim away. The fry feed on zooplankton, switching to larger invertebrates as yearlings. At about 9 to 10 cm (nearly 4 in.) they begin eating the fry of small minnows (Cyprinidae) and remain primarily piscivorous for the rest of their life.

As an apparent result of the extensive habitat manipulation and species introductions of the past century, the Sacramento Perch has steadily declined in numbers to the point where it now occurs in only two of its former native habitats: Clear Lake and widely scattered areas of the Delta where it occasionally turns up at the fish-screen facilities near Tracy. Luckily it has been successfully introduced into over two dozen natural and artificially created sites in California and in many more sites outside the state, where its popularity is based both on its relatively large adult size and on its ability to tolerate the high alkalinity of many western lakes, which most other fishes cannot. For this reason, populations have been established in states with such restricted habitats, such as Nevada, Utah, and Colorado,

A common trait shared by nearly all sites where the Sacramento Perch thrives today is a lack of serious competition from other sunfishes or similar species, especially with regard to its feeding niche. Its lack of aggression and vigor in pursuing prey, mentioned above, is especially apparent in small ponds and large aquariums. Where the Sacramento Perch coexists with Black Crappie *(Pomoxis nigromaculatus)*, Green Sunfish *(Lepomis cyanellus)*, or Bluegill *(L. macrochirus)*, it is usually driven out of the more lucrative natural or artificial feeding areas, resulting in poor growth and health when the food supply is limited. Even when some individuals grow beyond the average size of these panfish-size competitors, a spectrum of large, introduced piscivores such as Striped Bass *(Morone saxatilis)*, Largemouth Bass *(Micropterus salmoides)*, and White Catfish *(Ameiurus catus)* continue the feeding niche competition.

ANGLING NOTES: In reservoirs where populations have become established, such as Lake Almanor in Plumas County, Clear Lake Reservoir in Modoc County, and Crowley Lake in Mono County, you may still try your skill at catching this fine native game fish. Adult Sacramento Perch are lie-in-wait predators of the vegetated inshore zone. Their feeding niche is much like that of the adult Largemouth Bass or Green Sunfish. This species seems reluctant to move from a chosen haunt, and bait must be presented quietly at close range to avoid "spooking" it. A slow-moving night crawler, or a small minnow or minnowlike lure, worked through or along the edge of available cover, usually produces results. Keep in mind the lack of voraciousness in this fish's feeding behavior. It will often stalk a prey item and then hesitate for some time before grabbing it. Thus when using live bait, allowing it to pause here and there for a minute or so may turn out to be the best technique of all.

BLUEGILL ***Lepomis macrochirus***

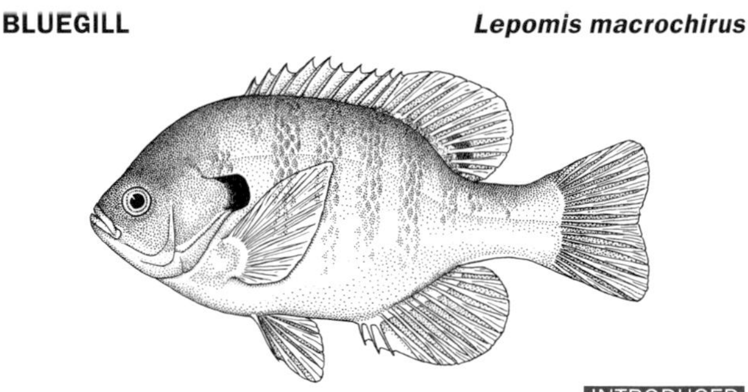

INTRODUCED

This species is now the most abundant centrarchid in California and the premier panfish throughout the United States (pl. 73). It has also been one of the main competitors for the ecological niche of the Sacramento Perch *(Archoplites interruptus)*. It is extremely prolific, capable of producing up to 50,000 eggs per female per spawning. This species exhibits the classic territorial nest defense and spawning of the sunfishes along with a few additional features. The nest construction sites are usually very crowded, which leads to almost nonstop patrolling and chasing by each territorial male. As spawning proceeds within such a melee, nonterritorial males, which exhibit none of the striking color of the dominant male, often come to cospawn with the females. This ensures a high rate of reproductive success that results in rapidly expanding populations in most habitats where it is introduced. Bluegills are also more catholic in their feeding habits than most sunfishes, and this flexibility further adapts them to variations in California habitats. In reservoirs they tend to utilize the offshore epilimnion water far more than other sunfishes, especially during the midday period.

The Bluegill's great success has led in some cases to the development of stunted populations. Fishes are different from most other vertebrates with respect to overpopulation. Whereas mice and humans experience large die-offs when their populations

Plate 73. Bluegill.

greatly exceed the existing food supply, most fish species adjust simply by growing less. Bluegill biomass for a given habitat, therefore, may be realized in very different ways. At one end of the scale would be an angler's dream: a biomass of 500 kg represented by 1,000 Bluegills, each weighing 500 g (about 1 lb). In a highly stunted population, that same biomass would contain 10,000 fishes each weighing 50 g (0.1 lb) This syndrome has been repeated numerous times in California lakes and ponds into which the Bluegill has been introduced.

The only real solution to the problem is adequate annual cropping by anglers or sustained predation by piscivores such as the Largemouth Bass *(Micropterus salmoides)* to hold the population within the carrying capacity of the habitat. Unfortunately, such cropping is rarely achieved. In larger California lakes, there never seems to be enough Bluegill fishing to keep up with the production. The problem is further compounded when a population of large fishes in a particular lake begins to exhibit stunting. If anglers spread the word that such and such a lake holds only small panfishes, angler visits to the lake drop just when cropping of its fish population is especially needed. The problem is even more vividly demonstrated in a small farm pond where the long-standing recommendation of planting Bluegills and Largemouth Bass has been followed. The basic biological concept is sound: the adult Largemouths should feed on the excess young Bluegills, keeping the latter's population below the pond's carrying capacity and producing large, catchable fishes of both species. However, most ponds surveyed eight to 10 years after stocking turn up thousands of small but reproducing Bluegills and only a few giant Largemouth Bass. It appears that mushrooming Bluegill populations consume all available food, including Largemouth Bass eggs.

The preference for stocking Bluegill and Largemouth Bass stems from a long-standing recommendation by the California Department of Fish and Game. A glance at a 1998 informational leaflet on farm fish pond management published by the CDFG shows that this combination still heads the agency's list of stocking recommendations. In another section, however, the leaflet warns against Bluegill overpopulation and stunting, and points out that the Redear Sunfish *(L. microlophus)* is less prolific and not as prone to stunting. The best management tool for ponds is an annual seining and removal of adequate numbers in all age categories. This is more effective than fishing, because anglers often concentrate on catching the Largemouth Bass rather than the Bluegill.

This stunting phenomenon is not unique to the Bluegill and can be seen in species of many fish families. Because of its wide popularity and great reproductive potential, the Bluegill simply exhibits this syndrome more often than other forms. When well managed, it is one of the best small game fishes in North America. Many also feel that it is the best tasting sunfish, and second only to the outstanding Yellow Perch *(Perca flavescens)* among all panfishes.

ANGLING NOTES: In the early morning and late afternoon, this species is found in shallow inshore water, whereas during the day it schools in deeper midlake water. Light intensity appears to play a major role in these movements, and productive inshore morning fishing may be prolonged by a lingering cloud cover.

Conversely, the species becomes inactive immediately after sundown and will not bite. The most widely employed fishing method is the basic line, bobber, split shot, number 8 hook, and worm held well off the bottom in the shallow inshore water. During the day, try slow midlake trolling with a large worm on a single spinner-hook combination.

Bluegill normally spawn late April to early June in central California waters. You can check other areas for the onset of spawning by noting when inshore day water temperatures approach 20 degrees C (68 degrees F). Surface fly fishing when the fishes are over nests in shallow spawning areas is also very exciting and productive.

REDEAR SUNFISH *Lepomis microlophus*

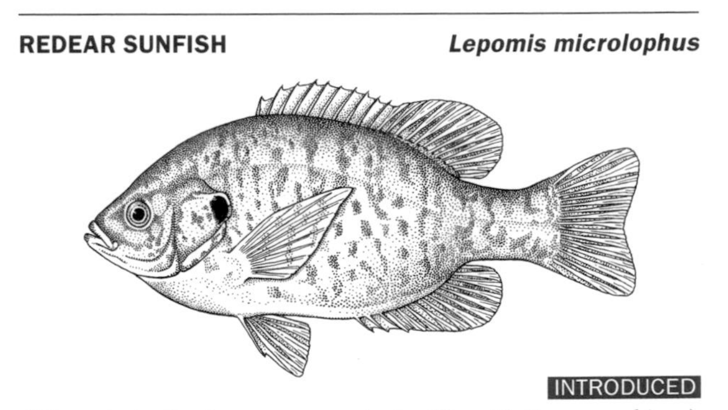

INTRODUCED

Although similar in appearance to the Bluegill *(L. macrochirus)*, the Redear Sunfish differs significantly in several aspects of its biology (pl. 74). It is a deep-water bottom feeder and takes a variety of benthic aquatic insect larvae and shelled invertebrates such as small clams and snails. By feeding outside the crowded inshore zone, it avoids competition with most other sunfish species. This may be one reason why in some lakes it grows significantly larger than the Bluegill. This species produces fewer eggs than the Bluegill and may not breed until three to four years of age, whereas the latter has been known to spawn as a yearling. The combination of its own special food resource and a slower reproductive rate has produced sizeable populations of large fishes in many California reservoirs. Its lack of dependency on a good inshore zone and its slower population growth often make it more suited for farm ponds than the Bluegill, and that trend has continued in recent years.

Plate 74. Redear Sunfish

ANGLING NOTES: The most important single factor to keep in mind when fishing for this species is that it is an offshore bottom feeder. Many of the deep inlets of central California reservoirs, such as Lake Berryessa, contain sizeable numbers of large redears, but if the area is fished in the normal Bluegill fashion with a worm well off the bottom, this fish will seldom if ever be taken. Worms, small crayfish, or pieces of clam or snail should be moved slowly and directly along the bottom. Patience also helps, as this species swims slowly in small schools while testing the bottom debris for food items. Once a school locates your bait, however, the action can be fast and furious.

PUMPKINSEED *Lepomis gibbosus*

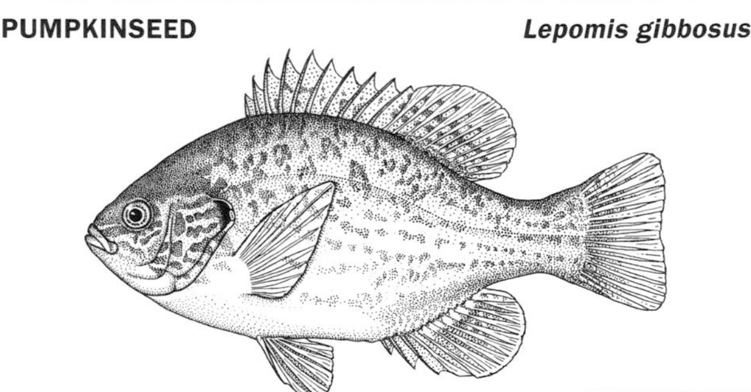

INTRODUCED

In both appearance and feeding habits, the Pumpkinseed and the Redear Sunfish *(L. microlophus)* are probably more similar than any other two sunfishes. In its native range east of the Rockies, the

Plate 75. Pumpkinseed.

Pumpkinseed is found in greatest abundance in the northern states, whereas the Redear Sunfish thrives in the South. In California this trend continues, with the main Pumpkinseed populations found in the northern, middle-elevation reservoirs. Like the Redear Sunfish, it is also a bottom feeder of invertebrates, especially snails. It forages mainly within the inshore zone, thus avoiding direct competition with the Redear Sunfish in waters where the two coexist. The reproductive rate for this sunfish is also somewhat low, with egg numbers averaging about 2,000 eggs per female per spawning, and it does not readily overpopulate a pond or small lake as does the Bluegill *(L. macrochirus).* The Pumpkinseed is therefore an ideal selection for middle-elevation farm ponds. Its bright colors also make it a favorite for the wild fish aquarium (pl. 75).

ANGLING NOTES: The Pumpkinseed is an avid snail feeder in the inshore zone, and it works over aquatic vegetation, logs, and pilings as well as the bottom substrate for this food. The standard worm bait, presented in the basic Bluegill fishing fashion, usually nets Pumpkinseeds in the proper habitat. Because this is normally an inshore fish, deep-water trolling or bottom fishing is wasted on this species.

GREEN SUNFISH *Lepomis cyanellus*

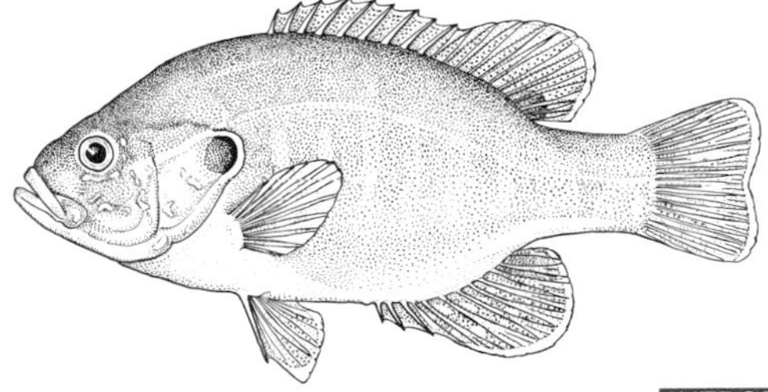

INTRODUCED

This is a sunfish of heavily vegetated backwater areas (pl. 76). Its tolerance for high temperatures (up to 38 degrees C, or 100 degrees F) and very low water-oxygen levels makes it well adapted to habitats where few other species can survive permanently. Green Sunfishes are very aggressive and will dominate a shallow-water area or a wild fish aquarium. It is one of the larger sunfish species and can attain a total length of about 30 cm (1 ft), and a weight of 1 kg (2.2 lb). Like the Bluegill *(L. macrochirus)*, however, it can rapidly overpopulate a shallow, well-vegetated pond, resulting in a stunted population of sexually mature fishes only 8 to 10 cm (3 to 4 in.) long. Spawning also follows the Bluegill pattern with accessory males often partaking in egg fertilization. It produces eggs only in moderate numbers (average of 6,000), but the exclusion of other species from a spawning area and good

Plate 76. Green Sunfish.

nest-fry care accounts for its ability to quickly overpopulate a small habitat. The young of this species feed on nearly all small invertebrates of shallow, well-vegetated water, whereas the adults are partially piscivorous. Like the Sacramento Perch *(Archoplites interruptus),* the adults are lie-and-wait hunters in dense-cover areas.

The Green Sunfish is a pioneer species that can quickly colonize warm, eutrophic habitats. I have seen it completely replace California Roach *(Lavinia symmetricus)* in intermittent creek pools after arriving there during winter flood conditions. It is also more susceptible to predation than the Bluegill, and thus large populations of Green Sunfish are not usually found in habitats with a substantial Largemouth Bass *(Micropterus salmoides)* population.

ANGLING NOTES: In fishing for this species, you must keep in mind its extremely territorial nature. In a slow-stream or lake-bay habitat where Green Sunfishes are present, prominent pieces of inshore cover are usually the domain of adult specimens. It is best to approach these from shore while being careful not to cast a shadow on the water. A worm or small minnow carefully lowered at the edge of the cover, or even between submerged branches, usually yields a fast strike. Deep-water fishing techniques should not be used for this species.

WARMOUTH *Lepomis gulosus*

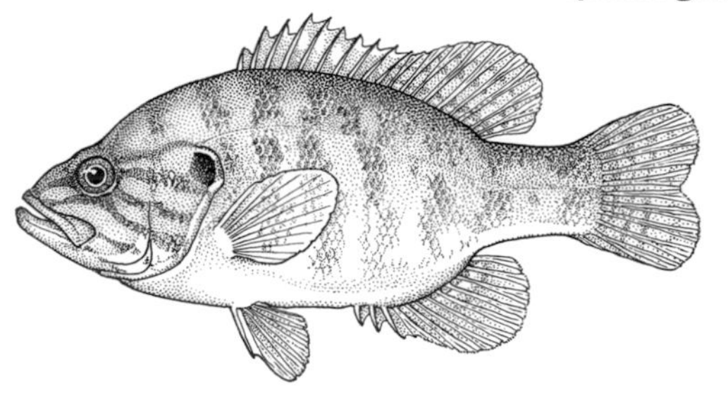

INTRODUCED

This fish is a small sunfish, rarely exceeding 28 cm (11 in.) total length and 0.5 kg (about 1 lb). It is closely related to the Green Sunfish *(L. cyanellus)* and similar in both body shape and behavior. Also like the Green Sunfish, it prefers heavily vegetated backwater areas. Young Warmouths feed on a wide variety of small invertebrates, and as they mature, small fishes and crayfishes become increasingly important in their diet.

The Warmouth (pl. 77) has not enjoyed the widespread distribution in California displayed by its cousin the Green Sunfish. It occurs only in the lower portions of the Sacramento and San Joaquin Rivers and in the lower Colorado River. It is scarce in

Plate 77. Warmouth.

the Delta, rarely seen at the fish-screen facilities near Tracy. On a three-day fish-collecting cruise on the lower San Joaquin River, my ichthyology class collected hundreds of specimens of several other sunfish species but only one Warmouth. That lone specimen was taken in a fyke net (large fish trap) set in one of the few heavily vegetated backwater bays we encountered. Angling methods for this species are very similar to those described for the Green Sunfish.

BLACK CRAPPIE AND WHITE CRAPPIE FOLLOW ➤

WHITE CRAPPIE and BLACK CRAPPIE — ***Pomoxis annularis* and *Pomoxis nigromaculatus***

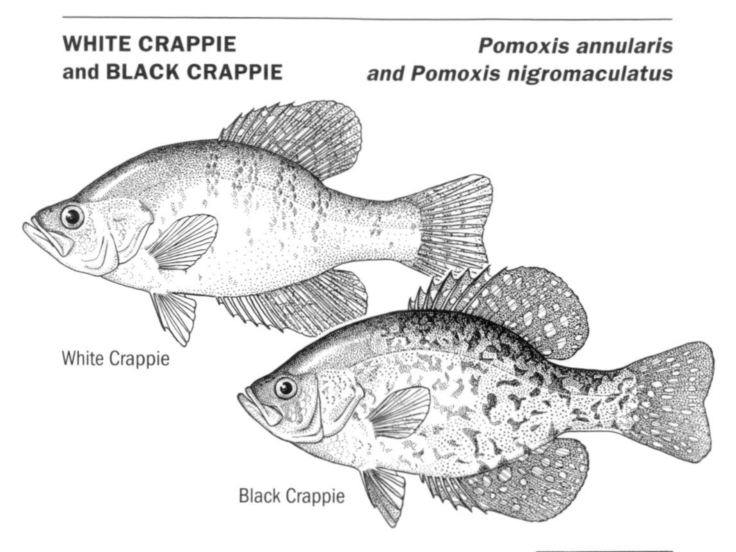

INTRODUCED

These species are discussed together because of their very similar nature and ecology. The two apparently resulted from a period of north-south reproductive isolation in their evolutionary history east of the Rockies. The White Crappie (above) is relatively scarce in northern U.S. lakes, whereas the Black Crappie (below and pl. 78) attains excellent growth in these waters. The latter species appears to fare far better in clear, cool reservoirs in California, whereas the White Crappie does better in warmer, turbid waters.

Crappies are far more pelagic throughout much of the year than sunfishes and usually spend their time in close-knit schools. As with all centrarchids, the schooling behavior temporarily breaks down during the nest-guarding and spawning period. The young are avid filter feeders with many fine, close-set gill rakers and surprisingly large mouths. A good standing crop of zooplankton is required for good crappie production. Under good conditions they may grow to about 20 cm (8 in.) by the end of their second year, at which time they begin to eat small fishes in addition to plankton. Full-grown adults may be 35 cm (14 in.) long and weigh over 1.5 kg (over 3 lb). The California sportfishing record for the White Crappie is 2.05 kg (4.6 lb). The large mouth of the crappie adapts these fish equally well to the adult

Plate 78. Black Crappie.

piscivorous habit and to zooplankton filter feeding, and in this respect it resembles the White Bass *(Morone chrysops).*

The reproductive potential of these species is extremely high, with large females producing in excess of 150,000 eggs per spawning. It is much rarer, however, to find a large lake full of stunted crappies than Bluegill *(Lepomis macrochirus),* so perhaps these forms experience high fry mortality, possibly from adults eating their young. Studies in Lake Chabot, Alameda County, showed the fry of these species to be particularly susceptible to bluestoning, which is the addition of copper sulfate to the water to control green algae production. Of the several species of fry that we exposed to lake water within a day after such treatment, only crappie fry were affected.

Before crappies disperse throughout a spawning area, they form large schools and engage in extensive surface filter feeding. You often hear anglers say that crappies are "running" at such and such a lake or stream, and it is usually this behavior to which they are referring. Crappies follow a standard centrarchid spawning pattern. Their fry are avid zooplankton filter feeders and continue this behavior through their third year, at which time they begin feeding on small fishes. For the rest of their lives, they utilize both foods. Given the extensive dependence of the fry on zooplankton, it is possible that introductions of lifelong planktonivores such as the Threadfin Shad *(Dorosoma petenense)* and Inland Silverside *(Menidia beryllina)* may hold this species in check in parts of its large range in California.

ANGLING NOTES: When going after crappies, remember their intense schooling habit. As mentioned above, this is most pronounced in spring just before the spawning season in April and early May and is characterized by large schools swimming close to the surface. Occasionally they ripple the surface as they chase small minnows or filter feed, and this is an excellent way to locate them. Streamer flies or a small minnow dropped into such congregations produce immediate action. After spawning, the schools move offshore and into deeper water but rarely exceed 5 or 6 m (16 to 20 ft) in depth. Midwater trolling techniques, such as described for Bluegill, work well here also, but with crappies a minnow as bait usually works better than a worm and can be moved faster. Another favorite haunt for postspawning crappies is in and around sunken brush piles. Crappies are attracted to such cover more frequently than other sunfishes, and when the situation permits, a "crappie hot spot" can be created by sinking a weighted brush pile in 2 to 3 m (6.5 to 10 ft) of water.

LARGEMOUTH BASS *Micropterus salmoides*

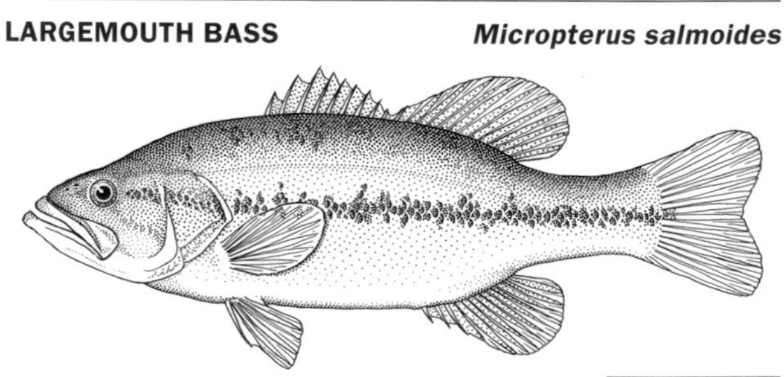

INTRODUCED

Without a doubt, the Largemouth Bass is the most popular warm-water game fish in North America, occupying the role of top piscivore in the vast majority of habitats where it occurs (pl. 79). It was introduced into California in 1891 and has since been found throughout the entire state except for higher elevations in the Sierra and the southeastern deserts. A southern variety, the Florida Largemouth Bass, was introduced in 1959 in response to demands for bigger bass, and most imports since are of that strain. Ironically, in many California habitats, these fishes do not appear to live up to their reputation for large size However, several southern California reservoirs such as Castaic Lake produced near-world-record Largemouths. These lakes were also regularly stocked with put-and-take trout, but the California Department of Fish and Game eventually noticed that only about 10 percent of the trout were being caught by anglers. Instead, the catchable trout stocking program here was in reality a huge bass feeding project!

Plate 79. Largemouth Bass.

The Largemouth Bass conforms to the nesting-spawning pattern already described for sunfishes. A large female may lay up to 100,000 eggs, but the average spawning is well under this number. The young are primarily zooplankton feeders and probably compete more heavily with crappie fry than with sunfish fry for this food source. Young bass tend to school inshore, becoming piscivorous at a surprisingly small size. It is not uncommon to find a 12 cm (about 5 in.) Largemouth Bass with a 6 cm (2.4 in.) relative wedged in its stomach. In such cases the tail of the prey fish actually protrudes from its captor's mouth while the head end is slowly digested.

This species is the largest of the Centrarchids. Specimens in the southeastern United States attain lengths up to 75 cm (29.5 in.) and weigh up to 10 kg (22 lbs). Large adults appear to occupy a shoreline feeding area, where they lie under plant or root cover and wait for prey. It is fair to assume that the lack of abundant shore cover in most California reservoirs limits the overall effect this piscivore exerts in controlling inshore forage-fish populations. As an apparent result of this lack of shore cover, it is not unusual to find schools of adult Largemouths frequenting the offshore waters of reservoirs, behavior rarely seen in its native habitat.

The aggressive, piscivorous behavior of the Largemouth Bass may have been a major factor in the decline of many native cyprinids such as the Thicktail Chub *(Gila crassicauda)*, which may have been pushed to extinction by the combined loss of its backwater habitats to river channelization and the combined presence of the Largemouth Bass and the Striped Bass *(Morone saxatilis)*. Now the Largemouth Bass appears to be declining in some reservoirs where introduced planktonivores such as the Threadfin Shad *(Dorosoma petenense)*, Wakasagi *(Hypomesus nipponensis)*, and Inland Silverside *(Menidia beryllina)* are numerous. Even though the Largemouth Bass becomes piscivorous at an early age, adequate food for first-year fry may be the ultimate key to success or failure in a given area.

ANGLING NOTES: In California and much of the United States, particularly the southern states, the Largemouth Bass is the darling of the nonsalmonid sportfishing world. Indeed, it possesses all of the criteria for this honor: It grows large, it is a piscivore, and it exhibits an aggressive, voracious appetite. When the angler presents a live bait fish or one of the thousands of highly crafted artificial lures designed to resemble a bait fish, the results are often positive.

During the past several decades, the Largemouth Bass has been the focus of a unique innovation in freshwater sportfishing since the invention of the level wind fishing reel: the competitive bass-fishing tournament. These tournaments began in the early 1970s as small, local events usually held once a year, dispensing modest, donated prizes for the winners. When the prize began to take the form of cash, however, the concept of competitive tournament fishing took off. The 1981 winner of the Western National Bass Tournament took home an impressive purse of $53,000. Two decades later in 2001, several "professionals" boasted earnings in excess of $300,000, not including additional "little" prizes such as boats, motors, trailers, and off-road vehicles donated by big-name companies that view sportfishing enthusiasts as an important part of their clientele. Today there is an ever-growing group of professional Largemouth Bass anglers whose work year consists of touring large tournaments in their region and, if successful, going on to the national competition. Granted, the money does not begin to compare to professional football, baseball, and hockey, or the payoffs common in major golf and tennis tournaments, but this sport is still in its infancy. Who knows how high the stakes might go, and you do not have to be a "pro" to play. Amateurs and professionals compete together. The only defining difference between the two is that the amateur does not quit his or her day or night job, but the pro usually does.

Several aspects of tournament bass fishing are unique among professional sports. It is the only professional competition in which men and women compete on an equal basis and directly against each other. The only other similar co-ed competitions are equestrian events, which are usually amateur, and horse racing, which fields an occasional female jockey. There is also no age limit. Old-timers and young upstarts alike have an equal chance at the prize money. At one recent competition, I witnessed one member of a two-person team competing from a wheelchair. Indeed, this new twist on the age-old sport of fishing seems to have no barriers.

But what about the poor fish? There must be a loser here somewhere. Not so—this is a true win-win situation. Aside from the discomfort of being dragged out of the water and held in a live box for a few hours, each fish, after being weighed, is carefully returned to the water, usually via a long chute similar to a human water slide. Why are tournament anglers so careful with their catches? You guessed it: Points are deducted for each dead fish they bring in. Fish release is also important because as more tournaments are held each year, often many at the same lake or river segment, such sites could soon run short of big bass. These anglers are very proficient at their sport, and even a small contest with only 50 contestants would remove up to 500 large, breeding-size fishes from a site if the "bring 'em back alive" rule was not enforced.

Competitive fishing is conducted from modern, open boats, usually in the 14- to 18-ft range and powered by relatively large outboard motors, the latter necessary in order to get to your favorite fishing spot(s) and back within the allotted time. Sometimes that may be all day, or for two time periods on two consecutive days. Smaller tournaments usually run from dawn to noon. Tournaments normally set a five-bass limit and a 13-in. minimum size, with the highest combined weight determining the winner. Most contests also award a "big fish" prize independent of the high-poundage winners. Cash prizes and assorted fishing gear usually go to the top 10 places. As you might imagine, the "weigh-in" of your catch is the moment of truth and the high point of the day; after that you can only wait in nail-biting anticipation until all other competitors have weighed in. Many tournaments are team affairs, and like doubles partners in tennis, winning teams tend to stay together, often wearing caps and shirts displaying their sponsor's logo. The current team entry fee for small tournaments in California is around $100. At this time, tournament Largemouth Bass fishing has yet to make it onto primetime television, but given its continuous rise in popularity, I would not count it out.

To the vast majority of Largemouth Bass anglers, of course, a good day on the river or lake fishing for this top sport fish is reward enough. You can make consistently good catches by keeping a few basic points in mind: First, this species demonstrates a particularly strong bimodal activity curve with strong feeding peaks at both dawn and dusk. In reservoir lakes during their midday activity slump, large bass move from their lie-and-wait inshore positions to deeper offshore bottom sites where apparently they remain inactive most of the time. Where shore cover exists, work it well with lures during the prime inshore periods. The proper choice of lures has always been a problem for bass anglers, and the reason probably lies in the highly variable food preferences this species exhibits. In one year Largemouths seem to feed heavily on Bluegill *(Lepomis macrochirus)*, but when Bluegill become scarce, the Sacramento Blackfish *(Orthodon microlepidotus)* may be the next season's choice. Studies in Clear Lake, Lake County, have confirmed this year-to-year shift in natural food preferences of the adult bass with the result that rapid lure switching has become popular among reservoir anglers.

The midday inactivity period has also been successfully exploited in many reservoir lakes by a technique known as jigging, in which a rubber lure, usually a worm with a weighted head, is hopped or bounced along the bottom in the offshore area. Likely, bass strike at such lures simply because the lures invade their resting territory and not because they resemble a food item. Whatever the reason, jigging, especially in late spring, produces exciting midday fishing.

Trolling for Largemouths in reservoirs has also been successful, especially for medium-sized fishes that tend to school offshore in the middle of the epilimnion. In contrast to the slow-moving trolling techniques recommended for smaller centrarchids, Largemouth Bass lures are usually moved rapidly. Sometimes a method known as zip trolling is successful. It employs specially designed lures that avoid breaking the surface when moved rapidly.

One type of Largemouth Bass fishing that is generally overlooked or prohibited in some California fishing districts is summer night casting in shallow inshore water. Here the casting skill of the angler is paramount, as the lure must be position near inshore cover by only moonlight, starlight, or both for illumination. Noisy, splashing surface lures with weed guards on the hooks are normally preferred. It is probably not the surface noise of the lure as we hear it, but instead the water pressure waves received by the fish's lateral line that help to produce a strike.

SMALLMOUTH BASS *Micropterus dolomieu*

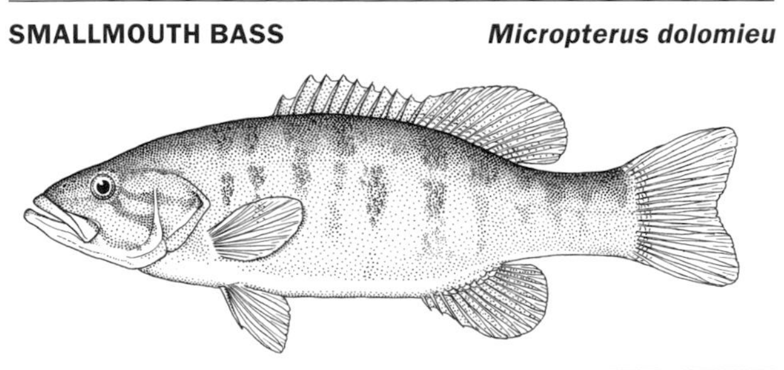

INTRODUCED

In its native range east of the Rockies, the Smallmouth Bass is primarily a cool-water stream-dwelling form that avoids direct competition with its warm-water, lake-dwelling largemouth relative (pl. 80). The Smallmouth Bass range extends into Canada, far beyond that of the Largemouth Bass *(M. salmoides).* In California, Smallmouth Bass has adapted well to middle-elevation lakes such as Lake Shasta and Clair Engle Lake. Some of its best potential riverine habitats, however, have been dammed to construct the many low-elevation reservoirs in the state. In Lake Berryessa, Napa County, for instance, this species is found primarily in the backed-up waters of Putah and Pope Creeks, whose main drainages were flooded to form the lake.

Aside from the Smallmouth Bass's preference for cooler,

Plate 80. Smallmouth Bass.

moving water, its biology is similar to that of the Largemouth Bass. The adult Smallmouth, however, appears to be less piscivorous and instead relies more on insect food in the streams that it occupies. The full adult Smallmouth is about half the size of the same category of Largemouth Bass. The California record Smallmouth Bass weighed in at 4.1 kg (9 lb), whereas the state record Largemouth Bass, a trout-fed specimen from Castaic Reservoir, registered 9.9 kg (21.8 lb).

ANGLING NOTES: Ardent anglers usually prefer the Smallmouth Bass for its exceptional fighting characteristics when taken on medium or light tackle. In general, the angling techniques used in lakes for Largemouths also work well for this species. In the stream habitat, however, modified trout-fishing techniques are far more successful. Large surface or streamer flies usually produce well. Numerous stomach contents studies of stream-dwelling smallmouths have shown a great affinity for crayfishes. Live crayfishes and good rubber substitutes often produce large catches. The Smallmouth Bass has a strong preference for the rubber lure, perhaps because it tends to mouth its prey before taking it. Trailing rubber lures, especially those combined with a spinner, are a consistently successful bait, no matter what the habitat.

SPOTTED BASS *Micropterus punctulatus*

INTRODUCED

The Spotted Bass is another native of the Mississippi River drainage stocked extensively in the Central Valley and associated foothills during the 1930s. Populations from early stockings became established only in the Cosumnes and Feather Rivers and their reservoirs. A second round of stocking took place in the 1970s, and many additional populations have since been established. In Millerton Lake, Fresno and Madera Counties, this species is viewed as the number one sport fish. Population shifts from Largemouth Bass *(M. salmoides)* to Spotted Bass are also occurring in Lakes Shasta and Berryessa. Its success in California reservoirs appears to be due to a tendency to spawn in deeper water and on more open substrate than the Largemouth Bass. Given the highly restricted and fluctuating inshore zone of most western reservoirs, this behavior may indeed be advantageous.

Like other bass-type centrarchids, the adult Spotted Bass is a piscivore and switches its food preference from invertebrates to fishes as it matures. It is short-lived compared to related species, seldom surviving longer than about five years and lays comparatively few eggs, averaging 8,000 per female. Spotted Bass spawning is similar to that of other sunfishes except that one male and one female compose a monogamous spawning pair. This species is also smaller than the Largemouth Bass. The northern subspecies, released in the first introductions, rarely exceeds 3 kg (about 6.5 lb), but one specimen taken during a California Department of Fish and Game electrofishing survey in Lake Perris, Riverside County, weighed 4.4 kg (9.6 lb). More recently the state has stocked an Alabama subspecies, which may reach 4 kg (8.9 lb). It is not the size of this fish, however, that has attracted a anglers in recent years, but its ability to provide an excellent fight on light tackle. Angling techniques for this species are similar to those used for Largemouth Bass, but with a greater emphasis on fishing the offshore waters.

REDEYE BASS *Micropterus coosae*

INTRODUCED

This is the smallest of the centrarchid bass, rarely exceeding 40 cm (about 16 in.) and 3.5 kg (about 8 lb). It is also slow growing, often taking up to 10 years to attain maximum size. The Redeye Bass is an upland stream species native to the southeastern United States. It occupies an ecological niche similar to that of trout and is highly insectivorous, but it does not thrive in lake or wide-stream habitats and competes poorly with other bass species.

Redeye Bass was stocked in a number of middle- and low-elevation rivers and streams in the early 1960s. Breeding populations failed to establish in some of these sites, but others were quite successful. In a recent electroshock survey of the south fork of the Stanislaus River above Fivemile Creek, 78 percent of all fishes collected were Redeye Bass. Such success should be greeted with restrained enthusiasm, however, because the Sierra foothill streams contain some of the least-manipulated native fish populations in the state, and the dominance of an introduction such as the Redeye Bass may have very undesirable results. Some stretches of the Cosumnes River now contain essentially nothing except this species. Indeed, this sanctioned introduction has been far more efficient in ridding our foothill streams of native fishes than any method yet devised.

PERCHES (Percidae)

The perch family is prominent in the upper Midwest and eastern United States, but California contains only two introduced species. In North America this family comprises three somewhat diverse groups: the pikeperch, consisting of the Walleye *(Stizostedion vitreum)* and it smaller cousin the Sauger *(S. canadense);* the yellow perch, of which the Yellow Perch *(Perca flavescens)* is best known; and the darters (subfamily Etheostomatinae), a very successful group of small, bottom-dwelling fishes of over 110 species. The Walleye, often referred to as the "walleye pike," is the largest member of this family, attaining lengths of over 1 m (over 39 in.) and weights up to 11 kg (25 lb). As an adult it is a deep-water piscivore and a top game fish in the northern Midwest. In states such as Minnesota, the yearly activities of many anglers center around this one species, including winter when anglers fish for it through holes in ice-covered lakes. It is also considered one of the best-tasting fish in the world, which has led to a long-standing commercial fishery in the Great Lakes. With such admirable qualities, the Walleye was a prime candidate for introduction into California waters. So far, however, attempts to establish breeding populations have been unsuccessful for reasons not yet understood. The Yellow Perch and one darter, the Bigscale Logperch *(Percina macrolepida),* have been introduced successfully into California habitats via two very different routes discussed below.

A brief discussion of the word "perch" is in order, because it is perhaps the most misused and misunderstood common name among North American freshwater fishes. The American College Dictionary gives as its first definition the correct ichthyological usage, that is, "a spiny-finned, freshwater food fish of the genus *Perca,*" but then goes on to add a second interpretation: "any of various other spiny-rayed fishes of the same and other families." The "other families" part of this definition has led to much confusion over the years. Ask an angler in the northern Midwest what a perch is, and the answer will be "the Yellow Perch." The same question put to an angler in the southern part of Texas, however, might be answered with a detailed description of what sounds like a Bluegill *(Lepomis macrochirus).* Of course, if you ask the average California angler (who has probably not fished the few sites in this state where the Yellow Perch has been planted), you are likely to receive an account of the Tuleperch *(Hysterocarpus traski)* or the many marine members of the "surf perch" family. No amount of commentary will solve this problem, but perhaps being aware of the many interpretations of the word "perch" will lessen the confusion.

YELLOW PERCH ***Perca flavescens***

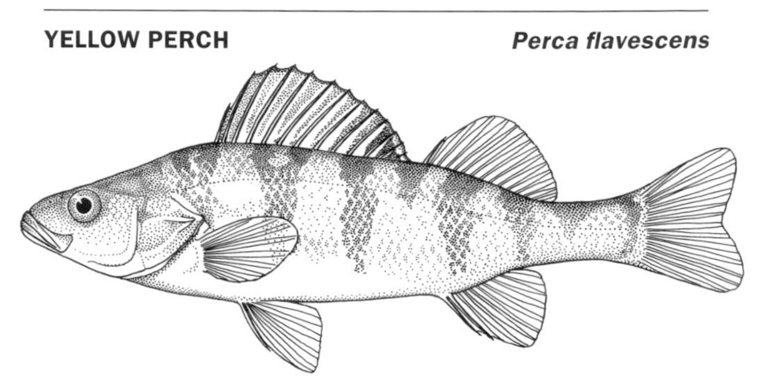

INTRODUCED

The Yellow Perch is a major sport fish or panfish throughout the Great Lakes region, and in many lakes it dominates the small-game-fish scene (pl. 81). Introduced in 1891 into the lower Feather River in Butte County, California, it spread to the Delta, where even more were eventually introduced. Here it became quite abundant but then began to decline in the 1930s. Today the Yellow Perch has vanished from Delta waters for reasons not yet known. This is unfortunate for the Delta angler, because many, including this author, consider the Yellow Perch to be the best-tasting fish in North America. Bay Area residents may investigate this claim for themselves by fishing an isolated introduced population in Lafayette Reservoir in southwestern Contra Costa County. In Iron Gate Reservoir, Siskiyou County, it feeds extensively on small snails. Larger Yellow Perch are piscivorous and take small fishes when available.

The Yellow Perch is a spring spawner and lays long strands of eggs on aquatic vegetation in shallow inshore areas. It reaches minimal catchable size (15 cm [6 in.]) by the end of its second year, and under favorable conditions may double that size by the

Plate 81. Yellow Perch.

end of its fourth year. In the Midwest, these big specimens are known as "jumbo perch" and are highly prized. A jumbo is usually any perch over 30 cm (12 in.) and 0.5 kg (1 lb), a size rarely attained in California. Unfortunately, this fine fish is also subject to stunting and overpopulation, following much the same course as the Bluegill *(Lepomis macrochirus)*. In shallow lakes of the Great Lakes region this problem is quite common. Dense inshore vegetation provides excellent cover for the unusually high number of eggs (up to 75,000 per female per spawning). The result is an enormous number of stunted Yellow Perch that dominates the entire lake. Fortunately, this has not occurred in the Klamath River and its reservoirs, where this species now lives. Pl. 81 shows a 20-cm (8-in.) specimen I captured in Iron Gate Reservoir.

ANGLING NOTES: Small perch are among the easiest fishes to catch, and in the Midwest, where they abound around lake piers, they are frequently a young person's first hook-and-line prize. This same fast action can be found here in the shallow waters of the Klamath River near or amidst the aquatic vegetation of Iron Gate Reservoir, Siskiyou County. These fishes are masters at cleaning large hooks, and the use of a number 8 or 10 hook can reduce rebaiting.

Catching large Yellow Perch can be difficult without an understanding of their basic behavior. This species is one of the most school-oriented of North American game fishes, and you rarely find an individual, especially an adult, away from the main school. Midday locations of schools are usually random, although Yellow

Perch seem to distinctly prefer deep water just off a shallow ridge or sandbar. To locate such ridges, a precise knowledge of subterranean topography is helpful. A good source of information is the old topographic maps of areas now flooded by reservoirs. By referring to the map topography and checking it against estimates taken with a sounding line or electronic depth-finder, you can sometimes locate an extensive underwater ridge that will prove well worth the effort on future trips. The adult Yellow Perch is pelagic and piscivorous, especially during the day. Unlike lake-dwelling salmonids with the same preference, the Yellow Perch rarely enters the thermocline, preferring instead the middle to lower reaches of the epilimnion.

At dusk the school moves to shallower inshore or sandbar areas, where the fishes will actively bottom feed until they settle on the bottom at sunset and all activity ceases. Usually the best summer fishing comes during the first or last hour of daylight as the school "gets up" or "goes to bed."

Worms and small minnows seem to work equally well as bait. The action is fast once a school is located and starts biting. Do not waste time baiting hooks or stringing fishes, because once the school moves on, you may not find another that day. To speed things along and increase the catch, use a live net rather than a stringer and provide a can of worms for each angler.

BIGSCALE LOGPERCH *Percina macrolepida*

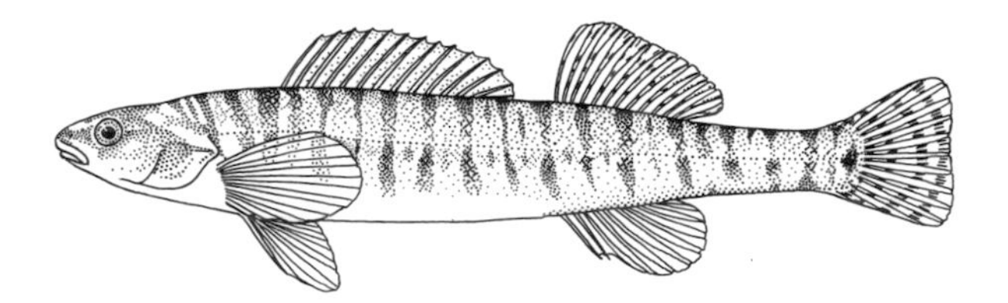

INTRODUCED

The Bigscale Logperch is a member of a subgroup of the perch family (Percidae) known as the darters, a very successful group of small, bottom-dwelling fishes in its native range east of the Rockies (pl. 82). In the Great Lakes region alone are some 20 species of darters. Fishes in this group lack a functional swim bladder and thus have permanent negative buoyancy, which allows them to occupy a bottom-feeding niche. On stream and lake bottoms they are quite sedentary and may remain for hours without a change in position. When they move, it is via a series of quick pectoral fin movements with short periods of rapid sinking between each. Darters are incapable of continuous swimming for even moderate distances. The food resource on the bottom of freshwater habitats is usually the richest in the entire ecosystem, and thus the darters occupy a lucrative feeding position. They appear to take all forms of bottom invertebrates, plus fish eggs.

Plate 82. Bigscale Logperch.

Their motionless posture on the bottom along with good cryptic coloration allows them to escape much of the predation that small pelagic fishes normally experience. In all ways, the darters are very successful fishes, and herein lies the problem with the Bigscale Logperch in California.

In 1953, the U.S. Fish and Wildlife Service stocked several large ponds at Beale Air Force Base, Yuba County, with Largemouth Bass *(Micropterus salmoides)* and Bluegill *(Lepomis macrochirus)* to provide recreation for base residents. They obtained their fishes from a river in Texas that also supported a thriving population of Bigscale Logperch some of which were included in the shipment of bass and thus brought into California, where apparently nobody bothered to check the shipment. During the next two decades the Bigscale Logperch spread throughout the lower Sacramento–San Joaquin Delta and it has further expanded throughout the San Joaquin Valley and into the Sacramento Valley and southern California reservoirs to the south.

The effect of the Bigscale Logperch on other fish species has yet to be studied carefully; however, their presence must be viewed as a further drain of the bottom-invertebrate food resource of these waters and therefore as potentially harmful to native species, especially the sculpins (Cottidae), whose general mode of existence they closely mirror.

One small benefit of this fish is its availability to Californians for freshwater aquariums. The adult averages 10 cm (4 in.) and is quite colorful compared to many small native species. Its bottom position in the aquarium helps to balance the overall distribution of fishes, and its unique swimming method is interesting to observe, especially when it darts after a food item and consumes it in one gulp.

CICHLIDS (Cichlidae)

The cichlid family currently contains 1,300 species, and this number continues to rise as more and more are described. Cichlids are found throughout the tropical and subtropical regions of the world, and many are familiar to North Americans through the aquarium fish trade. They have a high reproductive and growth rate and exhibit intricate spawning and parental behavior, which is perhaps their most attractive feature to tropical-fish fanciers. A few large species are also familiar to many Americans because they are now one of the few cultured food fishes in this country. These are marketed under the genus name *Tilapia* and in a short time have become a standard in fish markets. Their taste and preparation are discussed in the cooking section. Four species now inhabit southern California as a result of introductions authorized by the California Department of Fish and Game and by fish-farm escapes. Where two or more are introduced into the same habitat, they often produce fertile hybrids, making California cichlid taxonomy a challenging discipline.

MOZAMBIQUE TILAPIA ***Oreochromis mossambicus***

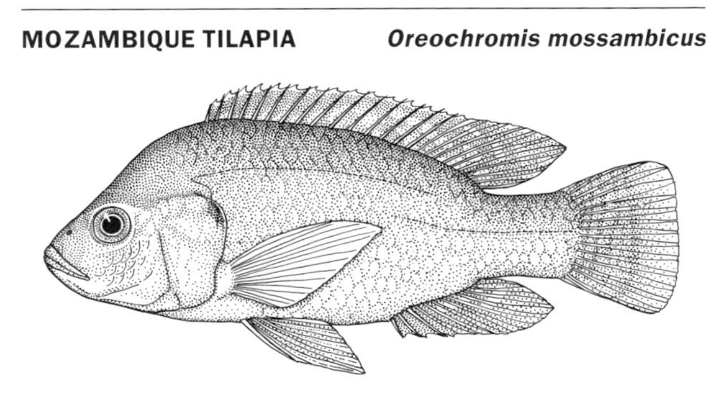

INTRODUCED

This species is native to nearly all major rivers in East Africa. The Mozambique Tilapia is a mouthbrooder, a term referring to its unusual method of protecting its egg and fry (pl. 83). After a female spawns in the territorial nest area of a male, she picks up the eggs and milt in their mouth, and that is probably where most of the eggs are fertilized. The female then retreats to a secretive place

Plate 83. Mozambique Tilapia.

in the aquarium, which is usually a ceramic flowerpot lying on its side, and incubates the eggs in her mouth for about 12 days. This behavior not only offers the embryos maximum protection but also ensures them an adequate oxygen supply from the water circulating through the mouth cavity. After the fry are released from her mouth, the female continues to guard them for another week or so. This high degree of protection of both the embryo and the fry allows these fishes to successfully compete with species offering little parental care to eggs and young.

Rapid growth is another characteristic of this group. At its optimal temperature of 30 degrees C (86 degrees F), the Mozambique Tilapia grows about 1 to 2 cm (about .5 in.) per month depending upon individual pond conditions. In freshwater ponds this species can grow to 16.5 cm (6.5 in.) in 18 months. In the Salton Sea, where its tolerance to high salinity makes it one of the few surviving species, some individuals grow to 32 cm (12.5 in.) total length in just one year. This ability for rapid growth has produced a long history of cultivation of cichlids dating back to 2000 B.C. in Egypt. Present-day cultivation began in Java in 1939, and in many parts of the world it is one of the main items sold in fish markets.

In 1971, the introduction of Mozambique Tilapia into southern California was authorized by the California Fish and Game Commission for the dual purposes of controlling aquatic vegetation in irrigation canals and providing further sportfishing. However, food preference studies, some of which were available before the introductions, show the Mozambique Tilapia to be omnivorous, feeding on a variety of aquatic invertebrates and even small fish species, which in California has unfortunately included the endangered Desert Pupfish *(Cyprinodon macularius).* Failure of this species to significantly deplete canal vegetation seems to be in agreement with these findings.

With respect to sportfishing, this species has about the same advantages and disadvantages as the Bluegill *(Lepomis macrochirus).* It is a scrappy fighter but can rapidly overpopulate a pond or canal with stunted individuals. In areas where the two species occur together, the Mozambique Tilapia outcompetes the Bluegill. Thus we have the potential for secondary interactions between California freshwater fishes, whereby introduced species such as the Bluegill, which already have displaced the native Sacramento Perch *(Archoplites interruptus),* are now open for similar displacement by the Mozambique Tilapia. The hope that such competition will be restricted to southern California is based on the apparent inability of the Mozambique Tilapia to withstand water temperatures below 12 degrees C (54 degrees F) for extended periods. This limitation theoretically eliminates them from central and northern California waters. In 1999 an unscheduled field experiment tested this theory when the winter water temperature of the Salton Sea dropped below 12 degrees C and produced a massive die-off of this species.

The ongoing shuffling of the genetic deck of the Mozambique Tilapia in California could, however, produce some exceptions to this rule. Indeed, one may have already occurred. In the early 1990s, anglers began catching Mozambique Tilapia in Mormon Slough, a small backwater of the San Joaquin River in Stockton, San Joaquin County (pl. 8). My farm is located about 33 km (20 mi) south of Stockton, and it is a rare winter when we do not have to break the ice on the horse-watering troughs once or twice. Given that the Stockton site is an urban slough surrounded by concrete structures that could produce a heat-sink effect, the apparent establishment of this species here may be just an isolated occurrence. However, if this indeed represents a first-stage adaptation to cool water temperatures, the rapid sexual maturity and high fecundity of this species could literally spawn a strain of northern California tilapia that would have a devastating impact on both the native and introduced complements of resident fishes.

REDBELLY TILAPIA *Tilapia zillii*

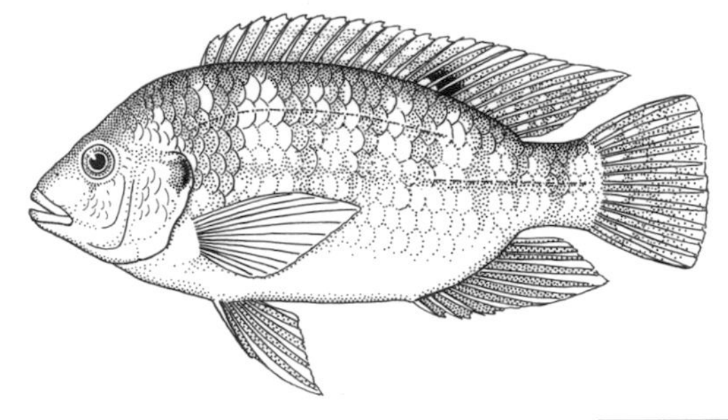

INTRODUCED

Like the Mozambique Tilapia *(Oreochromis mossambicus),* the Redbelly Tilapia was introduced into the irrigation canals of the Imperial Valley in 1971 by authorization of the California Department of Fish and Game. Unlike the Mozambique Tilapia, the Redbelly Tilapia appears to be almost completely herbivorous and has been used elsewhere in the world for canal weed control. It is native to West Africa and may attain lengths up to 35 cm (13.8 in.). The Redbelly Tilapia does not exhibit mouth brooding but instead constructs nests much like those of the Bluegill *(Lepomis macrochirus)* and other sunfishes (Centrarchidae), where it lays up to 6,000 eggs and guards them through the hatching and yolk-sac stages.

Its ability to control aquatic weeds has been mostly successful, leading to further promotion of this species in southern California. What seems to have been overlooked, however, are studies indicating that this species can tolerate lower water temperatures (8 to 10 degrees C [45 to 50 degrees F]) than the Mozambique Tilapia. The implications are far-reaching, because this lower thermal range mirrors that of most of the Delta region and surrounding reservoir lakes in winter. The Redbelly Tilapia has already been introduced into isolated ponds in Napa County; given how fish often mysteriously move from place to place, this is the ichthyological counterpart of keeping a loaded gun in the closet when company comes to visit. This species could very well produce the second-phase competition with northern California fishes theorized for the less thermally tolerant Mozambique Tilapia.

On the brighter side, the Redbelly Tilapia appears to be declining in southern California, perhaps because the entire population is descended from only three breeding pairs, a situation that eventually produces genetic and adaptation problems for most species.

BLUE TILAPIA *Oreochromis aureus*

INTRODUCED

This native of northeastern Africa is another favorite of aquaculturists and is raised for food in the United States, Mexico, and South America. It is a mouthbrooder and exhibits a breeding behavior similar to that of the Mozambique Tilapia *(O. mossambicus)*. Like other cultured tilapia species, its main attribute is rapid growth to a marketable size. After two years it averages about 25 cm (10 in.) and eventually can reach 37 cm (14.5 in.) and 1 kg (2.2 lb). It also tolerates high salinity, an attribute for which most tilapia species are noted (up to 99 ppt). Far more pertinent to its presence in California, however, is its tolerance to low water temperatures—as low as 5 degrees C (41 degrees F). Given this ability it could colonize all of California's lower- and middle-elevation habitats, and it does appear to be an expert colonizer. During the 1970s the Blue Tilapia spread across a southern tier of states from Florida to the California border. Six years after its release into one Florida lake, a survey revealed that 93 percent of the total fish biomass in that habitat was represented by the Blue Tilapia, and in some Florida and Texas lakes, populations have reached over 900 kg (about 2,000 lb) per acre.

The Blue Tilapia came to California via the Colorado River as a result of releases by the Arizona Cooperative Fisheries Research Unit at various sites in southern Arizona. Presently, its range in the wild still remains the southern portion of the Colorado River. In 1992, however, the California Department of Fish and Game authorized its use in aquaculture enterprises south of the Tehachapi Mountains, with the provision that only processed (dead) fishes could be sold. Such a provision is a good first step toward controlling the spread of this fish throughout northern California. Unfortunately, it and other seemingly sound restrictions can never ensure that human nature will not one day intervene, allowing the Blue Tilapia to begin a long and most likely successful journey through the waterways of California.

NILE TILAPIA *Oreochromis niloticus*

INTRODUCED

The Nile Tilapia is a legal aquaculture species in Arizona and as an indirect result has made its way to the California side of the lower Colorado River. It is very similar to the Blue Tilapia *(O. aureus)* and freely hybridizes with it. As a result, the river supports a mixture of Blue Tilapia, Nile Tilapia, and Blue Tilapia–Nile Tilapia hybrids. The basic biology of the Nile Tilapia is also very similar to that of the Blue Tilapia, but the former can grow larger (up to 65 cm [25.5 in.]), which explains its preferred status in aquaculture industries in several states. In California the same restrictions placed on Blue Tilapia culture are also applied to the Nile Tilapia and hybrids. Studies are now needed on these fertile hybrids to see if they are even better adapted than their parents for life in the Golden State.

Plate 84. Hybrid Tilapia in fish market.

It would be comforting to conclude the cichlid section on an encouraging note, to the effect that all is under control, and we may look to the future presence of tilapia species only at aquaculture facilities in this state. Unfortunately, this is wishful thinking. There will always be a small number of people who, if given the opportunity, will introduce new fish into new habitats.

In the case of tilapia, the opportunities certainly exist. As of 2005 California had 23 registered fish farms where you can buy one or more of the species discussed here. Even more convenient places to obtain live fishes are the hundreds of fish markets that sell live tilapias, and many of these are in northern California where the fear of introduction is greatest (pl. 84). Of course cichlids such as those discussed in this section have long been a favorite of tropical-fish keepers—until they get too large for their aquarium.

The tilapia time bomb is ticking, and eventually its explosion will be felt throughout the Golden State.

VIVIPAROUS PERCH (Embiotocidae)

Given the extensive coastal river system of California and the comparative ease with which some species such as trout and Striped Bass *(Morone saxatilis)* are able to move back and forth from freshwater to saltwater, you would expect to find a large number of California freshwater fishes of direct marine origin. This is not the case, however, and such marine-derived species are scarce. When they do occur, they are usually represented by only one or two species of an extensive complement of inshore or estuarine forms. Perhaps the best example of this rare marine-to-freshwater adaptation is seen in the viviparous perch family, a marine group with only one permanent freshwater form, the Tuleperch *(Hysterocarpus traski)*. Another essentially marine species, the Shiner Perch *(Cymatogaster aggregata)*, makes periodic excursions into freshwater and is therefore also included in this book.

TULEPERCH ***Hysterocarpus traski***

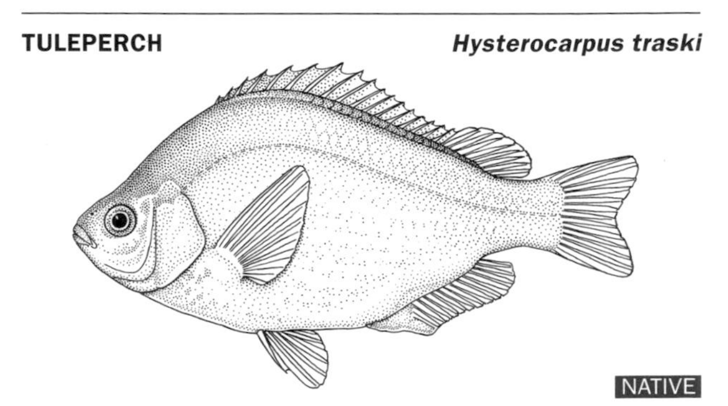

NATIVE

The Tuleperch is the only completely freshwater member of a complex of 20 viviparous perch species, which are similar in size and shape to sunfishes (Centrarchidae) such as the Bluegill *(Lepomis macrochirus)*. Some viviparous perch are inshore marine species that frequent estuaries during the breeding season, and perhaps it was from such an ancestor that the Tuleperch evolved.

This is one of the smallest members of its family, rarely exceeding 23.5 cm (9.3 in.) total length (pl. 85). Like other viviparous perch, it has a small mouth with flexible lips for picking small invertebrates off the substrate. The mouth also contains both normal teeth on the jaw bones and a set of pharyngeal plates similar to the pharyngeal teeth of the cyprinids. These are used for crushing the shells or exoskeletons of its invertebrate food.

Plate 85. Tuleperch.

My ichthyological colleague, Dr. John Hopkirk, has made the taxonomy of the Tuleperch more challenging by describing three subspecies: one from Clear Lake *(H. t. lagunae)*, one from the Sacramento–San Joaquin system *(H. t. traski)*, and a third from the Russian River *(H. t. pomo)*. Viviparous perch also have three body patterns: unbarred, narrow-barred, and broad-barred. These are not necessarily subspecies indicators but do occur in different proportions in each group.

By far the most interesting aspect of viviparous perch biology is their method of reproduction. In the broad sense of the word they are livebearers, but unlike the livebearer family (Poeciliidae), the viviparous perch are truly viviparous, which means that the embryos obtain their nourishment from the body fluids of the mother, just as in mammals. In contrast to most mammals, which develop a placenta for exchange of nutrients and waste products with the blood supply of the uterus, the viviparous perch have greatly enlarged ovaries and associated fat bodies that produce copious amounts of nutrient-rich fluid to nourish the embryos. Oxygen and carbon dioxide are exchanged through the fin vascular network, primarily that of the dorsal and anal fins. These fins are greatly enlarged and contain a dense capillary network in the developing embryo; they press up against the vascularized ovarian wall in the same manner in which the capillaries of the mammalian placenta lie next to the uterine circulation. This amazing system produces large, fully developed young. Unlike most fish fry, which are really larvae that drift with the current and are open to a variety of hazards, the young viviparous perch is actually a small adult that, in most cases, achieves sexual maturity soon after birth.

As in mammals, the key to the production of such young is the maternal-embryo nutrition and exchange system, which permits more extensive growth and development than a retained yolk mass could sustain. One apparent drawback would seem to be the reduced number of young. A fully grown female Tuleperch rarely gives birth to more than 80 young, and in a young female the number may be as low as 20. However, these newborn Tuleperch are comparable in size and development to yearling forms of egg-laying fishes, and on this basis the actual reproductive success is quite good.

Another interesting Tuleperch trait is delayed fertilization. Mating takes place from midsummer through early fall. Although the anal fin of the male is not highly modified to form a gonopodium as in the livebearers, the fin does bear a fleshy glandular structure that apparently improves the successful transfer of sperm from the vent of the male to that of the female. The breeding pair assume a V-shaped position with their anal fins forming the apex of the V. Breeding does not result in immediate fertilization, however; instead, the sperm is stored until midwinter when the actual fertilization of the eggs takes place. A female may also mate with several males and thus produce a genetically diverse set of young. Once fertilized, the eggs develop for approximately four months, and the young are born in May and June, when food and water temperature conditions are the most favorable for growth.

The Tuleperch is still moderately common in the lower Delta, the Sacramento and Russian Rivers, and Clear Lake in Lake County. In addition, it has been introduced into a number of small lakes and ponds, such as Lake Merced in San Francisco, where it is doing well. Its efficient reproductive method is perhaps the major factor by which it continues to survive in the face of severe competition from a seemingly unlimited array of introduced species. The Russian River subspecies, a CSSC, appears less numerous than several decades ago. The Hwy. 128 bridge over this river has been one of my long-standing fieldtrip stops where it was almost certain that a class could net and release Tuleperch. In recent years, however, the little side-pool areas there have failed to produce any. This subspecies differs from the others because it lives only two years, about one-third the life span of the other groups in this family. Thus, with an ongoing decline in water quality as a result of expanding agriculture (vineyards) upstream, the combination of low offspring numbers and a short life span may prove disastrous for this isolated group.

ANGLING NOTES: The Tuleperch receives no mention in the California Sport Fishing Regulations. No doubt it is caught occasionally by anglers and mistaken for a species of sunfish. As the key in this book points out, the row of large scales on the base of the dorsal fin is a good characteristic by which to recognize this species. As its name implies, the Tuleperch is a fish of the inshore cover, where the majority of small invertebrates can be found. Thick tule beds offer excellent habitat for this species. Its small adult size prevents it from ever becoming a widely sought-after game fish; however, its preference for the shoreline and its ability to thrive in small lakes and ponds make it a good fish for young anglers, especially in city park ponds with limited species variety. The Common Carp *(Cyprinus carpio)* and the Goldfish *(Carassius auratus)*, which inevitably find their way into such habitats, offer little competitive threat to the Tuleperch. A very small hook (number 10 or 12) should be used for this species because of its very small mouth. As for taste, the entire viviparous perch family is noted for its delicious flesh, which should be treated similarly to sunfishes. Should Tuleperch fishing ever become popular, seasons should be set to coincide with the postbirth season, as a heavy take of pregnant females in a small pond could greatly reduce the population.

SHINER PERCH FOLLOWS ➤

SHINER PERCH *Cymatogaster aggregata*

NATIVE

This is one of the most abundant members of the viviparous perch family. Although it normally occurs in marine bays and lagoons, it occasionally moves up coastal streams for varying periods of time. During a fall class field trip to the Albion River in Mendocino County, we caught nothing but yearling Shiner Perch in one lower river segment, and according to the "tip-of-the-tongue salinity meter," the water was totally fresh. This ability to penetrate and utilize lower river areas allows a marine species to

Plate 86. Shiner Perch female giving birth.

occasionally exploit a habitat with reduced feeding niche competition and predator presence. This latter advantage may be of particular importance because it is usually the smaller first-year fish (about 6 cm [2.3 in.]) that use these areas. Midge larvae are one of the main freshwater foods available to such migrants. In the marine habitat, both young and adults feed extensively on zooplankton and a variety of bottom invertebrates.

Breeding and birthing in the Shiner Perch is similar to that in the Tuleperch *(Hysterocarpus traski).* As mating time nears, the male loses his normal silver with yellow bars coloration and takes on an overall dark hue. Reproductive events take place in shallow backwaters in midspring (pl. 86). The number of young born ranges from six to 20, depending on the size and therefore the age of the female. Breeding and fetal development follow the pattern described for the Tuleperch.

MULLET (Mugilidae)

This is a marine family of fast-swimming, closely schooling species, some of which occasionally forage in freshwater rivers. Mullets have long been an important food fish for humans and are pictured in ancient Egyptian art. Because of their habit of entering streams, especially as juveniles, mullets may have been the first cultured food fish. In some species, or populations within species, the growth from juvenile to young adult may take place during long stays in freshwater, which is a mild form of catadromy.

STRIPED MULLET ***Mugil cephalus***

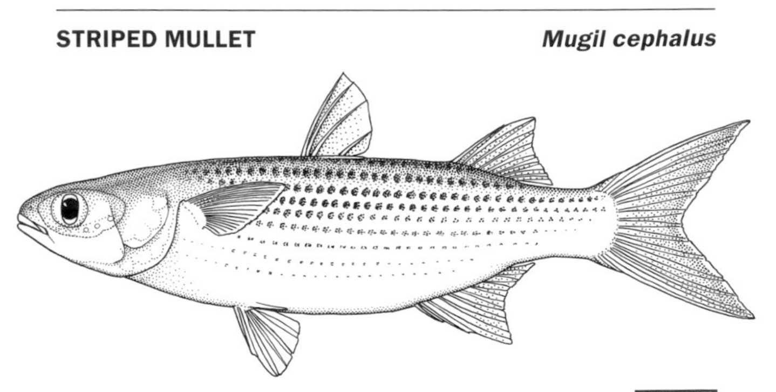

NATIVE

The Striped Mullet is found around the world in tropical and subtropical waters. It cannot withstand temperatures below about 15 degrees C (59 degrees F) and in California is therefore restricted to the lower Colorado River and southern coastal streams. Some populations pursue a catadromous pattern, living part of the year in freshwater and migrating as adults from the lower Colorado River to the Sea of Cortez during winter to spawn. The eggs hatch at sea, and during their first year of life fry move into the estuary of the Colorado River. Low water levels and actual drying up during parts of the year in the lowest reaches of the Colorado River have greatly curtailed this behavior.

The Striped Mullet was once a dominant species in the Salton

Sea. Canals from the Colorado River gave the species access to this inland saltwater habitat during the first half of the twentieth century. A large commercial fishery existed for this species in the Salton Sea from 1915 to 1921 and again from 1943 to 1953, but in the 1950s a flourishing sport fishery for introduced marine species brought commercial netting to an end. The canal links to the Colorado River were also altered, making the existing Salton Sea mullet an essentially landlocked population. Although adult Striped Mullet can tolerate salinities over twice that of seawater, the Salton Sea population failed to spawn and gradually died out. This spawning failure has been attributed to a lack of suitable temperature and depth conditions in this inland sea.

The Striped Mullet feeds on bottom detritus and gleans its nourishment from some of the smallest available food items, such as diatoms and bacteria. It takes organic debris into its mouth and grinds it between plates in the pharyngeal region, ejecting gravel and sand through the mouth while swallowing the soft organic material.

Because of its feeding behavior, the Striped Mullet is extremely difficult to catch with hook and line. The California Sport Fishing Regulations therefore, permit several alternate methods, including bow and arrow, and dip net. The latter is the method most commonly used and offers a considerable challenge because mullet are fast swimmers and excellent jumpers. Perhaps my most exciting seining experience was on an ichthyology field trip to Bahia de los Angeles on the Sea of Cortez. I decided to introduce the students to night seining, so with lanterns and flashlights in hand, we set a large beach seine in a sandy-bottom inshore area. We had the net about halfway to the shore when suddenly the water erupted in front of us as a couple hundred big Striped Mullet decided they did not want to become scientific specimens. Until that time, we actually did not realize we had anything in the net. Most fishes sound when encountering a moving seine net and try to swim under the weighted bottom or lead line. But, as we soon learned, mullet do not exhibit such behavior. Instead, they started jumping over the net, sometimes two feet or more above the surface, level with our heads and lights. The jailbreak was almost a complete success, and I came back with just one specimen of this species, which is still in our museum collection today.

GOBIES (Gobiidae)

Plate 87. Ventral view of the fused gobid pelvic fins.

The goby family is large and comprises inshore, mostly tropical marine fishes. They are small, bottom-dwelling carnivores that rest on the substrate or cling to the face of rocks with the aid of a suction cup formed by the fusion of the pelvic fins (pl. 87). The inshore marine areas and estuaries of California contain six marine and euryhaline native goby species, most of which are quite small. Two recently introduced species, the Yellowfin Goby *(Acanthogobius flavimanus)* and the Shimofuri Goby *(Tridentiger bifasciatus)*, have spread rapidly throughout the lower Sacramento–San Joaquin drainage system.

YELLOWFIN GOBY *Acanthogobius flavimanus*

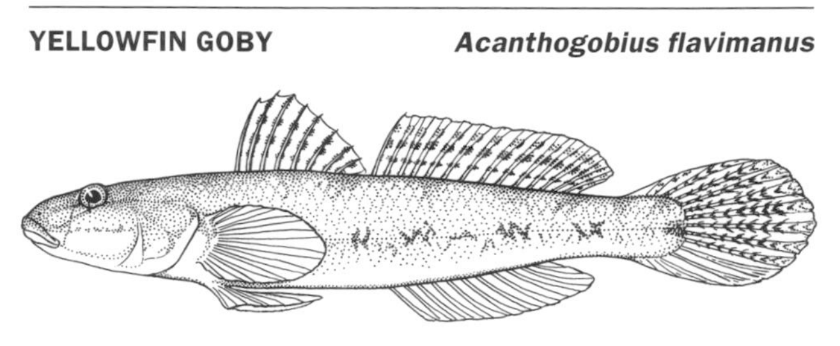

INTRODUCED

The Yellowfin Goby is a native of the inshore marine waters of Japan and China. It was first noticed in the lower Delta in 1963 (pl. 88). Its method of introduction is unknown, but four possi-

Plate 88. Yellowfin Goby.

bilities exist. One is that it was brought to California as fry in the water ballast of ships. Another is that adults were introduced, either as rejects from the home aquarium or possibly as escapees from small, private imports of live specimens intended as food. A third possibility is that the species arrived in California with spat of the introduced Japanese oysters. In addition, they are considered a delicacy in Japan, and such highly prized foods have a way of following their admirers to new lands. Gobies are particularly adapted for withstanding long periods out of water if kept moist, which makes this fourth possibility one to be considered. Whatever the case may be, this species has spread rapidly and, with the aid of the Delta-Mendota Canal, has established a landlocked, breeding population in San Luis Reservoir, Merced County.

The Yellowfin Goby is a bottom feeder on a wide variety of invertebrates and small fishes. In aquariums it exhibits a lie-in-wait feeding behavior, in which prey are captured after a short chase or dash as they swim near the waiting goby. This particular feeding niche is not well exploited in California freshwater habitats by other fishes, and this may be the key reason behind the species' rapid spread.

Unlike the small, native, marine gobies of California, this species may attain total lengths up to 27 cm (10.6 in.). Fishes of this size are three-year-olds, which is the maximum life expectancy for the Yellowfin Goby. Reproduction begins when a male constructs a Y-shaped tunnel in mud-sand substrate, with two openings positioned at the tips of the upper two arms of the Y.

He may use broken bottles or other hollow debris with dual openings that grace the floors of our freshwater habitats. The dual openings of the nest chamber provide a water circulation system, and the chamber itself affords protection to the adhesive eggs, which are placed on the underside of the chamber roof. The male usually guards the nest throughout the one-month hatching period. The newly hatched fry are free-swimming zooplankton feeders; when they reach a length of 1 to 2 cm (about 0.5 in.) they begin their bottom predatory role.

Within a decade of its "discovery" in California, the Yellowfin Goby had firmly established itself in the lower Delta system. The question now is not "What can we do about this?" but instead, "What can we do with it?" because, like other successful exotics, it appears to be here to stay. One recent use is for live bait, especially for Striped Bass *(Morone saxatilis)* fishing. The Yellowfin Goby is apparently mouthed and swallowed far more easily than the Pacific Staghorn Sculpin *(Leptocottus armatus),* a standard live bait species, because the it lacks the sharp preopercular spines of the Staghorn Sculpin, which are positioned outward from the head region upon disturbance. Striped Bass anglers are rapidly learning that hooks baited with Yellowfin Gobies produce strikes, whereas hooks in the same area baited with Staghorn Sculpins remain untouched. This discrimination implies a learning process on the part of the Striped Bass whereby it may recall the scent or taste of its prior painless meal and select the Yellowfin Goby again when the choice is present. As a result, commercial bait dealers in the lower Delta are anxious to obtain this species.

Another possible use of this new resident is as a food fish. Its reasonably large size and flaky meat rank favorably with a number of established edible species. The Yellowfin Goby has long been a delicacy in Japan, and this preference could well develop here in California. Perhaps the greatest drawback to harvesting at this time is this species' success in avoiding seine and trawl nets, possibly by means of its burrowing instinct. This species might lend itself well to pond rearing, however, where periodic harvests could be made by draining.

SHIMOFURI GOBY *Tridentiger bifasciatus*

INTRODUCED

This is a goby from Japan and Southeast Asia that began to appear in the lower San Francisco Estuary in the mid-1980s (pl. 89). The source of this introduction is not known, but ship-water ballast taken on at an Asian port and released in the lower Delta is suspected. It may have gone unnoticed before because of confusion with several very similar small Asian gobies all grouped under the name Chameleon Gobies. It took no less than an emperor to unravel this taxonomic confusion: Akihito, Emperor of Japan and a world expert on gobies. We now know that the Shimofuri Goby is a freshwater goby with a moderate tolerance for brackish water (up to 19 ppt), whereas the similar appearing Chameleon Goby is a marine species with little tolerance for freshwater.

While the ichthyologists were sorting out this introduced goby puzzle, the Shimofuri Goby was finding many freshwater habitats in California to its liking. It rapidly spread throughout the San Francisco Estuary, a phenomenon well documented by Dr. Scott Matern of the University of California at Davis, and then appeared in southern California reservoirs, having traveled via the California Aqueduct. Most of its long-range dispersal is by passive larval movement instead of active adult migration. Perhaps the prime reason for its rapid establishment in many areas is its feeding niche: freshwater hydroids (tiny jellyfishes) and exposed barnacle body segments. No native freshwater species occupies such a feeding niche, most likely because both of these prey items are themselves introductions into California freshwater and brackish-water sites.

The Shimofuri Goby is a small goby, with mature adults rarely exceeding 10 cm (4 in.) (pls. 89, 90). Like many other gobies, it lays its eggs on the underside of the roof of small cavities, including many human-created crevices. Popular nest sites attract sev-

Plate 89. Shimofuri Goby, female.

Plate 90. Shimofuri Goby, male.

eral spawning females, and the resident male may end up guarding several thousand eggs. If cavity nest sites are scarce, the aggressive males may prevent the smaller Tidewater Goby *(Eucyclogobius newberryi)* from spawning in some areas where they now occur together. Individual Shimofuri Gobies, on the other hand, continue to spawn throughout spring and summer, flooding the

San Francisco Estuary with very small larvae. Natural and artificial currents carry them to more new areas as the expansion of this pioneer species continues.

TIDEWATER GOBY *Eucyclogobius newberryi*

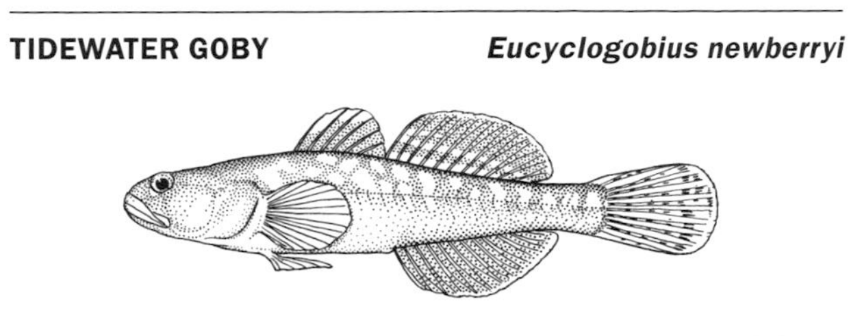

NATIVE

PROTECTIVE STATUS: FE in Orange and San Diego Counties.

This is a much smaller species than the Yellowfin Goby *(Acanthogobius flavimanus)*, rarely exceeding 5 cm (2 in.). It is a native California fish and prefers semiclosed estuaries or lagoons of small coastal streams low in salinity. Populations are found in such habitats along the entire California coast, apparently isolated by stretches of marine water between lagoons, where they feed on small bottom invertebrates, especially tube-dwelling species.

The spawning behavior of this fish is unique. It begins in the normal goby manner with the male constructing a spawning burrow. From this point on, however, the more brightly colored female is in charge, and once she decides that a particular male's burrow is where she would like to raise a family, she guards it against all other females in an aggressive "territorial male" manner. If the chosen male is not ready to breed, he simply plugs the burrow entrance and leaves his ardent suitor (the female) to continue defending the site. When finally ready to spawn, he removes the burrow plug, and both fishes enter the spawning chamber. This is no quick romance, as is the case with most fishes, for once within the nuptial tunnel, the male again plugs the door and the couple remains there for up to three days. At the end of what can best be termed the honeymoon, the female lays up to 1,000 eggs on the roof of the chamber and the male fertilizes them. The marriage now consummated, the burrow plug is removed, the pair exits, and the male replugs the entrance to protect the embryos. The pair now go their separate ways but do not seek out other mates during that spawning period. The male remains at the spawning chamber, which he eventually reopens to fan the developing embryos. Once the fry leave the nest, he constructs another spawning chamber. Repeated spawnings are necessary because the Tidewater Goby operates on the one-year plan: grow up fast, spawn several times, and die after one year.

Because of ever-increasing manipulation of coastal stream flows by humans, this species faces potential habitat destruction in the near future. In many cases, segments of lagoon populations are found in upstream pools off the main channel, and these groups may not survive if the lagoons continue to be altered or destroyed. The delisting in 2000 of Tidewater Goby populations north of Orange County as Federal Endangered species is certainly not going to help this situation.

SCULPINS (Cottidae)

The sculpins are primarily a marine family with 42 species present in tide pool and inshore habitats along the California coast. They are scaleless, bottom-dwelling fishes with large mouths and dorsally protruding eyes. The absence of a swim bladder allows these fishes to remain on the bottom because of their negatively buoyant state. The freshwater species of sculpin in California are relatively small and occupy a bottom-invertebrate feeding niche. Although the freshwater sculpins of California are unusual in that their species numbers have not been increased in recent years by eastern introductions, they have by no means escaped competition from exotics. This is especially true in the Delta region, where the Yellowfin Goby *(Acanthogobius flavimanus)*, Shimofuri Goby *(Tridentiger bifasciatus)*, and the Bigscale Logperch *(Percina macrolepida)* continue their rapid spread.

California freshwater sculpin identification is challenging. Unlike with most other families, a quick glance at the species illustrations will not do, although the renditions of this family are accurate down to the last fin ray. It's best to study the illustrations and written descriptions with the specimen close at hand, and knowing the geographic location where the sculpin in question was taken is also very helpful.

PRICKLY SCULPIN *Cottus asper*

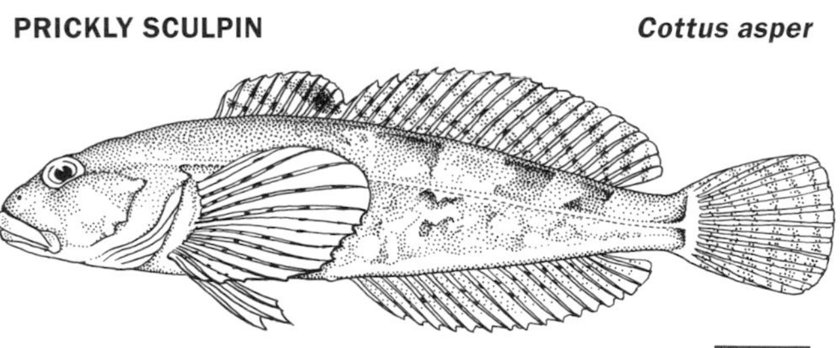

NATIVE

The Prickly Sculpin is one of the most widely distributed freshwater sculpins in California. Because its basic life history is similar to that of other species, it serves as an example for the entire group.

This small fish rarely exceeds 20 cm (about 8 in.) in freshwater

Plate 91. Prickly Sculpin.

and is found throughout a wide range of lower-elevation freshwater habitats of California. Like other sculpin species, it normally goes unnoticed by the average observer because of its excellent protective coloration and habit of hiding under bottom objects during the day (pl. 91). At night it actively feeds on a variety of bottom-dwelling invertebrates, particularly insect larvae. Viewing the abundance of bottom-living aquatic insect larvae allows you to appreciate the feeding opportunity, or niche, available to bottom-dwelling fishes. Doubtless, such an opportunity promoted the original invasion of freshwater habitats by marine forms. Another source of food in the upper reaches of coastal creeks and rivers is salmonid eggs. Although this may at first seem alarming to anglers, remember that these species evolved together over a long period of time, and the sculpins are probably taking those excess eggs that never become buried in the gravel. The mouth of an adult Prickly Sculpin is large enough to handle small fishes. I recently added a relatively large specimen to a large aquarium of smaller fishes whose numbers were soon noticeably reduced.

During the spring spawning season the male is territorial much like male centrarchids and cichlids. Here, however, the comparison ends, for instead of building an open-water nest, the male sculpin scoops out a depression underneath a rock and escorts a female to the nest, where she lays her eggs on the underside of the rock or nest ceiling. After spawning, the male drives the female from the nest site. A single female lays up to 11,000 eggs, but because a male may spawn with several females, a nest may contain up to 30,000 eggs. The fry are carried downstream after hatching. After several weeks they assume an adult body form and settle on the bottom. As they approach sexual maturity, they gradually move to preferred spawning sites.

Although the Prickly Sculpin is the most widely distributed California freshwater sculpin species, its range continues to be reduced by the many small barriers placed in coastal creeks and streams. Although a small check dam installed across a creek poses no significant barrier to most fish, the Prickly Sculpin body and fin complement is not made for jumping, so it is often deprived of any further upstream movement.

RIFFLE SCULPIN *Cottus gulosus*

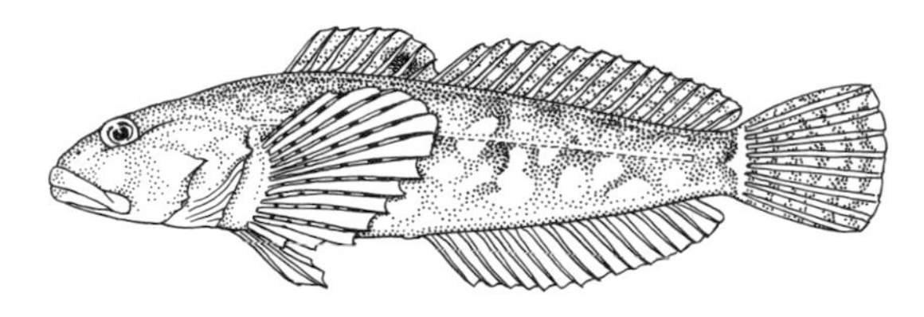

NATIVE

The Riffle Sculpin occurs throughout the Sacramento River Valley and in a few coastal rivers and streams. It appears to avoid direct competition with the Prickly Sculpin *(C. asper)* by occupying the cool, riffled, upper reaches of these streams to which this appropriately names species is well adapted. It wedges between rocks or under sunken logs and waits for bottom invertebrate prey to come its way. Spawning is similar to that of its downstream cousin, but the Riffle Sculpin is far less prolific, with each female producing only several hundred eggs. Despite this difference, the Riffle Sculpin appears to be abundant in many habitats. It is nearly always found with Rainbow Trout *(Oncorhynchus mykiss)*, which require the same high-quality stream-headwaters habitat.

PIT SCULPIN *Cottus pitensis*

NATIVE

The Pit Sculpin is confined to the faster-water areas of the Pit River system in northern California and appears to have evolved from the Riffle Sculpin *(C. gulosus)*. The latter is absent from most of the Pit Sculpin's range. Its feeding and reproductive habits are very much like those of the Riffle Sculpin and Prickly Sculpin.

COASTRANGE SCULPIN *Cottus aleuticus*

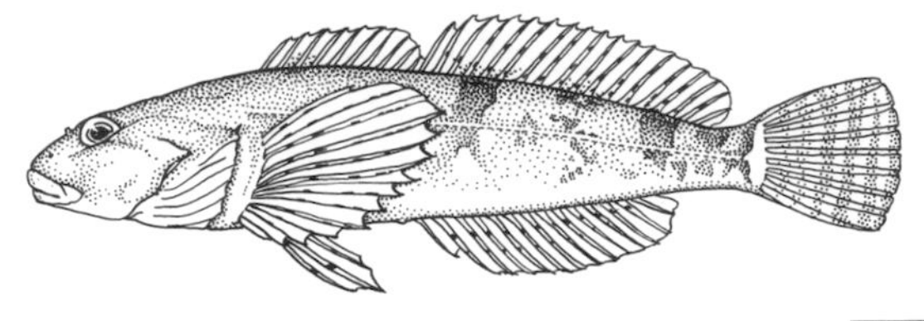

NATIVE

This is another small sculpin of coastal streams from Morro Bay northward to the Aleutian Islands. It is sympatric with the Prickly Sculpin *(C. asper)* in many areas. The Coastrange Sculpin seems to prefer fast-water riffle areas but may also be found in the quiet backwaters near the mouths of the coastal streams they inhabit. It feeds and reproduces in the same manner as other sculpin species. Like the Riffle Sculpin *(C. gulosus)*, it is less prolific than most other species, with the female producing about 1,000 eggs. The Coastrange Sculpin also migrates downstream to spawn at or near the stream estuary. This presumably is necessary to ensure that its larvae have access to zooplankton, which are not found in the fast-flowing upper reaches of streams.

RETICULATE SCULPIN *Cottus perplexus*

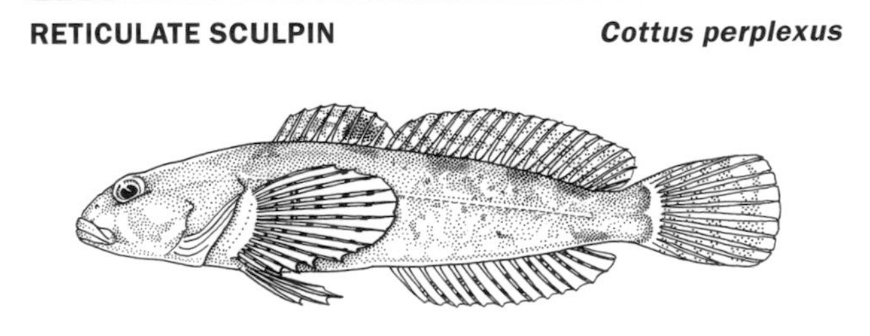

NATIVE

PROTECTIVE STATUS: CSSC.

This small sculpin is abundant in Oregon and Washington, but in northern California it is found only in the small California portion of the Rogue River. The Reticulate Sculpin appears to occupy an ecological niche similar to that of other small sculpins for the Klamath and Rogue River drainages and exhibits a spawning pattern similar to that of the Riffle Sculpin *(C. gulosus)*.

MARBLED SCULPIN *Cottus klamathensis*

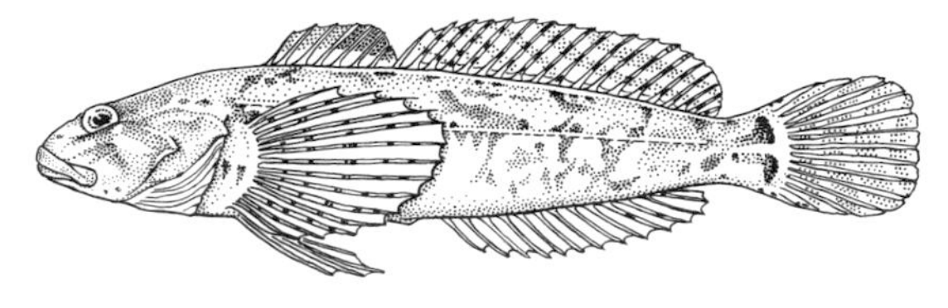

NATIVE

The Marbled Sculpin comprises three subspecies: the Upper Klamath Marbled Sculpin *(C. k. klamathensis)*, the Lower Klamath Marbled Sculpin *(C. k. polyporus)*, and the Bigeye Marbled Sculpin *(C. k. macrops)*. The latter occurs in the Pit River, along with the Rough Sculpin *(C. asperrimus)*, a designated State Threatened species. Although the Bigeye Marbled Sculpin has not been accorded threatened status, it receives similar habitat protection by virtue of coexisting with the Rough Sculpin. Where both species occur together, the Bigeye Marbled Sculpin tends to occupy clear, slower-moving waters and muddy bottoms. It spawns in areas with relatively flat rocks, laying its adhesive eggs in round clusters on the rock undersides.

ROUGH SCULPIN *Cottus asperrimus*

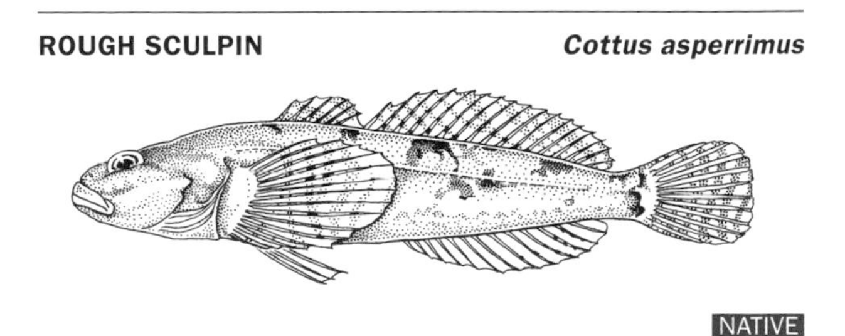

NATIVE

PROTECTIVE STATUS: ST.

The Rough Sculpin exists in cool, spring-fed tributaries of the Pit River near Burney, and in Hat Creek and Fall River. It was one of the first fishes to be protected under the 1970 California Endangered Species Act because of its low numbers and the numerous direct and indirect impacts it has sustained. These impacts include incidental loss from rotenone poisoning for Brown Trout *(Salmo trutta)* elimination and stream siltation resulting from excess logging and cattle grazing. This slow-growing species is also the smallest freshwater sculpin in California, with the mature adult rarely exceeding 8 cm (3 in.).

PAIUTE SCULPIN *Cottus beldingi*

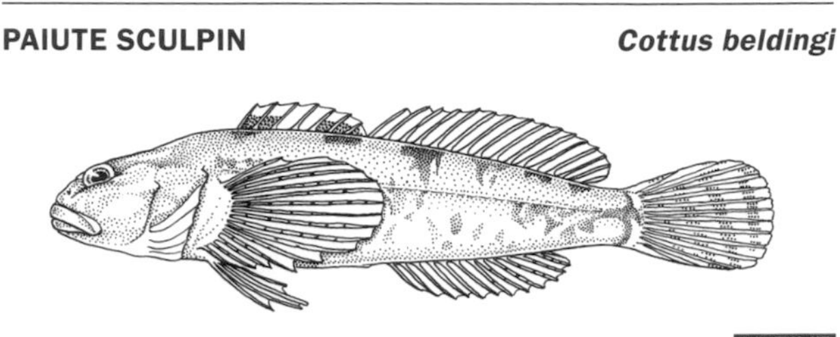

NATIVE

The Paiute Sculpin is the only California freshwater sculpin inhabiting the eastern slopes of the Sierra. It occurs in high numbers in numerous lakes and streams, including Lake Tahoe. It feeds on a variety of bottom invertebrates, many of which are outside the taste and size preferences of adult trout. The Paiute Sculpin appears to be a preferred food of trout in many areas and thus is an important link in the food chain in these Sierran habitats. This is the least fecund of the freshwater sculpins, with the female laying fewer than 200 eggs. Posthatching behavior of the fry also differs from that of most species in that the young remain in the protected nest site until the yolk sac is absorbed. Such behavior may make up for its low egg count.

PACIFIC STAGHORN SCULPIN *Leptocottus armatus*

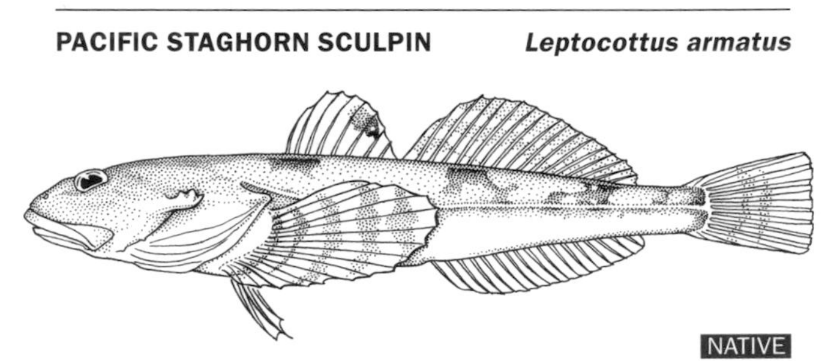

NATIVE

Many who have observed this species in marine bays and estuaries may wonder why a book about freshwater species includes it. The reason is that although most individuals of this species live in

Plate 92. Pacific Staghorn Sculpin.

seawater or highly brackish water, small numbers occasionally occupy freshwater sites. Occasionally, I collect this fish near the confluence of the Sacramento and San Joaquin Rivers, and it can be found in the lower portions of many coastal streams. This species' behavior may actually represent a first stage of freshwater invasion through which ancestors of freshwater sculpins have already passed.

The Pacific Staghorn Sculpin (pl. 92) is the largest of the sculpins found in freshwater in California, with older specimens attaining lengths up to 22 cm (about 8.6 in.). Despite its tendency toward piscivorous feeding habits in marine waters, this species feeds primarily on invertebrates when in freshwater. Although it is present in coastal streams year round, it breeds only in bay areas where the salinity is much closer to seawater than to freshwater. Females lay an average of 5,000 eggs, and the young begin their bottom-dwelling life in the bays before they move into freshwater.

A technical point concerning the name of the Pacific Staghorn Sculpin: the "horn," which is the term used to describe the long, scalloped projection on the operculum, is incorrect because stags (male deer) have antlers, not horns. A biologically correct substitute name would be "Antlered Sculpin." Do not look for a name correction just yet, however, because the California Department of Fish and Game still refers to deer antlers as "horns" in its hunting regulations booklet.

RIGHTEYED FLOUNDER (Pleuronectidae)

This is one of three marine flounder, or flatfish, families in California marine waters. Their unusual body plan is acquired a few weeks after hatching. Up to that time they are normal appearing pelagic fry. One eye then migrates to what becomes the dorsal side, and the anal fin and dorsal fin enlarge, nearly covering what are now the two respective edges of the body. These fins serve as the main propelling force for swimming, except when a sudden dash of speed is needed and the caudal fin briefly comes into play, performing as a standard outboard motor. Once the larval metamorphosis is complete, flounders take up residence on the bottom, where they feed on a variety of invertebrates and some fishes. Although this family has no freshwater species, one local marine species, the Starry Flounder *(Platichthys stellatus),* spawns in the lower reaches of the Delta and coastal streams, where the young spend the first portion of their life.

One brief taxonomic note: although the Starry Flounder has been placed in the righteyed flounder family, many individuals who apparently have not been informed of their taxonomic fate have both eyes on the left side of the body. Therefore, the best identifying characteristic for this one member of the righteyed flounder family, is the bold banding of the dorsal and anal fins.

STARRY FLOUNDER *Platichthys stellatus*

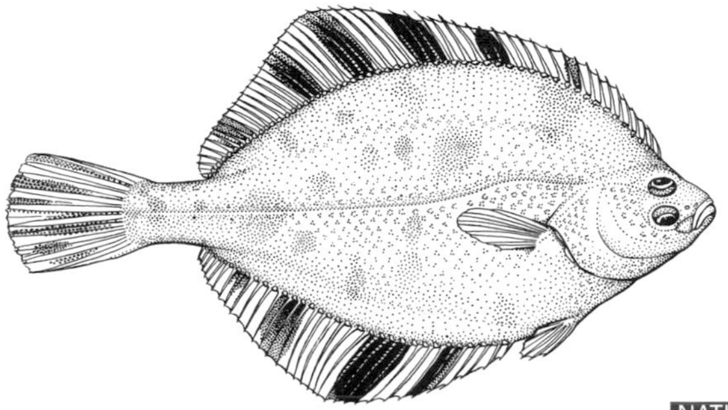

NATIVE

This species presents another interesting marine-freshwater puzzle. In the inshore marine and bay areas of central California it is an abundant and popular sport fish, well adapted to its saltwater habitat. However, small individuals, ranging from 1 to 15 cm (0.4 to 6 in.) total length, are quite common in the lower portion of the Sacramento–San Joaquin Delta and coastal streams from Santa Barbara northward. It has been found in coastal streams as far inland as 120 km (72 mi) and in the California Aqueduct in the Antelope Valley, Los Angeles County. These deeper inland excursions often take place during drought periods when tidal influence can extend farther up streams. Once beyond the 15-cm (6-in.) length, they migrate to marine water. The adults that occasionally are taken in freshwater appear in poor condition in contrast to the healthy-looking young, a fact suggesting that an irreversible change in the osmotic regulatory system may occur after the marine migration.

The adaptive significance of the freshwater site for early growth may be centered around the rich supply of bottom-dwelling invertebrates in these lower river areas. Competition for such food may be far less severe here than in comparable inshore marine zones. In freshwater, large aquatic insect larvae not found in marine water make up a large portion of the diet.

Angling for this species in freshwater is not recommended because of the small size in this environment, but do not overlook fishing for the adults in San Francisco Bay and along the California coast. Indeed, this is one of the best-tasting of all the Pacific Coast flounder species. As for the juveniles in streams and rivers, freshwater aquarium keepers can add a new dimension to their tanks with one or two of these most unusual fish. A number 10 or 12 hook baited with a bit of worm and bounced along the bottom can usually secure an aquarium-sized specimen. The species adapts well to fish-bowl life and grows rapidly when fed well. Flounders also exhibit some of the most dynamic color changes of any fish, and a "flounder tank" with purposely arranged bottom areas of light and dark gravel makes an instructive addition to the biology classroom (pls. 93, 94).

Plate 93. Starry Flounder, dorsal.

Plate 94. Starry Flounder, ventral.

PIKE (Esocidae)

In North America the pike family consists of the Muskellunge ("muskie") *(Esox masquinongy)*, the Northern Pike *(Esox lucius)*, and three species of pickerel. In Europe and the British Isles, however, it is represented only by the Northern Pike. These are lie-in-wait piscivores that occasionally take other small vertebrate prey. They are masters at hovering motionless in shallow, well-vegetated inshore waters, using the slow movement of the pectoral fins for stabilization. When a small fish passes by, they burst forth with amazing speed made possible by the long caudal peduncle, large caudal fin, and the posterior position of the dorsal and anal fins. The mouth of genus *Esox* contains many sharp, conical teeth that help it to successfully seize its prey fish at the end of such a dash.

This voracious piscivorous behavior is consistent with the qualities that make large sport fishes popular throughout the world, and in many states and provinces in the eastern United States and southern Canada, this family is the angler's favorite. Indeed, if you happen to visit the town of Hayward, Wisconsin, you will be greeted by a statue of a Muskellunge several stories high through which you can climb a stairway to an observation balcony in the open mouth. To the best of my knowledge, this is the largest statue of any organism (including humans) in that state.

As with other popular eastern U.S. sport-fish species such as the Largemouth Bass *(Micropterus salmoides)* and the Striped Bass *(Morone saxatilis)*, several early attempts were made to introduce the Muskellunge, Northern Pike, and one pickerel species into California. These attempts all eventually failed, probably because of the great loss of shallow, well-vegetated inshore habitat throughout the Delta during that period for levee construction and river channelization and because of the lack of similar habitat in the many newly built, steep-sided reservoirs. At this same time, other large, introduced piscivores (Striped Bass, Largemouth Bass) were getting a firm fin-hold in these new or modified habitats, which they have not as yet relinquished. Sometime in the 1980s, however, an unsanctioned introduction of Northern Pike occurred in Frenchman Reservoir in Plumas

County. At this writing, this premiere eastern game fish appears to have finally achieved California citizenship, albeit after a controversial and expensive eradication campaign by the Calfornia Department of Fish and Game.

NORTHERN PIKE *Esox lucius*

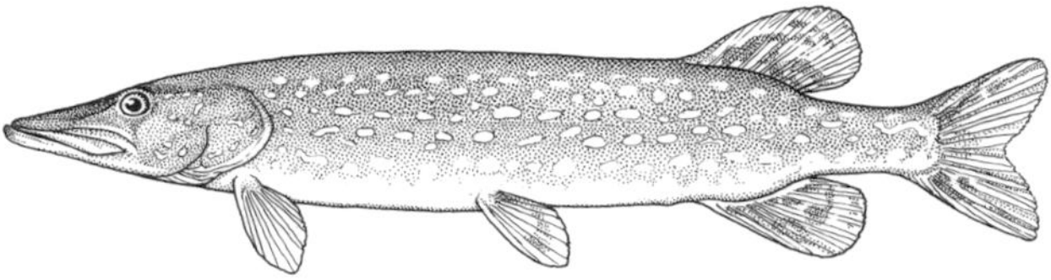

INTRODUCED

Along with various salmon species, the Northern Pike is one of the best-known freshwater game fishes in the world (pl. 95). Unlike the centrarchid basses or the large ictalurid species, the "northern," as it is often called in the United States, occurs throughout most of Europe, Russia, and the British Isles. In England, Scotland, and Ireland, it is simply referred to as the "pike," and over the centuries a wealth of folklore and highly embellished "fish stories" have grown up around it. Its large adult size is certainly one of its most impressive features. A three-year-old northern has an average length of about 50 cm (20 in.), which is already an acceptable size for many anglers, especially the younger set. The North American record is 110 cm (43 in.) and 14.2 kg (31.2 lb), which is dwarfed by the world-record Northern Pike taken from a German lake in 1985 that weighed in at 25 kg (55 lb).

The other major attraction of this fish to anglers is its manner of feeding, which, of course, dictates the method used to catch it. The Northern Pike is the premier ambush piscivore. Its elongate, loglike body with its dappled cryptic coloration blends in perfectly with its preferred feeding area: the well vegetated inshore area of a natural lake or stream. Certain areas of the littoral zone edge seem to be favored by large pike, and it is the angler's challenge to deduce where these are and then place the lure, usually a weedless spoon, accordingly. When successfully done, the next thing you usually see is the aquatic vegetation beyond the lure parting as the northern makes its rush at the "prey." For many anglers, this is the most exciting moment in freshwater fishing.

Northern Pike abandon their normal solitary foraging exis-

Plate 95. Northern Pike caught by ice fishing.

tence in spring when they aggregate at shallow, well-vegetated spawning areas. A female lays adhesive eggs that are fertilized by up to three accompanying males before attaching to an aquatic plant surface. The newly hatched fry also adhere to the vegetation for up to two weeks before swimming free. Such adhesion features for eggs and fry are unusual in lake species. Northern Pike fry feed on zooplankton and, like Largemouth Bass *(Micropterus salmoides)* fry, turn piscivorous during their second year. From this point on, the Northern diet consists mainly of fish species frequenting the littoral-zone edge, including many of the young sunfish species such as Bluegill *(Lepomis macrochirus)*. In well-balanced Midwestern lakes, Northern Pike likely help prevent overpopulation and stunting of Bluegill and Green Sunfish *(L. cyanellus)*. However, the Northern Pike's feeding strategy is not well suited to pelagic species such as young Kokanee Salmon *(Oncorhynchus nerka)* and Striped Bass *(Morone saxatilis)*, and therefore I believe that some of the recent hand-wringing about its reappearance in California is not soundly based.

ANGLING NOTES: The recent reappearance of the Northern Pike on the California sportfishing scene is in itself a fish story worth retelling. In 1891 the Northern Pike and the Largemouth Bass, two of the most popular large game fishes east of the Rockies, were brought to this state by the U.S. Fish and Game Commission. Both are voracious top piscivores and share a position at the top of the food chain in thousands of eastern and Midwestern lakes, where they occur together. One century or so later, the

Largemouth Bass is the darling of most California freshwater anglers, some of whom compete for hundreds of thousands of dollars in annual prize money at bass-fishing tournaments. Conversely, the California Department of Fish and Game has recently spent a similar sum in an unsuccessful attempt to eradicate the Northern Pike from a reservoir in Plumas County. In the early 1990s, preceding this effort, the Northern received some of the worst "bad press" ever given any fish, including the White Shark *(Carcharodon carcharias)*. "On the verge of catastrophe? Game officials fear predatory pike could decimate fishes in Delta" was the lead to a May 26, 1997, front-page story in the *Stockton Record*. It went on to quote the director of the CDFG's Region 2, who called the Northern Pike "the mountain lion of the river," and the founder of the California Striped Bass Association, who asserted that "these toothy creatures are voracious feeders and operate like a wolf pack."

The CDFG statement was at least more accurate in that both the Mountain Lion *(Panthera concolor)* and the Northern Pike do indeed employ a solitary, ambush-type feeding strategy that precludes running or swimming in packs. Overlooked once again, however, was the basic biological fact that with only a few exceptions (Kokanee Salmon and sturgeons [Acipenseridae]), all large California game fishes are voracious predators, which is the attribute that so endears them to the angler. However, logic did not seem to matter much as "pike paranoia" ran rampant. These and many other statements of nonfact were part of a newspaper, radio, and television news blitz apparently intended to rally support for a $2 million CDFG proposal in the late 1990s to "poison out" Davis Reservoir, into which the Northern Pike had migrated from nearby Frenchman's Lake, the site of the recent unsanctioned introduction. The proposed poison was rotenone (a vasoconstrictor) with a small amount of trichloroethylene (a known carcinogen) added for good measure. Opposing the plan were the residents of nearby Portola, a town whose drinking-water supply comes from Davis Reservoir. Despite Portola's rigorous protest, a modified and ultimately unsuccessful, version of this plan was carried out, which was described to me by one attending game warden as a "first-class fiasco." In the end, the Northern Pike is still with us, and, most importantly, so are the good people of Portola!

Given that from the beginning, public comments from many

amateur and even a few professional ichthyologists have been notably one-sided against *Esox lucius,* I attempt here a more balanced assessment in a question-and-answer format:

1. *Is the Northern Pike a top-notch game fish?* Yes. Millions of North American and European anglers agree.
2. *Will it "decimate the Delta" as the newspapers warned?* Probably not. The inshore zone of most of the Delta is packed so tightly with tule stands that it is hard to imagine even a moderate-size Northern Pike using this area for "lie and wait" foraging cover. Furthermore, many fish species in the Delta region are similar to those found in many eastern habitats where the Northern Pike is present, and all, including many species of native minnows and suckers, are doing very well. Most long-standing piscivore introductions into the Delta (Largemouth Bass, Striped Bass, Channel Catfish *[Ictalurus punctatus]* and White Catfish *[Ameirurus catus]*) exert sustained predatory pressure on most other species there, and one more introduction will not likely make much difference.
3. *Do we need yet another introduced species in Delta waters?* No. The line against further introductions should have been drawn a long time ago. Fortunately, the CDFG now carefully screens all proposed introductions.
4. *Is it possible to manage the Northern Pike within well-suited "pike lakes" and prevent its further spread with well-designed outflow screens, downstream backup trap basins, and rigid enforcement of a live-fish possession law?* Yes, and it is encouraging to hear that the CDFG is considering a containment management plan. I believe that many anglers in this state who grew up fishing for Northern Pike and miss the challenge and good eating that this species offers would be very willing to pay a reasonable lake entrance fee. The fee could fund (1) information pamphlets describing the potential ecological consequences of illegally stocking any fish species, as well as the fines imposed if caught doing so; and (2) random inspections at boat launch ramps for illegal species transported in ice chests and car trunks. Such a plan would cost far less than another large-scale eradication effort and would help to accommodate more of the California fishing public.

You can, by the way, legally fish for the Northern Pike in Davis Reservoir but must immediately kill (by beheading) any northern you catch, notify the CDFG within 24 hours after the catch, and retain both the head and body in the refrigerator for collection by a CDFG employee (2005 California Freshwater Sport Fishing Regulations, paragraph 5.51). Given the fine taste of baked Northern Pike, however, may I suggest that a little angler psychology be inserted into this requirement, whereby the head (for species confirmation), the innards (for feeding and reproductive data), and some scales from the midline (for aging) be neatly packed in a plastic bag and frozen for later collection. The body, however, should be the angler's to take home for a fine pike dinner!

OBTAINING AND OBSERVING LIVE CALIFORNIA FRESHWATER FISHES

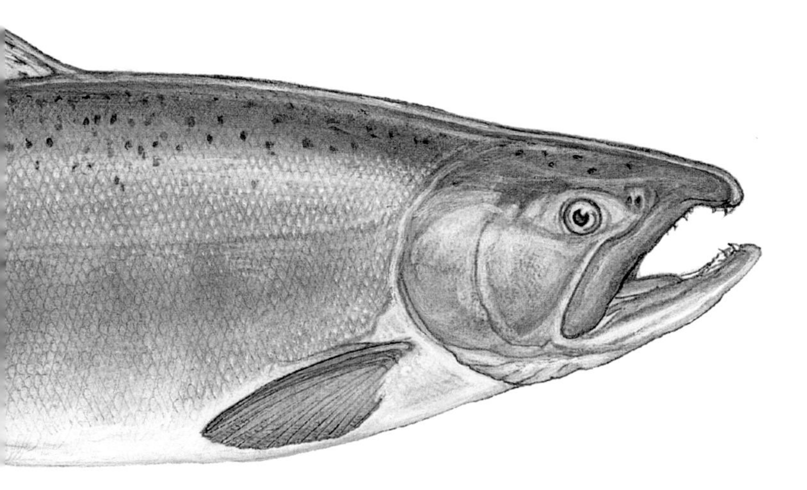

REGULATIONS AND METHODS

Fish collecting spans several fields of interest. Biology teachers at all instructional levels may wish to acquire a preserved collection, a live collection, or both of wild freshwater fishes for their classroom, often with the students participating in the process. Freshly killed fishes also may serve as excellent dissection specimens, whereas those preserved in 50 percent (100 proof) alcohol after an initial treatment in 37 percent formalin solution can be exhibited in the classroom for many years to come. Fresh dead specimens are also the basis of the fish printing art form, which is also a good classroom project (see "Photography and Making Fish Prints"). Certainly the most exciting collections are those for a classroom or home wild fish aquarium. Many people also enjoy wild fishes in their garden ponds.

Of course, a number of regulations exist pertaining to the collecting, transporting, and possession of wild California freshwater fishes. The first to consider are those set forth in the state and federal endangered species acts and additional protective listings by the California Department of Fish and Game (CDFG). These categories include endangered, threatened, fully protected, and species of special concern. Fish species with one or more state or federal protective designations are noted in the key to species, species accounts, and checklist in this book. In order to possess an endangered, threatened, or fully protected species, you must qualify for and obtain a permit from the U.S. Fish and Wildlife Service, the CDFG, or both. For species of special concern, a Memorandum of Understanding is required by the CDFG. For all such permits, the applicant must demonstrate a sound, scientific need for obtaining one or more such species and must possess a biological background that adequately qualifies him or her as a valid collector.

A second permission required for collecting wild fishes is the CDFG Scientific Collecting Permit. Of special importance is that it allows qualified persons who wish to establish a wild fish aquarium or pool to transport live fishes from a field collecting site. The 2005 California Sport Fishing Regulations contain a provision (1.63) that states, "Live fish taken under the authority of a sportfishing license may not be transported alive from the water where taken." The Scientific Collecting Permit can grant such permission for specifically requested species. Biology teachers, park naturalists, and interpretive center guides can usually obtain this permit. Provision 1.63 does not preclude you from collecting fishes for preserving or printing, as long as they are killed at the capture site and possession and size limits, where applicable, are adhered to.

Note that the majority of freshwater fish species in California are neither special status species nor "sport fish" as listed in the sportfishing regulations. For instance, the commercial bait minnows that are legally sold and transported in California (Golden Shiner, Red Shiner, and Fathead Minnows *[Notemigonus crysoleucas, Cyprinella lutrensis,* and *Pimephales promelas]*) all make excellent and, during their breeding season, colorful aquarium fishes. Also available are a number of wild bait fishes that can be legally captured for bait and, depending on the Sport Fishing District, often transported from the capture site. These include two species of goby, the Pacific Staghorn Sculpin *(Leptocottus armatus),* Threadfin Shad *(Dorosoma petenense),* Sailfin Molly *(Poecilia latipinna),* and, in the Colorado River District, "sunfishes." There are also many small species, both native and introduced, that do not fall under any of the current collecting and transportation regulations and would do well in a classroom or home aquarium.

Another way to obtain a variety of freshwater fishes, including small specimens of game fish species, and legally transport them to your aquarium or pool, is to purchase them from a registered freshwater aquaculturist. Currently, we have over 160 in the state, and a list, by county, of their names, addresses, phone numbers, and species they sell is available from the CDFG in Sacramento. Their offerings include some of the more striking centrarchids such as Green Sunfish *(Lepomis cyanellus)* and Black Crappie *(Pomoxis nigromaculatus),* several catfish and bullhead species, the Sacramento Blackfish *(Orthodon microlepidontus),* Striped Bass *(Morone saxatilis),* several carp varieties, and even White Sturgeon *(Acipenser transmontanus)* and Paddlefish *(Polyodon spathula),* an ancient species from the Mississippi River.

The Scientific Collecting Permit

The CDFG issues the Scientific Collecting Permit only to persons associated with aquariums and museums, and to college and university personnel engaged in research, teaching activities, or both with freshwater fishes or other wildlife species. In 2003, the permit cost $45.25, and it must be renewed every two years. At renewal time, you must report the number of each species taken during the previous year and their disposition. The permit allows collecting by a variety of methods and covers all students or assistants under the immediate personal supervision of the permit holder. This is an excellent means of obtaining specimens for a laboratory freshwater aquarium.

Capture and Transport Methods

Needless to say, the standard fishing gear must be modified in order to catch aquarium-size fishes successfully. The hook size should not be more than a number 10, and often a number 12 or 14 is needed. (Remember that hook size decreases as hook size number increases.) You should also file off the barb, because the objective is to hook the fish while making the smallest possible hole in the lip membrane area. Use the lightest possible monofilament line along with a very sensitive fly rod. Small bits of worm will usually tempt any little fish into biting. Without the barb, however, you cannot afford to "play" your trophy but instead must pull it directly out of the water. Much initial injury can be caused to the fish by handling, which often results in the removal of much of the protective mucous coat. When possible, the best technique is to lower the fish directly into a pail of water on shore and slacken the line so that it may swim free of the hook.

Small fishes not covered by any of the aforementioned restrictions can be captured by several types of small nets and commercial or homemade minnow traps. The latter all rely on the fact that fishes and other small animals will move readily through a funnel from the large end to the small end, but not the reverse. Large-diameter, fine-mesh dip nets are effective for catching fishes confined to shallow pools. Small seine nets usually need two persons to operate, and the pond or creek bottom must be free of large objects or dense aquatic vegetation, otherwise the fish will escape under the weighted "lead line." All of these items can be purchased at a well-equipped fishing tackle store.

Once the decision has been made to keep the specimen, use a soft hand net to transfer it to a transport container. Plastic bags supported by a strong cardboard box serve this purpose well. Products such as Novaqua also aid greatly in fish transport. Novaqua is a synthetic mucus that, when added to the transport water, coats and protects those areas of a fish's body where the natural material may have been rubbed off. Aeration of the transfer tank may be accomplished by small, battery-powered aeration units, which are available at modest cost from larger fishing tackle stores. Before leaving the collecting site, take the local water temperature. This should be approximated within 2 to 3 degrees C (3.6 to 5.4 degrees F) in the transfer tank en route. When collecting from cooler waters in summer, an ice-filled plastic bag submerged in the transfer tank can make the difference between live and dead fish upon arriving home. Once successfully home, begin the thermal acclimation program if the temperature difference between your aquarium and the collecting site is more than 9 degrees F.

THE WILD FRESHWATER FISH AQUARIUM AND POND

The Wild Freshwater Fish Aquarium

A home aquarium stocked with wild fishes is a refreshing change from the tropical fish tank and is often more instructional (pl. 96). By visiting a natural habitat instead of the fish store, you may view the ecological situation firsthand and obtain a natural complement of fishes and aquatic vegetation. Finally, you have the excitement of collecting wild fishes for an aquarium. To successfully stalk a school of minnows (Cyprinidae) in a shallow pool with dip net or miniature fishing gear takes skill equal to that of the craftiest Steelhead *(Oncorhynchus mykiss irideus)* anglers. The following suggestions should be helpful to those interested in starting a wild freshwater fish aquarium.

Water Quality

Water quality is perhaps the most important consideration in preparing a wild fish aquarium. Most tropical fishes come to your home in some preconditioned state. They have been exposed to foreign water conditions at the aquarium store, and those individuals that could not cope never made it to the sale counter. Freshly caught wild fishes, on the other hand, are commonly expected to make water quality adjustments instantaneously, often with disastrous results. Fishes are especially susceptible to small amounts of antibacterial agents, such as chlorine, in the water. City water should be aged for at least three days under continuous aeration before aquarium use. The best practice is simply to obtain all your aquarium water from the habitat where you collect your fishes. Large plastic bags supported within strong cardboard boxes may be used for this purpose and also serve as excellent fish transporters. Do not be too concerned about moderately cloudy pond or lake water, as a good filter

Plate 96. Juvenile Bluegills in a classroom aquarium.

system normally will clear it. Any of the better commercial filter units does an adequate job as long as the aquarium is not too crowded. For heavily populated tanks or those with rich algal growths, periodic treatment with a dolomite filter system is recommended. Most filter systems also provide adequate aeration, but again, crowded situations should be given supplemental air.

Thermal Acclimation

One major difference between the tropical and the wild fish tank is that the latter is not heated. Indeed, by placing the aquarium in a cooler portion of the house you may greatly assist wild fishes in their thermal acclimation process. As mentioned in the fish physiology section, the metabolic rate of fishes is adjusted to a given water temperature and needs time to readjust to a new thermal state. This is especially true for cold-adapted species, such as most suckers and trout. When brought from the wild directly into a room-temperature aquarium, they are forced into a metabolic high that they have great difficulty sustaining, even with adequate aeration. With proper acclimation or adjustment to the aquarium temperature by increments of 2 to 3 degrees C (3.6 to 5.4 degrees F) each day, most species can usually be conditioned to room-temperature water. If a collector wishes to specialize in cool-water fishes, an aquarium situated on a shaded outdoor porch can serve well. When positioned in front of a window, it can be fully appreciated from inside without taking up room space. Extensive collections of cold-water species may require a refrigerated tank.

Feeding

Although adult wild fishes consume a wide variety of foods, the diet of small species and fry of larger forms is quite uniform: it is based on zooplankton and small insects, with plant matter as a standard side dish. This is essentially what small tropical fishes thrive on, and the better commercial fish foods, especially the freeze-dried types, are most adequate. The expense of this type of food may be reduced by alternating it with meals of chopped earthworm or mosquito larvae.

Feeding is a most instructive time in the wild fish aquarium because it allows the various species to exhibit their individual adaptive feeding techniques. For example, many cyprinids will make rapid passes at the surface to pick off a food morsel. This behavior reduces the time they spend in the vulnerable surface position, where they have a good chance of becoming food items themselves. This sort of feeding pattern usually produces the small concentric ripples so often seen at dawn and dusk in quiet water areas.

In contrast, many centrarchids, particularly the Bluegill *(Lepomis macrochirus)*, wait until food begins to sink to the bottom and then peck away at it with the same staccato movement that produces those tantalizing movements of the fishing bobber. Suckers lie quietly on the bottom while other species above them feed frantically on the newly introduced food. Once several food particles reach the bottom, the suckers quickly come to life and begin working over the substrate with their ventrally opening mouths. A sucker is an important fish to have in an aquarium because it consumes food that would otherwise lodge in bottom crevices and pollute the water.

Aquarium Decor

To fully appreciate wild aquarium fishes, provide them with an aquarium decor that best suits their behavioral needs. Very few species prefer vacant, brightly lighted tanks with no retreat areas.

Several large rocks, tilted to provide a cavelike retreat, can help create a seminatural condition, as can waterlogged root segments and aquatic plants. The latter should be approached with some caution, because they are many times a preferred food of wild species. If this turns out to be the case, and you cannot afford to replenish the vegetation continuously with more fish-store purchases, artificial plants are a good second choice. No matter what decor you choose for the aquarium interior, the top of the tank must be extremely well covered. Many wild fish species are excellent jumpers, and some can even plane along the surface and slip out through any small space between the cover and aquarium wall.

Species Selection

A number of factors enter into your choice of wild species for the home aquarium. One of these is individual preference, and on this we can make no comment; however, certain species and families are far more adaptable to aquarium life than others. Much of this adaptability is based on physiological adaptations to the natural environment. Even with good filtration systems, many home aquariums are more characteristic of backwater, shallow areas than of midstream or midlake conditions. We would therefore expect to see species such as the Green Sunfish *(Lepomis cyanellus)*, Brown Bullhead *(Ameirurus nebulosus)*, or Threespine Stickleback *(Gasterosteus aculeatus)* thriving in home tanks, whereas species such as the Threadfin Shad *(Dorosoma petenense)*, Inland Silverside *(Menidia beryllina)*, and Rainbow Trout *(Oncorhynchus mykiss)* may not survive without a good filter and aeration system. The following list contains those species and families that have proven particularly suitable to the average home wild fish aquarium:

Most centrarchids, particularly the Green Sunfish, Bluegill, and Largemouth Bass *(Micropterus salmoides)*
Most ictalurids, particularly the Black Bullhead *(A. melas)* and Brown Bullhead
Most cyprinids, except those mountain species that have oxygen and temperature needs similar to those of trout
All livebearers (Poeciliidae)
All killifishes (Fundulidae)

Both the Yellow Perch *(Perca flavescens)* and the Bigscale Log Perch *(Percina macrolepida)*
Young Tuleperch *(Hysterocarpus traski)* and Shiner Perch *(Cymatogaster aggregata)*; the adults do not do well in captivity
Sacramento Sucker *(Catostomus occidentalis)* and those other suckers that do not require exceptionally cool water
Threespine Stickleback
Prickly Sculpin *(Cottus asper)*
Cichlids, where transport laws permit
Introduced gobies (Gobiidae)
Young Starry Flounders *(Platichthys stellatus)* from freshwater habitats

The Balanced Wild Fish Aquarium

Balance implies several things. Perhaps the most important is balance between species size and food habits. For instance, young Largemouth Bass make attractive aquarium fishes. They are also voracious piscivores and begin feeding on small fishes when they are only 8 to 10 cm (3 to 4 in.) long. Thus an aquarium that starts off as a Western Mosquitofish *(Gambusia affinis)* and Largemouth Bass tank will soon become monospecific in favor of the latter. This does not suggest that you should not keep Largemouth Bass, but that they should be matched with other species of similar size.

Sometimes one species will not eat another but will still cause much unrest in the aquarium as a result of its very aggressive behavior. Large Green Sunfish (8 to 10 cm [3 to 4 in.] and up) normally exhibit such behavior and should be avoided in favor of smaller (3 to 4 cm [1.2 to 1.6 in.]) individuals.

Color plays an important part in any selection process. The color plates in this book give you some idea of what to expect from wild fishes. Although we have some species in California that rival most tropical species, the majority of fishes from temperate latitudes are rather subtle in coloration. Do not be fooled by what you see upon first viewing a specimen in a bucket or jar after capture. Fishes change their color hues remarkably fast, and what may appear to be a washed-out, drab specimen at first glance will often take on a lovely coloration after a day or two in the aquarium.

Dominance may also play a role in coloration. An extreme example is found in the pupfishes, where only territorial males achieve a rich, metallic blue coloration. Because most aquariums are smaller than the average-size pupfish territory, usually only one dominant male patrols the bottom area in its handsome colors; all other males and females in the tank retain a drab coloration. Should the territorial male be removed from the tank or die, a new, brightly colored male will appear from out of the pack, so to speak, within a day or two.

Perhaps the most pleasing balance of all comes from obtaining a natural complement of species for your tank. By concentrating on one habitat and bringing into captivity a representative sample of those forms that occur together in the wild, you can start off on your first experiment in fish husbandry on the best possible footing.

The Freshwater Fish Pond

The other way to observe freshwater fishes literally in your own backyard is in a small pond. In recent years these have become very popular additions to home gardens or general landscaping, and a wide variety of preformed plastic and fiberglass basins are available at reasonable prices. You can also make your own basin by lining it with a variety of impervious plastic or rubber-based sheet products. For an even more natural-looking pond, a liner material called Bentomat can be used. This is a thin fabric blanket that is impregnated with bentonite, a clay-based material that swells greatly when wet and seals the bottom of an earthen pond basin. Several inches of soil are added over the liner, and then aquatic vegetation is planted.

For many years the McGinnis family has enjoyed what is perhaps the simplest of all small ponds to install and maintain: the round, galvanized stock watering troughs that are usually 2 ft high and range in diameter from 3 to 10 ft. They are available at most farm supply stores. We find that the 6- to 8-ft range placed above ground works very well. These come with a threaded pipe outlet on the tank wall just above the bottom that permits easy drainage when needed. We attached a garden hose to this outlet and then water the surrounding garden beds by gravity feed with fertile "fish water." A stock water tank float valve unit clamped to the top rim of the tank maintains the water level when watering plants in this manner or when evaporative loss requires some water input.

Most of the aquatic plants in our two ponds are ones we have collected in the field, and these grow in large submerged pots of soil with a light gravel layer on top to keep light organic material from floating out. Different sizes of concrete blocks and bricks are used to position the various plant containers at the appropriate distance below the water surface. As for the fish species, just keep in mind the suggestions for a community aquarium. I find that wild freshwater fishes behave far more naturally in a large outdoor pool than in even a good-sized aquarium. They often become wary of human presence and are best observed by standing back from the tank edge. If closer observation is desired, nearly all fish species soon learn to recognize the hand that feeds them and before long may be waiting as you approach for your daily viewing. Food can range from a variety of commercial dried flakes and pellets to small earthworms and live brine shrimp. The presentation of several food types at once will reveal the feeding niche divisions within your species complex. With all of the electronic home entertainment products competing for family members' attention today, a freshwater fish aquarium or pool is a refreshing departure from the norm and is often a bit more interesting than much of what the popular media have to offer.

FISH PHOTOGRAPHY AND FISH PRINTING

The hobbies of angling, collecting fishes for the home aquarium, and observing fishes in the wild can be greatly enhanced by making a permanent record of your subject. The three methods presented here can be explored with equipment and materials found in many homes. In addition, teachers of all grade levels can use cameras and camcorders to bring their fish-viewing experiences to the classroom, and fish printing is an art form that has already found its way into grade schools, high schools, and colleges throughout the state.

Freshwater Fish Photography

For many years the standard for close-up photography has been the single-lens reflex (SLR) 35-mm film camera. At the time of this writing, however, the digital point and shoot camera revolution is well under way, and several companies now sell digital backs to replace the "old" SLR film backs. Film SLR cameras with a macro zoom lens or extension tubes coupled with a standard lens continue to produce good fish pictures, but the ability to instantly review images from the digital camera can be especially advantageous when the subject is a fish in continuously changing light. The new (2005) models of cell phones may also be used to photograph fish in the field. One can then send such pictures to oneself via email and print them later at home.

The film-based SLR photos in this book were taken using two methods, and both rely on a small aquarium (five gallons is fine). The first method requires an additional piece of loose glass cut so that it just fits into the aquarium lengthwise. This piece of glass is the key to success in this type of photography because it acts to restrict the fishes to the front of the tank. The area behind the glass can be used as background, which in some cases may be quite simple. Pelagic or midwater fishes such as silversides or shad are often best photographed with only a blue or green piece of paper taped to the outside back wall of the tank. Because the sharp focus is always on the fish, the texture of the paper is lost and only the hue is visible. The picture of the Inland Silverside *(Menidia beryllina)* (see pl. 65) has such a colored-paper background. In most cases, however, a background of aquatic vegetation is more appropriate. Aquatic plants collected from the site where the fish was taken are the most technically correct and often the most pleasing. Only those portions of the plants that touch the movable inner glass wall are usually in focus, presenting the pleasant, diffused background seen in many of the color plates in this book.

Although the majority of fishes are best photographed in a pelagic or swimming attitude, others are better shown near or on a substrate. This is particularly true for bottom-oriented groups such as the catfishes, gobies, and sculpins. When photographing these species, add substrates such as gravel or sand both in front of and behind the inner glass wall. The substrate directly under the fish will be in sharp focus while that behind the inner glass will fade out of focus, giving a sense of depth to the picture. The quality of the water also affects the appearance of the background. Clear water brings out the background objects more strongly, even if they are out of focus, whereas cloudy pond or lake water greatly diffuses the background while still permitting a sharp image of the fish.

Good lighting is perhaps the most crucial of all aspects of fish photography. Whether you use the sun or artificial light, the aquarium or lights must be adjusted so that the angle of the light path is about 45 degrees to the front face of the tank. Positioning the camera at 90 degrees, or perpendicular to the front glass, should eliminate reflection of the light source. Even with proper alignment, the metal portions of the camera and tripod may reflect from the front glass. Eliminate this by positioning a large piece of stiff black paper on the front of the camera with just the lens protruding through a precut hole.

Now you are ready to photograph a fish: your tank is at the proper angle to the light source, the background is set to your liking, and your snug-fitting inner-glass wall is positioned about 2.5 to 5 cm (1 to 2 in.) behind the front wall of the aquarium, creating a space to receive the fish. You can photograph the fish in one

Plate 97. The author photographing fish in a field aquarium.

of two settings: swimming freely in the restricted zone, or gently held against the front glass by moving the inner glass forward as the fish passes through the field of focus. The widest point on most fish species is the area around and behind the pectoral fins; thus, gentle squeezing holds the fish motionless but still allows it to breathe, because pressure is not being applied to its operculum. A thin stick can be useful for final positioning of the specimen. The second method assures that the entire length of the fish is in focus, and it is normally the best choice for "textbook" pictures in which all of the body topography must be clearly seen. On the other hand, the first method can produce a more realistic posture and attitude. With practice and patience, you can achieve rewarding results using either method and in the process add a new dimension to your appreciation and enjoyment of wild freshwater fishes.

I prefer to take fish pictures at the side of a stream or lake whenever possible (pl. 97). A freshly captured fish seems to have more color and alertness than one that has been held in a bucket. If the latter is the only option, use a dark holding container. Fishes respond to the lightness of their surroundings by contracting the pigments in their skin color cells (chromatophores). What appears as a beautiful specimen in the net or on a small hook when captured may look very different when retrieved from a white bucket a few minutes later. If you have an established freshwater fish aquarium in which a number of species are well adjusted to the surroundings, merely alter your light source

Plate 98. Camera aquarium set up for shallow underwater photography.

and camera/tripod distance, and then wait for the right fish to appear in the right area. The picture of the Amargosa Pupfish *(Cyprinodon nevadensis)* (see pl. 61) and several others were taken this way.

For taking pictures in the fish's natural habitat, a technique developed by my colleague Dr. Chris Kitting works well. He reverses the previously described method by putting the camera in a small aquarium and then photographing the fishes outside of it, in their natural habitat (pl. 98). This, of course, works only in relatively shallow water, but photos at greater depth may be obtained by using a tall (rather than long) 5- to 10-gallon aquarium. Fill several socks with fine gravel or sand and place them on the bottom of the aquarium to counter the buoyancy of the tank and to allow you to lower it to a point where the top is still safely above the water surface. Stabilize the entire unit by attaching Styrofoam blocks below the rim on both of the long sides. Further protect the camera from water by attaching with duct tape a vertical skirt of plastic sheeting around the top of the aquarium. Once all is in place, the wading photographer needs only to wait until a fish is in the estimated field of view before pressing the shutter button. The auto-focus feature of modern SLR cameras and digital "point-and-shoot" cameras makes this technique possible.

This method also works well in shallows where fishes have congregated for spawning, such as the guarded nest sites of sunfishes (Centrarchidae), and salmonid spawning aggregations. The picture of the brilliant Kokanee Salmon *(Oncorhynchus nerka)* male in this book (see pl. 28) was taken by Chris with this method. A well-vegetated inshore area with abundant juvenile fishes is also a productive place to employ this technique, and you can often bait in small schools with bits of worms or live brine shrimp. When purchasing an aquarium for this use, choose one with a clear glass bottom so that you can use it from a boat or while wading in deeper water to photograph straight down. A self-adjusting flash unit coupled with a zoom lens will usually allow you to fill the view screen with your subject.

Freshwater Fish Videography

A good-quality camcorder is now a part of many California households, and each year more models have the close focus and low-light capabilities that are perfect for filming fish. Either the home aquarium, with its free-swimming fishes, or the partially sunken aquarium will produce very good results. The fact that the photographer does not have to be literally right on top of the camera also allows for more freedom and natural behavior of the subjects in the wild. Quiet, clear, creek pools that house schools of salmonid parr, cyprinids, or suckers are well suited to videography. Normally the towering human form at poolside is enough to send fishes into deep, underbank cover. However, once the intruder retreats, all the foraging and schooling activity begins again. Leave the camcorder inside the aquarium in a safe, shallow location and pointed toward the main activity area of the pool; the auto focus and exposure features will do the rest.

Even the most accomplished diver would be hard pressed to adequately recount all that happens during schooling, feeding, or spawning activities. But an hour-long video record can be analyzed and reanalyzed, providing a fine source of raw data for projects ranging from science fair entries to Ph.D. theses, with the assurance that the presence of an intruder has not modified the subjects' behavior. In addition, the more sophisticated digital camcorders allow you to choose that exact moment when all the most desired visual aspects come together and then to print it through your computer, just as you would a digital still camera shot.

Fish Printing ("Gyotaku")

As exciting as fish photography is for many people, nothing is more satisfying than graphically creating their own fish images. Unfortunately, very few of us are talented enough to produce fine illustrations with pen, ink, and brush. But do not despair. For all of us "artistically challenged" persons, there is an alternative: "gyotaku"! This art form was developed in Japan in the mid-nineteenth century and gained popularity in the United States during the second half of the twentieth century. It is really an ichthyological form of the ancient art of wood-block printing, where the raised portions of a carving on a flat wood surface are covered with ink and then pressed to paper to create a precise but reverse image. In "gyotaku", the relatively flat surface of the fish body, with its many fine, raised points such as scale edges, opercular striations, and fin rays, substitutes for the carved wood-

Figure 9. *Piscivore and Prey:* A gyotaku created by the author that depicts a Sacramento Perch and three Golden Shiners.

block. When done correctly, the finished product is a precise mirror image of the subject, accurate in every respect from its actual size to the number and shape of the scales along the lateral line (fig. 9). In addition to scientific accuracy, fish printing permits a degree of creativity rarely available in photography. You can produce that special fish-interaction scene or body attitude that has always eluded your camera. Here is how to begin.

The Subject(s)

Freshly caught, dead fishes are always the best specimens because of their body firmness, but fishes frozen soon after capture and thawed later also work well. Species with relatively flat bodies and large scales, such as the centrarchids, are best to start with, but with practice, even round-bodied, scaleless fishes such as ictalurids can be successfully printed. If you wish to begin with "store bought" fishes, choose the freshest available (see "Finding a Fresh Freshwater Fish"). For printing, you need the entire, uncleaned fish, which is rarely available at the local supermarket. Instead, visit an Asian fish market and browse the wide selection of whole fishes, usually displayed on open trays of crushed ice.

Materials

You will need school-type modeling clay, a 1-in.-wide paintbrush (either bristle or foam plastic), a small jar of black poster or acrylic paint, paper towels, and paper or cloth on which to print your fish. This last item is by far the most important and may to a large degree determine the success of your project. Because only one California freshwater fish is really flat (the Starry Flounder *[Platichthys stellatus]*), the printing material must be flexible enough to conform to the irregular body and fin surface without crinkling or folding, so that all portions of the fish body make contact. "Gyotaku" masterpieces by the old Japanese masters and more recent ones by American artists such as Christopher Dewees of Davis, California, are usually printed on expensive, hard-to-find paper made from specially processed tree bark. For the beginner, who may need to do repeat printings of a fish before achieving the desired effect, basic newsprint, available in large tablet form at most art supply stores, is a good medium to start with. I have found that fine, tightly woven fabric, such as muslin, broadcloth, and cotton sheeting, works very well because it conforms readily to the fish surface. For instance, I successfully used muslin for a print of a Sacramento Perch *(Archoplites interruptus)* pursuing Golden Shiners *(Notemigonus crysoleucas)*. Whether you use paper or fabric, you must first test the paint for blurring, or bleeding, on the material. Acrylic paint is least likely to do this.

Printing

First, dry the fish thoroughly with paper towels and place it on a newspaper-covered surface opposite the direction you want it to face on the print. Continue to blot dry the upper side and, if necessary, use a small fan or hair dryer with a cool setting to keep it dry. Fish skin continues to release mucus long after death, which will cause blurring or fuzziness on the print. Next, fashion small support blocks from modeling clay and place them under all fins except the pectoral. Spread each fin to its fully extended position and press it gently onto its supporting clay pad so that the rays on the underside catch in the clay surface and hold the fin in place. A few very small pins may be used here, but the heads must be somehow hidden so that they do not show up on the print. If you want the fish mouth open, prop it in the desired position with a toothpick inserted on the newspaper side of the jaw. There is a toothpick in the mouth of my Sacramento Perch print, but the paper never contacted it.

Once your fish is positioned to your satisfaction and as dry as can be, apply a light but complete coat of paint over the entire body except for the eye, starting at the head end and working to the tail. Avoid creating linear brush marks, which will often show up on the print. A plastic foam brush is less likely to create brush marks than a bristle brush, and a small foam paint roller works best of all. Leave the eye orbit blank for now. Next, paint each fin, being careful not to slop paint onto the clay pad beneath; wipe it off if you do. Finally, brush or roll the entire body once more but in the opposite (tail to head) direction and without adding more paint. This allows the scale edges to accumulate a bit more paint so that each prints clearly.

Now, for the moment you have prepared for: Carefully lay your printing material over the entire fish so that it contacts the fish surface cleanly with no wrinkles or folds. If you are using a fabric or very thin, floppy paper, a second pair of hands can be helpful to hold and lower the material, as you would a bed sheet. When the print material is on the fish, press gently over the entire body and fin surfaces, being careful not to move the paper or fabric in the process. A slight rubbing motion of the fingers ensures good ink transfer, but too much movement results in smears. Once you are satisfied that all portions of the print sheet–fish interface have made good contact, carefully peel, not pull, the paper or cloth off the fish, starting with the head end, and view your long-awaited print. Do not be discouraged if it does not look perfect. Often, a first print is needed to remove undetected moisture or mucus from the surface, and the second or third prints are the real masterpieces. You will have to repaint for each successive print, but take care not to use too much, because some is on the fish already.

If you wish to make a multiple print, that is, a print of more than one fish, let your first printing dry thoroughly and then set up the second fish in the same manner as the first. You must carefully align the successive images in the desired location, and here again a second pair of hands and eyes is helpful. In making my piscivore-prey print, I printed each of the four fishes separately, waiting for each to dry before printing the next.

When the fish or fishes are printed and dry, it is time to work on the eye or eyes. A fish's eye is one of its most noticeable features, however, it is covered by a smooth cornea so the details do not "print." Because you have left the eye orbit unpainted, however, an empty circle awaits your one chance at a personal artistic touch. Study the fresh fish eye carefully to note where the highlights occur, and then with an ink pen try to duplicate these on the print. You may want to practice a few times on some circle outlines before you draw on the print itself. Some fish have colored eyes, and it is perfectly within the wide lat itude of the "gyotaku" format to color the print eye accordingly.

You may also wish to color the entire fish and the background. The latter is best done by simply using a lightly colored paper or fabric that is easily masked by the printing paint. Pastel blues, grays, and greens are the usual favorites because they suggest the aquatic habitat. You can, of course, produce a monochromatic print by simply using a paint other than black. If you wish to reproduce the natural color of the specimen, colored pencil or pastel chalks rubbed on lightly produce a pleasing effect. If you wish to paint the fish, watercolors work well as long as the printing paint does not dissolve and smear with the brush strokes. Sometimes simply highlighting a species' most prominent color, such as the red wash on the side of a Rainbow Trout *(Oncorhynchus mykiss)*, produces the best color effect. Remember that once the print is made, the fish has done all it can, and you become the artist with all the freedom and inventiveness that goes with that title.

FRESHWATER FISH WATCHING IN THE WILD

The bird watcher, lizard watcher, frog and toad watcher, game mammal watcher, and even the insect and worm watcher all have one great advantage over the freshwater fish watcher: their subjects are available on land, often in broad daylight and in full view. Freshwater fishes, on the other hand, are normally seen out of their natural habitat only when they are removed by hook and line or by net. Only the freshwater aquarium enthusiast has a continuous opportunity to view fishes in their habitats, and then only in restricted and modified habitats. Nonetheless, a variety of sites around the state of California provide the opportunity to observe fishes in their natural habitats. Perhaps the most spectacular sights are to be seen at the salmon and Steelhead hatcheries, located at or near dams on California's great coastal rivers. For example, at facilities such as the Iron Gate Hatchery on the Klamath River, Mad River Hatchery on the Mad River, and Nimbus Hatchery on the American River, adult King, or Chinook, Salmon *(O. tshawytscha)* and Silver, or Coho, Salmon *(O. kisutch)* may be seen leaping up the steps of fish ladders or thrashing through shallow diversion streams during fall spawning runs. The Warm Springs Hatchery near Geyserville in Sonoma County is a good place to observe hatchery spawning of Steelhead beginning in late December and continuing through April. Besides offering one of the most spectacular wildlife sights in North America, these fishes are excellent subjects for the wildlife photographer, who can often get within a few meters of them.

Many state hatcheries are, of course, not devoted to salmon and Steelhead culture, and even salmonid hatcheries have adult fishes for only a few months of the year. However, all hatcheries are excellent places to view large numbers of young and catchable-size fishes, and many offer tours of the facilities that explain the art of fish propagation. The names, locations, and contact information for fish hatcheries in California are listed at the end of this book. Calling before you visit is recommended to learn when

Plate 99. Chinook Salmon viewing through fish ladder windows at the Feather River Hatchery.

tours are offered and, in the case of the salmon hatcheries, when the peak of the run is expected.

Several of the larger hatcheries now host weekend events. The Oroville Salmon Festival is held at the Feather River Hatchery in late September and features tours of the facility, artificial spawning demonstrations, and underwater viewing chambers where you can observe Chinook Salmon as they make their way up the fish ladder (pl. 99). For details, call 800-655-4653 or visit www.oroville-city.com/chamber on the Web.

The American River Salmon Festival is held at the Nimbus Hatchery and Lake Neotoma in Rancho Cordova in early October. It features over 30 activities and demonstrations, all featuring the king of salmon, the Chinook. For information, call 916-361-8700 or visit www.salmonfestival.net on the Web. A third event is the Anderson Return of the Salmon Festival, held at the Coleman National Fish Hatchery in Anderson, Shasta County. Again, salmon viewing, facility tours, and children's activities are featured. For information and dates, call 530-225-2300.

You do not have to go to hatcheries, however, to see a salmon spawning run. In fact, the most visible run is perhaps that of the Kokanee Salmon as they move from Lake Tahoe into Taylor Creek each fall to spawn and then die. The best time to view this spectacle is mid-October to mid-November. Taylor Creek crosses

Hwy. 89 between Cathedral Road and Fallen Leaf Road, just below Fallen Leaf Lake. The best viewing can be found by taking Fallen Leaf Road south to the national forest campground and from there hiking in to the creek. For a different view of these fish and other mountain stream species, turn north on Fallen Leaf Road to the Eldorado National Forest Visitor Center, which features a stocked pool with an underwater viewing room that provides in-close fish watching at its best; and yes, it hosts a festival too. For information, call 530-225-2300 or visit www.fs.fed.us/r5/ltbmu/ on the Web. In the greater San Francisco Bay Area, Lagunitas Creek in Samuel P. Taylor State Park, Marin County, offers good creekside sites from which to view Coho Salmon spawning runs from December through January.

At a different scale, in a vastly contrasting habitat, are the pupfish observation areas of Death Valley National Monument. Each of the two sites—Salt Creek, in the Mesquite Flat area, and Cottonball Marsh, in the northwest portion of the monument—contains an endemic species that bears the site's name (Salt Creek Pupfish and Cottonball Marsh Pupfish). Boardwalk facilities that meander through these areas permit close-up viewing and allow you to study the many facets of this unique desert habitat. Plenty of fish can be viewed in the warmer months of the year, when you may observe behavior such as territorial display and spawning (pl. 100). A spring sanctuary for a third species, the Desert Pupfish *(Cyprinodon macularius)*, has been established in Anza-Borrego State Park, San Diego County.

Often, the best fish watching happens unexpectedly, usually in calm-water areas. Bridges over small, quiet streams in the latter part of summer and early fall offer good vantage points. Half the fun of viewing under these circumstances is simply trying to figure out what species, or perhaps what families, you are seeing. Under such circumstances even the best drawings and photographs in field guides are of little use, and you must instead rely on a knowledge of basic body shapes as viewed from an upper angle, swimming behavior, and whatever unique light or dark marks may be present to make an identification, or at least a good guess. Centrarchids, for example, are relatively easy to recognize because of their deep body and prominent fin structure; however, correctly naming the species within this family is a different matter. A few have individual markings that may be seen at some distance. The terminal dark tail band of the Smallmouth Bass

(Micropterus dolomieu) stands out, as does the dark lateral line band of its cousin the Largemouth Bass *(M. salmoides)*. The dark opercular flap of the Bluegill *(Lepomis macrochirus)* is also prominent. On the other hand, you might be looking at a Redear Sunfish *(L. microlophus)*, or perhaps it is really a Pumpkinseed *(L. gibbosus)*. Needless to say, the sunfishes and many other fish groups pose a real challenge. Often, it is your knowledge of the geographic range, habitat preference, and general behavior of a species that ultimately leads to a correct identification. Of course, many sightings go unidentified, and perhaps those are the most exciting of all.

Plate 100. Pupfish watching at Salt Creek, Death Valley.

No list of fish-watching sites would be complete without mention of the newly relocated and remodeled Steinhart Aquarium at 875 Howard Street in San Francisco. This fine aquarium has long been known for its excellent marine and tropical freshwater fish exhibits. There is really no substitute for viewing a fish broadside, and the fine, natural-appearing backdrops behind the aquarium's specimens help to showcase the wide variety of protective coloration patterns that enable our native and introduced freshwater fishes to blend with their surroundings.

Perhaps the ultimate experience for fish watchers is to join their subjects in the aquatic habitat. This type of fish observation has long been popular in marine habitats because of the usually good clarity of seawater. In many freshwater habitats, however, the water is clouded by phytoplankton or silt, so although fish populations may be high they are not visible. Some California freshwater habitats do, however, offer excellent opportunities for viewing because of their clear water conditions. Mountain lakes and foothill and coastal streams are often worth exploring. In higher-elevation habitats a wet suit is desirable, and when used with few or no weights, a large-bore snorkel allows you to float effortlessly at the surface while viewing the fish activity below. Quiet, shallow bays, with their lovely backdrop of sunken dead trees and aquatic vascular plants, are the prime sites for viewing in lakes.

Surface floating is also productive in the still waters and deep pools of streams. Contrary to what you might think, a large, floating object in a small pool often attracts fish. Many small species avoid shallow, brightly illuminated water, preferring instead to carry out their activities in the shadows. Once the initial disturbance of your entry has passed, you may find that your midpool shadow, or even your immediate underside, has acquired a collection of small fishes that tag along as you float through their domain.

DINING ON CALIFORNIA'S FRESHWATER FISHES

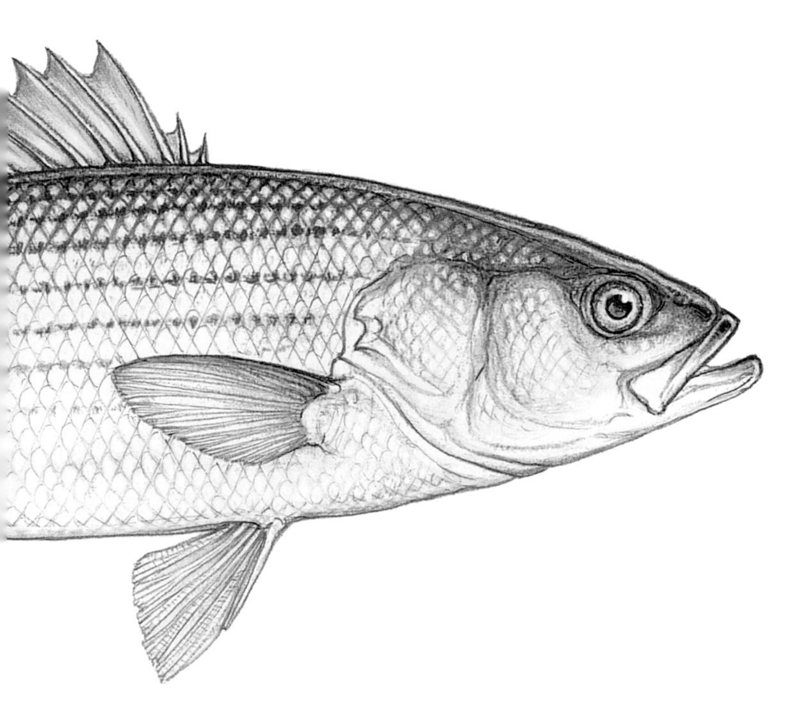

The Golden State offers golden opportunities to both those who catch and those who purchase freshwater fish for the table. In addition to its native salmonids, sturgeons, large minnows, and large suckers, California has been the recipient of most game fish species from the midwestern and eastern states. While such introductions have often adversely affected native species, they have combined with native game species to offer the California angler more choices than are available to those in most other states. Choices for the freshwater fish shopper are also increasing as fish farms become more numerous throughout the state, and California's large human population continues to attract more out-of-state producers.

With these facts in mind, I have devoted the last major segment of this book to guiding both angler and shopper through a series of steps and decisions aimed at producing many rewarding freshwater fish meals. The section covering the last and most important part of this process, the cooking of freshwater fishes, provides some of the many innovative fish-cooking techniques and recipes developed over the past several decades.

HEALTHY DINING ON FRESHWATER FISH

Fish has long been considered one of the healthiest foods in the human diet. It is one of the best sources of the proven heart-protecting omega-3 fatty acids. Fish is high in protein, and even oily fish are low in fat compared to other protein sources. The American Heart Association's advisory "Fish and Omega-3 Fatty Acids" says, "We recommend eating fish (particularly fatty fish) at least two times a week." The AHA recommends that adult men and nonreproducing women eat three or four 8-ounce servings a week, preferably of fatty fish, because of their greater amount of omega-3 fatty acids. The Department of Health and Human Services/Environmental Protection Agency's 2004 advisory says,

> Fish and shellfish are an important part of a healthy diet. Fish and shellfish contain high-quality protein and other essential nutrients, are low in saturated fat, and contain omega-3 fatty acids. A well-balanced diet that includes a variety of fish and shellfish can contribute to heart health and children's proper growth and development. So, women and young children in particular should include fish or shellfish in their diets due to the many nutritional benefits.

But there is another side to the story. Fish can acquire pollutants of various sorts from their habitats, and some of these materials have the potential to adversely affect humans who consume the fish. The biological basis for the problem of fish and toxic contaminants is shown in figs. 3 and 4. Many of the freshwater fish species preferred by both anglers and fish market shoppers are large tertiary consumers, feeding on smaller fishes and aquatic invertebrates. Prior to being eaten, these small prey species have acquired their own toxin loads from the various plant and detritus materials on which they feed, and these substances are then passed on to, and concentrated by, the larger predatory fishes which consume them. Thus, by the time a top piscivore such as a Striped Bass or Largemouth Bass arrives at the dinner table, a possibility exists that substances listed as being of concern by federal and state agencies or the World Health Organization will have accumulated in its body.

Given these basic truths concerning fish ecology and human behavior, just what should the California freshwater angler and fish shopper consider before going afield? One seemingly obvious answer would be to consult a published table listing levels of various contaminants in the species you plan to catch or purchase. However, lists and tables published by the World Health Organization and by several U.S. government agencies differ greatly in their assessment of the parts per million or billion in fish flesh at which each potentially toxic material might begin to cause damage. In addition, consumption advice is often given using only vague terms such as "large quantities" and "very frequently." The California angler receives an excellent guide upon purchasing a license: the California Department of Fish and Game fishing regulations booklet! Its "Public Health Advisories" section gives clearly written, easy to understand general and site-specific consumption guidelines. Most stores selling bait and tackle also distribute free copies of the fishing regulations booklet.

However, all guidelines must be adjusted to the situation. For instance, since the actual edible flesh of a whole fish with head and intestines intact is half or less of its total weight, the average angler probably does not have either the leisure time or the ability to catch enough fish week after week to provide even the generally accepted half-pound "safe amount" of fish flesh to every adult in the family. Nor is it considered a risk to eat more than a suggested amount at a meal or to eat several meals of fish in a row (as you might on a fishing vacation) unless you do it "often." Given the difficulty and expense of finding enough good fish to eat even as much as the AHA-recommended minimum of two eight-ounce servings of fish per week, most fish shoppers will no doubt simply continue to enjoy fish dinners as a way to vary the standard fare of chicken, beef, and pork.

The cautious angler's next question might be, "How can I find the places where the fish are safest?" In addition to the information in the regulations booklet, site-specific information is conspicuously posted on signs at most public boat launching sites and commonly fished river bank locations. For instance, the lower portions of the Delta and the lower portions of its tributary streams and rivers receive far more agricultural irrigation runoff water with its pesticides and herbicides than do the upper reaches of these waterways. Public boat launches and other fishing sites where contaminants in the fish caught may be of concern are conspicuously posted with specific information on these waters, as are similar sites throughout the state.

So, the solution here is to go well upstream to catch that fresh trout—or is it? Remember, we live in the state with the largest put-and-take trout stocking program in the world. This means that those fine-looking 12-to-16-inch (30-to-40-cm) rainbow trout in your creel most likely spent all but the last few days of their lives in hatchery runways eating "trout feed," essentially the same formula given other farmed fish. Up to 62 percent of the protein and 100 percent of the fat in this feed comes from fish meal, most of which is manufactured in Peru where it is processed from small marine filter-feeding fishes. Given that knowledge, the question of contaminants in the habitat now shifts from the essentially pure water of a Sierra stream to the Pacific coastal water of South America. Unfortunately, contamination figures for hatchery-reared trout are not readily available. However, salmon reared in marine cages are fed a "salmon feed," which like "trout feed" is based on Peruvian fish meal, and tests show that farmed salmon may contain up to 10 times more PCBs, dioxins, and pesticides than are found in wild-caught salmon. Similar loading of such toxins might be expected in some hatchery-reared trout.

Findings such as this are especially relevant to the fish shopper, since in California no game fish species caught in freshwater may be sold commercially. Thus, the normal fish market offerings of Channel Catfish, White Sturgeon, Largemouth Bass,

Striped and White Bass, Rainbow Trout, Atlantic Salmon, and tilapia all come from fish farms. Salmon lovers can obtain wild-caught salmon for at least part of the year, during the commercial salmon season when Chinook Salmon caught in coastal waters are readily available. During other times of the year, wild-caught Chinook and several other species of salmon and even Steelhead come to California markets from the Pacific Northwest and Alaska, and ardent fish shoppers soon learn the timing of these annual supplies.

Given the continuing lack of regulation of the U.S. and world marine fishing industry, we can expect to see more and more farmed fish on the market, most of which will be freshwater species. Besides filling the gap left by exhausted marine stocks, fish farming can be a very profitable business. The feed conversion ratio for most farmed fish species is 1.2 to 1. In other words, a fish adds another pound (0.45 kg) of body weight for every 1.2 pound (0.5 kg) of feed consumed. This is substantially better than the feed conversion ratio for chicken (approximately 2 to 1 at best) and pork (3.5 to 1), and way below that for beef at 6 to 8 to 1.

As with wild-caught fish, the question still comes down to what is and is not safe with respect to toxin concentrations. As just one example, both the U.S. Food and Drug Agency and Health Canada have currently set the tolerable level of PCB in fish at 2,000 ppb (parts per billion). In a recent analysis of small samples of farmed Atlantic salmon sold in the San Francisco Bay Area, salmon farmed in the Pacific Northwest had PCB levels averaging 29 ppb, 69 times less than the current safety threshold of 2,000 ppb. Clearly, a lot more research in this area is needed, including long-term fish consumption studies with laboratory mammals, and to date no such studies have even begun.

In the meantime, my best advice to the angler and fish market shopper is simply to keep alert for more and more study results, agency claims, and warnings as they come forth in future months and years. In the meantime, I hope you will continue to enjoy good fish dinners from your angling or shopping trips. However, if after reading this, you are now thinking: "Well, maybe I'll just give up fish dinners altogether and stay with chicken and 'the other white meat,' pork," you may want to think again. These animals require complete protein in their diets, much of which may be supplied by the same fish meal fed to farmed salmon, with the remainder of the ration derived from genetically modified grains and soybeans. Beef, sheep, and other livestock may also be fed fish meal or oils (fish products are not permitted in the feed of animals grown for certified organic meat). However, it will probably be a long time yet before we see toxin warnings posted at the supermarket meat counter!

How to Clean a Fish

WHEN TO CLEAN

Perhaps the first question should be: "When should you clean a fish?" The best answer here is: as soon as possible after the fish is dead. If the fish cannot be kept alive on a stringer or in a live box for the duration of the fishing trip, then it is best to clean them one by one as they are caught. The reasons for this are several. First of all, fish flesh tends to deteriorate much faster than other types of meat. Such deterioration can be greatly retarded by removing the flesh from the body as soon after death as possible and placing it in a cool place. Second, a fish should be bled immediately after death; again, this step is not possible if the fish died in the bottom of the boat several hours ago. Finally, there is the problem of the scales or skin, which get harder to remove with each minute after death. In addition, cleaning immediately helps prevent migration of parasites from the gut or internal organs into the flesh—though such organisms are killed when the fish is cooked.

The actual killing of a fish should be done as quickly and humanely as possible. The best method is to deliver a hard, sharp blow to the top of the head with a blunt object, such as a piece of pipe, and then cut into and sever the spinal cord from the top just behind the head. This will break all neural connections between head and body and will also deprive the brain of any further blood supply, since the dorsal aorta, which sends blood to the brain, lies just below the spinal cord. Do not, however, cut down any farther once you feel the knife pass through the spinal column, as the head presents a convenient handle with which to hold the fish during much of the skinning or filleting.

SCALE OR SKIN

To scale or to fillet and skin, *that* is the question! This is actually a simple matter of taste—literally. When a fish is scaled and then fried, the individual taste of that particular species is quite pronounced, since much of the flavor is contained in the oils of the skin. If you really relish and prefer the taste of Largemouth Bass over that of Bluegill, you should definitely not skin your fish. If, however, you dislike the somewhat pronounced taste of Largemouth Bass, skinning is the answer. In addition, if the presence of bones is undesirable, filleting should be added to the preparation.

If you choose just to scale your fish, select a large, rather dull knife or, better yet, a soup spoon from your kitchen drawer. Place the fish on a firm surface and hold it to this surface by placing the fingers of one hand firmly on the tail. Strong but moderately short fingernails are helpful in obtaining a purchase. Now, right ahead of the tail, scrape with short, quick strokes up toward the head (fig. 10, step 1). The scales should come off quite easily from a fresh fish and, if you are using a spoon, they will not fly up in your face. After each side is fully scaled, you will have to work over the areas on the dorsal and ventral surfaces between the fin groups to remove those few remaining scales. Next, remove the head by cutting down just behind the pectoral fins and then insert the tip of a sharp knife into the vent and slit the entire belly (fig. 10, steps 2 and 3). A few well-controlled scrapes of your knife in the body cavity should remove all of the internal organs. All that is left to complete the cleaning are cold running water and a few well-placed scrapes with the thumbnail along the underside of the backbone to remove the kidney tissue.

SKINNING

If you wish to skin your fish instead of scaling it, perform the beheading and cleaning functions as described above first. Then slit the skin from the vent backwards to the tail and also down the top line (fig. 11, step 1). Pass the knife on either side of the dorsal fin so that the skin is free from this major attachment area. At this point you may choose between two methods of skinning. One requires hot water, which may come from either the tap or a pot on the camp stove or fire. It should be just beyond the heat that your hand can tolerate for a few seconds. Hold the fish with a fork and let the hot tap water run over the skin on both sides for about 30 to 45 seconds each; or else hold each side of the fish just under the surface of the pan water for the same length of time. This heat treatment loosens the skin to the extent that it can eas-

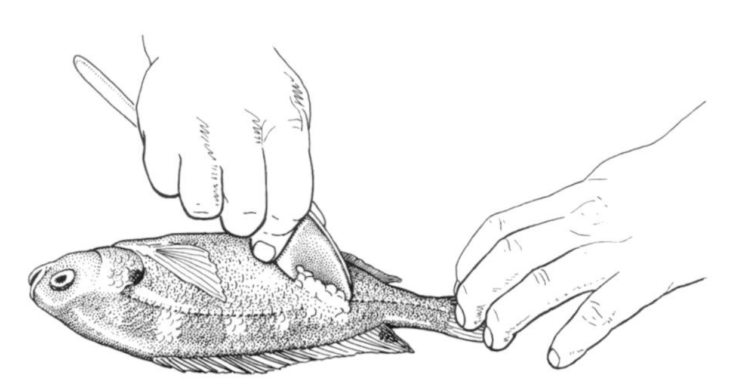

Step 1. Move spoon in quick strokes against scales.

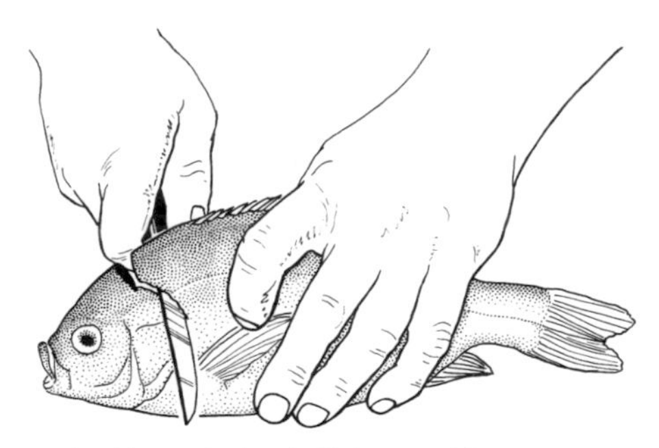

Step 2. Remove head by cutting just behind pectoral fins.

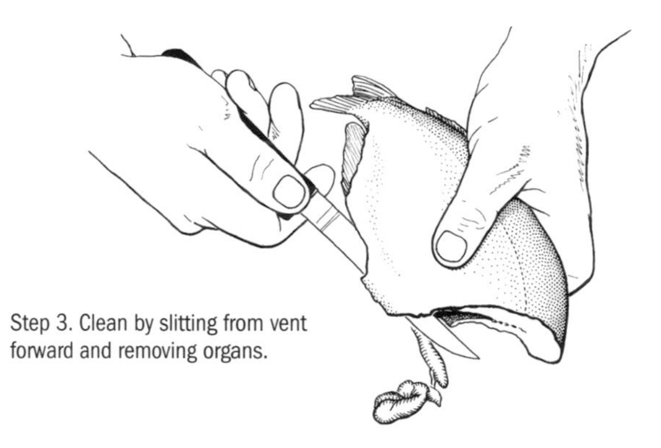

Step 3. Clean by slitting from vent forward and removing organs.

Figure 10. Scaling, beheading, and cleaning a fish.

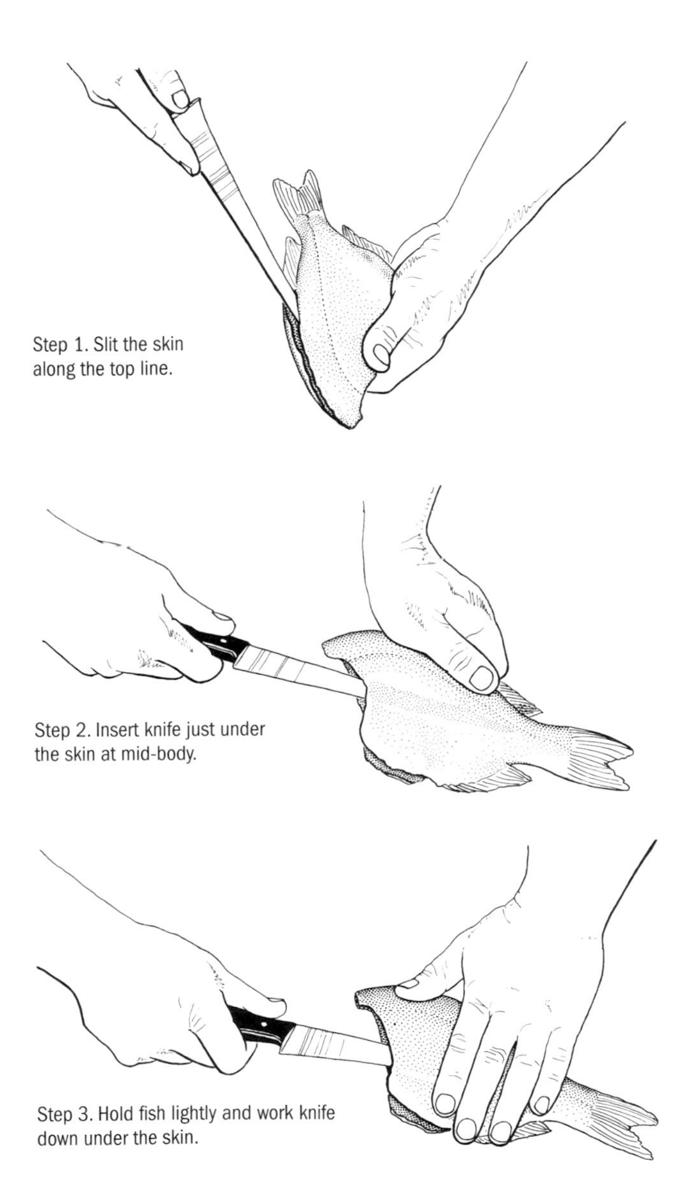

Figure 11. Skinning a fish with scales.

ily be peeled off using either the thumb and finger or a pair of fine pliers for a purchase on the skin edge. An added bit of heat treatment may be necessary for very tough-skinned fish.

If hot water is not available, you will have to skin entirely with a knife. As in filleting, the knife design is quite important. It should be at least 5 or 6 inches long, very thin, and very sharp. Most fishing tackle or recreational supply stores carry such knives at a reasonable price.

Again, begin with a beheaded fish whose skin has been slit as described under "Skinning." Insert the point of the knife under the skin anteriorly at mid-body, just beneath the lateral line (fig. 11, step 2). Continue to work the thin knife down the length of the body at midline just under the skin. You should be able to see the outline of the knife under the thin skin. Once the knife is inserted to its total length, place the palm of your free hand slightly on the upper side of the body and carefully work the knife under either half of the lateral surface while keeping the blade edge angled slightly upward towards the skin (fig. 11, step 3). This slight upward angle is very important because it prevents the sharp knife from cutting down into the flesh. The skin is much tougher than the flesh and can withstand a fair amount of pressure from the angled knife edge before it breaks. What does break away is the connective tissue that holds the skin to the flesh. When you reach the slit top or bottom area, free your knife, return it to the center line, and remove the other half of the skin. If the fish is longer than the knife, you will have to insert the knife point again at the midline under the remaining skin and repeat the same actions.

There are variations of this basic technique, and you may want to experiment with several before settling on one. All variations entail moving a thin, sharp knife under the skin at a slightly upward angle. One word of caution: the knife must be sharp to do the job properly, so be careful! Keep your skinning operation confined to the fish!

FILLETING

The decision to fillet can be made after the fish is either scaled or skinned. A good filleting knife should also be rather long, but may be a heavier and thicker instrument than the skinning knife. Because filleting involves cutting through all of the ribs near the vertebral column, the more delicate skinning knife should not be used, as it will be dulled in one filleting operation.

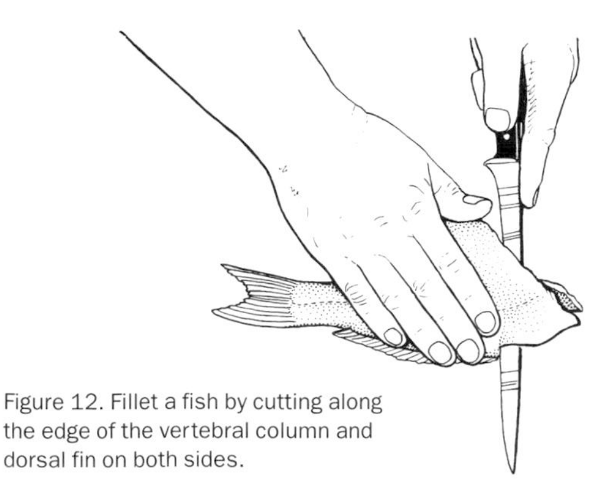

Figure 12. Fillet a fish by cutting along the edge of the vertebral column and dorsal fin on both sides.

To begin, again place the fish carcass on a firm surface and place the palm of your free hand firmly on the upper side. You will have to press down rather hard during the cutting, since it is only this pressure which holds the fish in position. Locate the vertebral column at the cut, anterior end of the fish, and begin your cut so that the knife blade passes just along the edge of the vertebrae and the dorsal fin (fig. 12). The portion of the knife nearest the handle will pass through the cleaned body cavity until you reach the vent, at which time it should pass just above the anal fin. Throughout the entire cut, keep the edge of the blade angled slightly downward toward the vertebral column. This will prevent the knife from cutting up into the flesh. If you angle downward too much, you will feel much resistance as the knife is forced into the spinal column instead of through the thin ribs. This resistance will tell you that it is time to turn your blade upward just a little before continuing to cut. Filleting differs from skinning in that you cannot see most of your knife moving through the flesh; thus filleting is much more a matter of feel and touch than of visual inspection. Once the flesh from each side has been removed, the only remaining bones are the ribs, which were previously severed from the vertebral column. To remove these, lay the fillets skin side down on the table and work your skinning knife just under the ribs, starting at the upper side of the body. In small fish, it is sometimes much simpler just to cut out the entire rib area, since the amount of flesh between the ribs and skin is very slight. This method, like the knife-skinning process, has a number of variations. You may

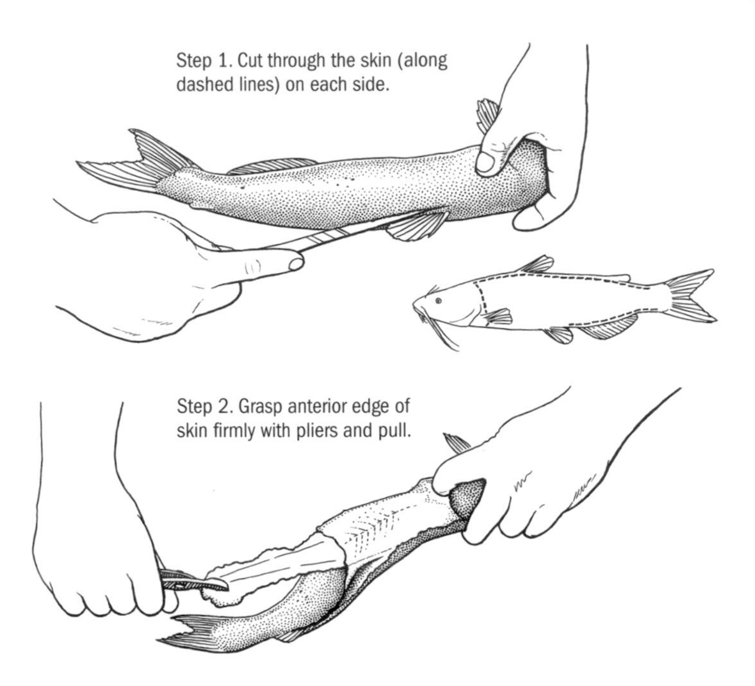

Figure 13. Skinning catfish and bullheads.

want to experiment. Don't be discouraged if your first few attempts leave too much flesh along the spinal column. This will be corrected with practice, and very soon you should be producing fillets just like those you see at the fish market.

Catfishes and Bullheads: A special set of instructions must be given for catfishes and bullheads, for although the filleting process is the same the skinning process is entirely different. First of all, do not begin by removing the head or cleaning the fish. Just kill it by the recommended procedures. Next, continue the cut that you made to sever the spinal cord completely around the fish, but this time just under the skin (fig. 13, step 1). At this point you will need another tool—a pair of pliers with tips that close firmly together. Hold the fish by placing your first two fingers just behind the spined pectoral fins while holding the head in your palm. With the pliers, obtain a good purchase on the cut edge of the skin along one midline of the body and slowly but firmly pull toward the tail (fig. 13, step 2). The skin of the ictalurids is much tougher than that of the scaled fishes, and it will usually pull free from the entire length of the body without tearing. If the skin tears, grab the torn end that still remains on the flesh and pull again. Then repeat the same process on the other side. Once the fish is completely skinned, remove the head, clean the body cavity, and, if desired, fillet. Except for lightly scaled fishes, such as the trout and salmon, the catfishes and bullheads are among the easiest of all fishes to prepare for the pan.

Caring for the Catch

If you take a fishing vacation or a fishing weekend, you may be eating your catch within a few hours or even a few minutes after you take it off the hook. Or you may travel a hundred miles or more to your fishing spot and drive home with your catch. The flavor of the fish you catch at your campsite can seriously degrade in the few hours between catching and eating, but a properly cared-for freshwater fish can be as good as if the fish just came out of the water. It all depends on how you treat the fish between the time you take it off the hook and the time it goes to the table. Follow the recommendations and detailed instructions for cleaning a fish above and always kill and gut your catch as soon as possible. If they are biting hard and fast, gut the fish during the first lull. Remember, if your fish are of a species to which a size limit applies, leave the heads on until they are safely home. Otherwise, suit yourself. Many panfish enthusiasts believe that the fish should be prepared with the head on—but head-on fish do take more room in the storage bags and in the pan.

Bacteria are the main cause of deterioration of fish flesh, and their multiplication is slowed by chilling the fish. If you are fishing away from your ice chest, gut and scale immediately and keep the fish cool by covering them in damp sacking or grass. Don't let them sit in the sun! The wise way to fish is to take a soft-sided insulated pack for your lunch, your drinks, and your fish. A few small gel-ice packs will do an amazing job of keeping things cold, and cans of chilled beverages help, too.

Cleaning your fish as you catch them can make the difference between a wonderful fish dinner and a mediocre one. A gutted fish will chill through faster, and fish flesh easily picks up the offensive odors generated by the multiplying gut bacteria. Scaling the fish will also help it chill and will remove some of the slime in which bacteria thrive. Onboard cleaning is strongly recommended. You can easily assemble a simple cleaning kit to use on a boat. These will do for a start: a cardboard box and/or newspaper for the cleaning table, a big spoon or a table knife with a serrated blade for scaling, and a knife for gutting. Your kitchen sink for rinsing is all around you! Use the top of a hard-sided ice chest or a boat seat as a table to support the cardboard or newspapers. A pair of inexpensive but strong scissors is handy for clipping fins. Don't forget to include plastic bags!

Place the cleaned fish in a plastic bag, seal it water-tight, and bury the bag in ice to chill. If you have caught a fish with stiff, spiny dorsal fins, such as Largemouth Bass, Striped Bass, or sunfish, clip the fins close to the body so they don't puncture the bag and let water in. When fish flesh absorbs water, bacteria soak into the exposed flesh with the water and deteriorate it fast, and bacteria multiply even at ice water temperatures.

If you don't want to clean your fish in the boat, you can keep many species alive for a short time on a stringer or in a live net—but trout and salmon will die from lack of oxygen. If you tow the live box or stringer with a motor boat, some of the fish will die, and nothing deteriorates faster than a dead fish soaking in warm lake water. An aerated onboard live box is a must if you plan to spend a day fishing from a boat and don't want to clean the fish as you catch them. Trout and salmon have a high oxygen requirement and may not live long even in an aerated live box, especially if the water warms up. Clean the fish as soon as you land. Many boat launch sites now provide fish-cleaning tables and even running water.

CUTTING UP FISH

Consider the size of the cooking pan or freezer shelf when deciding whether to leave fish whole. Some of the best meat is just behind the head and on the cheeks (some consider the eye a great delicacy also) and is lost when the fish is beheaded, but small fish often won't fit in the frying pan unless the head and tail are removed.

Fillets and Split Fish: A fillet is the boneless meat from the side of the fish. You can fillet almost any size fish, even small panfish such as bluegill, whose tiny fillets are delicious pan-fried. The skin helps hold the fillet together and keep it moist, and leaving the skin on fillets of good-tasting fish intensifies their flavor. Removing not only the skin but the fat belly strip from a fillet of a strong-flavored or very oily fish will reduce "fishiness." A split fish is like a fillet with ribs, sometimes held together by backbone. A fillet cooks quickly and evenly because either side lies flat in the pan, but split fish require more cooking time because the ribs arch away from the pan.

Steaks: A steak is a cross-section or crosswise slice of a whole fish, usually about an inch thick. Steaks usually include the backbone and ribs. Steaks can be cut only from a quite large fish but you can cut one steak from each of a catch of fish of borderline steak size to collect a meal of steaks, leaving the remainder whole as a "tail" or cutting it into fillets, split fish, or chunks. Remember that a steak need not be so large that it makes a whole serving. Two or three small steaks taste as good as a huge one.

Slashed Fish: "Slashing" is used to make whole thick-bodied fish (Largemouth Bass, Striped Bass, or large catfish, for instance) or thick pieces cook more quickly and evenly and to make the cooking sauce penetrate the fish more thoroughly. Slashing and grilling a bony fish such as a large nonprotected cyprinid, a sucker, or a Northern Pike makes it easy to deal with the bones.

During a fish collecting trip to the upper Amazon River several years ago, my guide prepared many excellent meals of equally bony large tropical fishes as follows. He first scaled and gutted each fish. Then, with a long, sharp knife, he scored the fish at right angles to the backbone, keeping the cuts shallow so that no bones were cut. These cuts were spaced about a half inch apart, creating strips of boneless fish meat still attached to the body. The fish were then grilled on both sides, and the crisply browned, succulent strips were peeled off with the fingers and eaten. Since then I have applied this technique to both the Sacramento Sucker and Sacramento Pikeminnow with good results.

"Slash and grill" is not only an excellent way to cook bony fish but a good way to cook any fish in a camp or outdoor situation, especially if you have forgotten the silverware! Slashing a fish also saves precious camp fuel, as the strips on the slashed fish cook more quickly than a whole fish would.

Chunks: Salvage the rest of a fish from which steaks or fillets have been cut by cutting out boneless cubes and chunks, which make excellent fish kabobs, tacos, curries, soups, and stews. Indeed, chunks are so useful that large morsels cut from farm-raised catfish are now a standard item at many supermarket fish counters.

STORING THE FISH YOU CATCH

A large catch or a catch of one very large fish presents a storage problem. To keep part of a large catch or a large fish for up to two weeks, you can hot-smoke the fish to eat hot or cold. Freezing is the most popular and most practical long-term preservation method, and a properly frozen fish can retain much of the quality of a fresh one.

FREEZING FISH

For good results, fish to be frozen must be

1. properly cared for when caught;
2. held chilled, preferably at 32°F to 34°F, until prepared for the freezer;
3. frozen as soon as possible after bringing the catch home;
4. completely protected from air in the freezer container;
5. frozen as quickly as possible once placed in the freezer;
6. held at no more than −2° (check with a freezer thermometer);
7. held for no longer than four months, preferably for only two months.

Small fish can be beheaded and frozen whole or split. Large fish should be cut into fillets or steaks for freezing. Once the fish is packaged, the packages should be spread out on a freezer shelf in one layer so that below-zero air circulates around all sides of the fish and freezes it quickly. For better protection and for convenience in finding small packages, group similar-sized packages into a large standard (twist-tie or zip-sealing) freezer bag and label the outer bag clearly with the kind of fish, the cut, and the date.

No matter how carefully fish is frozen a few areas of freezer burn, whitish or discolored spots caused by contact with air, may occur. Tiny spots may be ignored, but cut away any bite-sized patch before cooking.

Vacuum-Packing is one of the best and easiest ways to freeze fish at home. The sealing machine pulls air out with a vacuum pump, then heat-fuses the open end. If you have a large amount of fish to freeze, many freezer locker businesses will vacuum-pack a catch for you, using their powerful sealers and extra-heavy bags. Most offer "quick-freezing" at a very low temperature as well. You will pay for quick-freezing, so make sure the place is really quick-freezing the fish, not just throwing it all into a basket to freeze at standard locker temperature.

Double Wrap: This method of freezing fish pieces is almost as effective as vacuum-sealing. Fold plastic "shrink wrap" around the fish with enough overlap to make the wrap secure. Tape if it seems necessary, then arrange one or two layers of wrapped fish pieces in a zip-sealing freezer bag. Press as much air as possible out of the bag as you "zip," or put a drinking straw in the last small opening and suck air out as you complete the closure. The combined wraps give very good protection, but it is best to use heavy freezer bags for the outside layer. Again, don't forget to label the outer bag with date, species, and cut of fish.

SHOPPING FOR FRESHWATER FISH

As one marine fishery after another continues to collapse due to overfishing, farmed freshwater and anadromous fish will become more and more common in the markets and perhaps even the dominant fish to be found on the fish counter. Most states prohibit the selling of sport fish species and this also will drive the farming of more and more species of fish. At the present

Plate 101. A wise shopper selecting a fresh fish.

time, farmed Rainbow Trout cultigens, Atlantic Salmon, Channel Catfish, and several species of tilapia can be obtained in most markets. Farmed Grass Carp or Mirror Carp and newly farmed species such as White Sturgeon, Striped Bass, and Largemouth Bass are beginning to appear in specialty markets and can be expected to become more commonly sold. Several other nongame freshwater fish species such as Sacramento Blackfish and Common Carp are caught in the wild and legally sold in markets. There are also a few species that are protected in California but can be legally imported from other states. For instance, Coho Salmon and Steelhead from the Pacific Northwest may be sold in California if taken by Native Americans from their tribal lands.

Occasionally, you may find eastern or midwestern freshwater fish species in the markets, especially Walleye and several whitefish species, some of the very best-eating freshwater fishes. These are among the traditional white-fleshed freshwater fishes served during the Passover and Easter season, and they appear briefly in California markets at this time. Regardless of your religious affiliation, Walleye and whitefish are two types of fish you shouldn't pass up if you're lucky enough to find them.

Another most delicious freshwater fish comes all the way from Vietnam. "Basa" (a Vietnamese name despite the Spanish appearance) is a type of catfish farmed in the Mekong River. It is processed in Vietnam; the Basa in the fish market trays has been imported as IQF (Individually Quick Frozen, in nitrogen) fillets. Basa has the firmness of catfish and a pleasing flaky texture, without the musty or muddy flavors found in the eastern-farmed catfish or tilapia sold from California fish counters. In fact, Basa is so good and so competitively priced that when it was originally marketed in the United States as "catfish" or "Basa Catfish," sales of southeastern-farmed American catfish plummeted. Southern catfish farmers complained to their senators, the senators complained to the USDA, and the USDA enforced an old rule stating that only members of the North American catfish family, Ictaluridae, could be sold as "catfish," thus forcing the Basa vendors to revert to the much less recognizable Vietnamese name.

FINDING A FRESH FRESHWATER FISH

Finding a good fish to take home for dinner is not easy. Some supermarket fish counters have very acceptable fish but others are more appropriately described as "Death Row." There is also a tremendous range of quality among specialty fish stores. If you're lucky, you will find a market that generally has acceptable fish and get to know the owner or counter person. But no matter where you shop, you need to learn to recognize a fresh, well-cared-for piece of fish. Let's start with the most common case for freshwater fish, a whole gutted fish. Evaluate a good fish by the points on this checklist.

1. Look the fish in the eye. The eye of a truly fresh fish is rounded, clear, and glossy, not clouded, scummy, and sunken. *Note:* Gloss can be put on by the fish market, and probably has been, if every single fish in the case has a brilliant shine regardless of cut or species.
2. Look at the gills, if they have not already been removed. Like the eye, gill filaments reveal deterioration faster than muscle tissue and fish markets often remove them for this reason. The gill color varies somewhat with the fish but the gills should be bright red and the gill segments should be well defined and separate. Very pale pink gills or sunken, clotted-together gills are a bad sign.
3. Touch the fish; bring a disposable glove or a plastic bag to cover your hand if you wish. Service counter markets rightly prefer not to have the customers pawing the fish and will be happier with the bag procedure. They may even give you something to protect the fish from your germs. Give the fish a gentle push in the back flesh, using the tip of your finger. Does the flesh feel somewhat elastic and does the little dimple you made spring back quickly? If not, the fish may be several days old. A sticky, slithery feeling is another bad sign, indicating that the natural mucous covering of the fish has become a bacteria culture. If the fish is packaged, the tight-stretched wrapping may give a stale fish a deceptively firm feel, so pay careful attention to the other criteria.
4. Next, smell the fish. If you're at a service counter and the salesperson doesn't feel that this is within the rules, smell the hand or glove with which you were prodding the fish. Old fish smells like, well, old fish. Or it may reek of ammonia. Fresh fish has almost no odor or a delicate watery odor reminiscent of a cool mountain stream. By the time a fish has developed ammonia in the tissues, it is seriously old and nothing you can do will make it taste good. A fish with just a slight fishy

smell, but which has a good eye and is firm to the touch, may be perfectly good to eat. Surface moisture or the mucous coating can deteriorate a bit without really affecting the deeper tissues.

5. If you are choosing among packaged fish, as a last test nick a tiny hole in one corner of the wrap and sniff. No doubt the market hates this and I am not suggesting that you punch open every package, but if you're actually intending to buy the package you punch I consider it fair enough. Maybe finding punched and rejected packages will encourage the market to mind its fish better. If breaking open packaged fish bothers your conscience, you could purchase the fish, sniff before leaving the check stand, and attempt to return it if it is not satisfactory.

The freshness of a fillet or steak will have to be evaluated by other criteria. It's often easier to judge a cut fish than a whole one, because the skin of a whole fish can disguise the condition of the flesh.

1. Look at the texture. Most fish should have the muscle flakes well knitted together, not falling apart with big spaces between the flakes. A loose-flaked cut of fish might be old, or it might have been frozen and thawed for some time. Markets may not be willing to reveal that a piece of fish has been frozen, and there's no need to pay a premium for "fresh" fish unless it has never in fact been frozen.
2. Look at the color. Fish flesh darkens and dulls as it ages and though different species have different appearances, most fresh fish will look pearly and somewhat translucent, white or pink-shaded in the major muscle. *Note:* Farmed salmon is bright orange because it can legally be "dyed" by feeding carotenoid pigments. In California, farmed salmon must be labeled "farmed" and dyed salmon must be labeled "color added" or "dyed" as well. The color does not fade no matter how stale the fish is, so you have to judge its freshness by the other tests.
3. Smell and feel the fish if possible. If the salesperson doesn't want you handling the fish, ask that a small sliver be cut for your inspection.
4. If the fish is prepackaged, check the liquid in the container. Most fish is packed on one to several absorbent pads, but if there is a lot more juice than the pads can soak up, the fish has

been sitting around for too long and may have been frozen before packaging as well.

5. Finally, when you've found a possibly acceptable package, poke a hole in the plastic wrap and smell the fish.

TRICKS FISH SELLERS USE

There are many tricks for making old fish seem fresher, and some I have detected while shopping for fish are listed below.

1. Frozen fish is supposed to be labeled "previously frozen" in this state, but neither service counters nor package counters are necessarily to be trusted in this respect. They may sell thawed, previously frozen fish without the "previously frozen" label or even labeled as "fresh." Make your own decision based on the above tips for selecting fresh fish. Commercially frozen fish can be perfectly acceptable, but not when it has been sitting thawed on a fish counter for perhaps several days while being offered at fresh fish prices.
2. The shiny appearance of the fish laid out in a service counter case may be applied from a jar.
3. Some stores will use borax to neutralize the smell of old fish and to whiten and firm it. Borax has an alkaline, soapy taste if more than the tiniest amount is used, and the fanciest fish markets as well as the cheapest supermarkets have occasionally fooled us with fish so saturated with borax that it was inedible. Borax treatment for fish and shellfish is considered acceptable and even desirable in Asian cuisines, and it's quite probable that any extremely white piece of fish you see in an Asian market has been "boraxed."
4. Don't take a fish seller's word for when a fish was caught, where a fish was caught, or even what it is, until you know the seller well. Use your eyes, nose, touch, and judgment no matter what you are told about the fish.

Ask, "When did this fish come in?" and you will receive an invariable, "Oh, just this morning." I once asked this question on three successive days, pointing to the same fish at the same market but with a different person behind the counter each time, and miraculously the fish was always fresh that day!

Fortunately, I had a much different fish-shopping experience a few years ago. Imagine my delight when upon my first visit to a well-appointed fish market in Berkeley, California, I saw not only good-looking fish but a familiar face on the other side of the counter. It was that of Ted Iijima, a former student in my ichthyology course and now the proprietor of what I believe is the best fish market in the greater Bay Area or beyond. What makes it and a few others like it so good is that Ted knows where most of his fish come from and how they are handled between the fishing line or net and the loading dock of his store. For instance, he has one long-time supplier, a husband and wife team, who catch Chinook Salmon on their 22-foot boat north of San Francisco in the morning, clean them within 20 minutes after capture, and then deliver them on ice to his market that afternoon. By contrast, many larger salmon boats don't clean the fish immediately, and when they do clean them, they are tossed into an ice and brine slush where they remain for up to a couple of days or until the boat comes back to the dock.

I doubt if my lectures on the finer points of ichthyology had much if anything to do with Ted's success in the fish market business, but I do know that my weekly chats with him over his fish display counter have been a real education for me. I strongly encourage any frequent fish buyer to seek out that good market and then get to know the person or persons who operate it.

COOKING THE FRESHWATER FISHES OF CALIFORNIA

BEFORE COOKING

Most freshwater fish can, and should, be eaten as soon as possible after catching, but cuts of sturgeon and some other large fish may be tough unless they are held for a day or two. Keep fish in the coldest part of the refrigerator—usually the meat compartment. To keep fish fresher tasting, cut a lemon, squeeze slightly, and rub the cut side over the fish. You may smooth on a very few grains of salt or drops of soy sauce with the lemon. Place in one layer on a plate, cover lightly with plastic wrap, and refrigerate promptly. Hot-smoked fish can be stored in the meat compartment for up to two weeks if carefully handled. No cooked fish freezes well, including hot-smoked fish.

Plate 102. Brown (above) and Rainbow Trout (below) cleaned and ready for the kitchen.

Plate 103. Brown Trout pan-fried on a camp stove.

WAYS TO COOK THE CATCH

An angler can't predict how many fish or what species he will bring home, and shopping for fish is hardly more certain. Each of the "master methods" described below can be used to cook many kinds and cuts of fish. A few sample recipes which illustrate ways to use these cooking techniques follow most methods. The chart below matches master methods with types of fish to help you choose how to cook your fish. The "ratings" are necessarily subjective and tastes vary—you may find some low-rated combinations quite acceptable. Please note that a fish rated with a "0" for one method can have the highest (★★★) rating for another way of cooking. Some fish are untested—don't hesitate to experiment with combinations marked "?"!

COOKING YOUR FISH

The delicate flavor of a good fish deserves to be complemented with high-quality, fresh ingredients, and even muddy or strong-flavored fish may be made more palatable with assertive spices,

Ways to Cook the Catch

Fish type examples	White-Cook	Sauté	Sauté-Poach	Pan-Fry	Bake	Grill	Smoke/Pickle
Panfish: Bluegill, Sunfish, Perch	?0	★★★	★★	★★★	★★	★★★	★★★
Trout	★★	★★★	★★	★★★	★★★	★★★	★★★
Bullheads, small catfish (cold water)	?	?	★★★	★★★	?	★★	★★★
Salmon, Steelhead	★★★	★★★	★★★	0	★★★	★★★	★★★
Large, mild-flavored fish: Striped Bass	★★	★★★		★	★★★	★★★	★★
White Sturgeon	?	★★★		0	★★★	★★ 1	★★★
Large catfish (cuts)	?	★★★		★★★	★★	★★	★★★
Strong-flavored fish: Largemouth Bass	0		★ 1	★	★	★ 1,2	?
Carp, Pikeminnow, suckers from cold water	0	?	★★ 1	★	?	★★ 1,2,3	?
Cyprinids, suckers from warm water (don't try)	0	0	0	0	0	0	0

(1) Cook with sauce or flavoring vegetables. (2) Cook in foil. (3) Try "Slashing"!

herbs, and flavoring vegetables. All the ingredients used in this chapter are standards in my kitchen and most are carried by any supermarket. Whole spices are cheaper and fresher in the big discount supermarkets (look for words like "Pak," "Maxx," "Less"), because they cater to a diversity of ethnic groups. You can find many of the same foods that you would find in Mexican, Asian, Indian, or Near Eastern markets in the discount supermarkets. Look for natural yoghurts and cold-pressed oils in health food stores, specialty markets such as Trader Joe's, and supermarkets that emphasize natural foods.

Ghee and clarified butter may be the least familiar ingredients used in these recipes. These butters have had milk residues removed so that they withstand high cooking temperatures, store well, and remain semiliquid at room temperatures. The delicate flavor of a fish fried in pure butter can be achieved only by cooking in one of these special butters. It is simple to make a supply of either to keep on hand, and directions for making both are given at the beginning of the "Special Ingredients" section, below. You can buy ghee and sometimes clarified butter in discount supermarkets, specialty markets, and Indian and Near Eastern markets.

Cooked fish are fragile. Serve fish dishes in their cooking pans so that they arrive at the table whole and hot. Most techniques and recipes given here were tested in a heavy 7- or 9-inch frying pan, which will hold a pound of most cuts of fish plus sauce or flavoring vegetables—two servings for big eaters. The methods do not for the most part assume a specific quantity of fish.

For further exploration of fish-cooking techniques, and for specific recipes for additional fish dishes and salads and side dishes to serve with fish, Julia Child's brilliant general cookbook, *The Way to Cook,* has 80 pages devoted to fish and accompaniments. The recipes and techniques presented are marvelously uncomplicated. The techniques and recipes for pureed fish dishes in Volume I of *Mastering the Art of French Cooking* (Child, Beck, and Bertholle) are especially useful when dealing with a catch of large bony fish such as Northern Pike or carp, and the large section on sauces includes many fish sauces. Don't forget to check general cookbooks in your own collection. Irma Rombauer's *Joy of Cooking* ("Classic" editions), the first cookbook of millions of Americans, has many good fish recipes. Cookbooks have not kept pace with the collapse of marine fisheries and the accompanying increase in farmed freshwater fish in markets. Use the chart to adapt ocean fish recipes to freshwater fish. Freshwater fish is preferred in Chinese cooking, but few freshwater fish species have been available in general markets until recently, and many otherwise authentic Chinese cookbooks specify ocean fish substitutes. Recipes can now be re-adapted to use freshwater species. Again, use the chart to choose a species to fit Chinese cooking techniques.

At last we're ready to start cooking the catch! The first step in most cooking is to "seal" the surface of the food, and in the first method, white-cooking, surfaces are sealed by immersing the fish in boiling water.

Method 1: White-Cooking

In white-cooking, a fish or piece of fish is dropped into a large pot of boiling water, then allowed to cook in the slowly cooling water (pl. 104). The result is a firm fish with all its natural flavor intact. White-cooked fish is delicious hot, and since no fat is used in cooking, the fish can also be served cold or used as an ingredient in salads or pasta dishes.

Special Advice: Large trout and salmon tails are especially good cooked by this method, as are firm steaks, large firm fillets, and gravlax (see Method 6). Most fish can be white-cooked with or without skin. For small panfish or small filets, sautéing

Plate 104. The author's wife, Molly, and son, Ross, "white-cooking" salmon.

(Method 2) or pan-frying (Method 3) may be better. White-cooking won't conceal or dissipate "off" flavors. Cook strong or muddy-tasting fish such as Largemouth Bass or catfish that have been living in warm water by another method.

Before you begin, remove large or thick cuts of fish from the refrigerator and bring to room temperature.

YOU WILL NEED:

1. A large pot and lid, such as a spaghetti pot. It takes 4 quarts (1 gallon) of water to properly cook 1 to 1½ pounds of fish, and more won't hurt. A large or thick fish requires from 6 quarts to 2 gallons of water.
2. Salt for the water. A pure salt such as sea salt or pickling salt is best.
3. Seasonings for the water (see list below).
4. Spatulas, wide strainer, or plate for lifting the fish out of the water.

BASIC PROCEDURE

Bring to a full rolling boil in a covered pot:

at least 4 qt water

Add to the boiling water:

a big handful of sea salt
seasoning combinations

Slide the fish into the water and if you are white-cooking a fillet or steak, immediately turn the heat off, cover the pan, and let fish and water cool to serving temperature. A very thick piece such as a salmon tail needs a little more heat. Watch closely as you let the water come almost to a boil again, and turn off the heat as soon as a few tiny bubbles begin to rise. Let fish and water cool in the covered pan as above. That's all there is to it!

The timing for white-cooking is very flexible, as the fish can remain in its water for many minutes without overcooking. A fillet or steak may cook in only 10 minutes, while a thick whole fish or a thick piece of fish will take 30 minutes or more.

When ready to serve, put a serving plate near the pot and lift the fish out carefully, supporting as much of it as possible on two spatulas, a very wide or long spatula, or a plate. (A Chinese woven wire lifting basket is one of the best tools.) Slip the plate under the fish, hold it on the plate with a spatula, and tilt plate and fish together, lifting slowly while liquid drains from the plate. The water will usually have cooled enough so it won't burn you, but test by feeling the side of the pot. Slide the fish onto the serving plate, and again tip any liquid from the plate, or blot with paper towels, before garnishing and serving.

To serve cold: Let the fish cool in the broth, remove to a plate, cover, and refrigerate. White-cooking preserves gelatin elements, so that white-cooked fish is particularly firm and good served cold.

Rescues: You may find that a fatty, thick piece—for instance, a salmon tail or midsection—has not cooked through. Test a thick piece after about 10 minutes: stick a paring knife in a thick area and twist to open the flesh. If an inch or more of the inner meat is raw and translucent, reheat the water to a near boil, turn the heat off, and finish as before. You can make a deep slit along the thickest part to speed cooking. Or remove the fish from the pot to an ovenproof plate, slit, stuff the slit with butter, and finish in a low (250°F) oven. Avoid the problem in the future by using a bigger pot of water and wrapping a towel around the pot after it is removed from the heat to slow cooling. With experience, you'll have to rescue an undercooked fish only rarely.

SEASONINGS FOR WHITE-COOKED FISH

The heavily salted water is necessary to firm the fish. Add all seasonings except fresh dill to the water as you begin heating it.

1. *Scandinavian-style seasoning:* Peppercorns and dill. Add a heaping tablespoon of dill seed or dried dill to heat with the water or add several fresh dill sprigs or one or two flower heads with the fish.
2. *Chinese-style seasoning:* Smash an inch or two of fresh ginger root (peeling is unnecessary); add with peppercorns. Snip cilantro over the cooked fish and serve with one of the soy-ginger sauces (recipes below).
3. *"Crab-style" seasoning:* A good handful of premixed crab spice. Zatarain's Crab Spice, a flow-through bag of whole spices in a little box, is sold in many supermarkets. Cut the bag and use a handful of spice or briefly boil the whole bag, remove, and freeze in a plastic bag for re-use. Some fish take on a flavor somewhat like that of well-seasoned boiled crab when cooked with crab spice. The ingredients (peppercorns,

mustard seed, bay leaf, dried hot peppers, and a few bits of cinnamon stick and clove) could be also be added separately.

4. *Continental combination:* A handful each of bay leaf, carrot slices, peppercorns, celery sliced from the leaf end, parsley stems. Garnish with parsley.
5. *Lemon slices or peel:* Add to any combination above.

SAUCES AND CONDIMENTS FOR WHITE-COOKED FISH

The butter sauces below are made with plain salted butter.

1. *Melted butter:* Pour generously over a hot fish, which may be sprinkled with minced dill or parsley.
2. *Lemon butter:* For two to four people, heat half a cube of butter until it begins to brown, add juice of half a lemon, and heat together briefly. Serve in the pan and stir the mixture as it's dipped out. Keep warm over a candle warmer if you have one; you can then add more butter and lemon at the table.
3. *Lemon wedges* should always be served with white-cooked fish, alone or as a backup to other sauces.

Soy-Ginger Sauces for White-Cooked Fish

These low-calorie pan-Asian classics make a nice change from butter sauces. The variations (and names for them) found in different cuisines, different cookbooks, and different restaurants are innumerable, but the basis is the same: a tart liquid, soy sauce, and ginger. Toasted sesame oil or seed is a typical ingredient, and the sauce is often sweetened. By choosing different assortments of ingredients from the suggestions below, you can make at least two dozen different sauces reflecting the styles of several countries! But taste as you add and don't add so many contrasting ingredients that the flavor is muddled. Soy-ginger sauces keep for weeks if refrigerated, but add fresh cilantro or other herbs at serving time.

For a starter dip sauce, mix:

½ cup tamari or other Japanese soy sauce
¼ cup rice wine vinegar, or to taste
1 inch fresh ginger root, peeled, grated
[optional: sugar to taste]

For enhancements, add:

a few drops Chinese-style toasted pure sesame oil
(or 1 tsp pan-toasted sesame seed)
1 clove of garlic, smashed
a few cilantro leaves
finely cut scallions or chives

Vary the character of the sauce by changing the basic ingredients. Try a light-colored Chinese soy sauce, or replace all or part of the vinegar with citrus juices (lime, lemon, orange, sour citrus, or any combination). Substitute sherry or Chinese Shao-Sing wine for part or all of the tart liquid for an approximation of the "Special Sauce" or "Special Wine Sauce" served in dim sum houses.

Serve any of your soy-ginger sauces in a pitcher or in individual "bit dishes" for dipping. Plain mayonnaise is a traditional sauce for cold fish and there are innumerable recipes for flavored and enhanced mayonnaises to serve with fish. It's easy to "enhance" a mayonnaise: for example, scissor dill, chives, or other fresh herbs into an unsweetened mayonnaise. Tartar Sauce (see "Universal Accompaniments," below), another enhanced mayonnaise, is very easy to make, and delicious.

ACCOMPANIMENTS FOR WHITE-COOKED FISH

Serve light side dishes with white-cooked fish. Many accompaniments can be prepared while the fish is cooking, as white-cooking takes so little of the cook's time. Boil small peeled potatoes and shake with butter in the Scandinavian style or serve rice or a small pasta like orzo or tiny shells. Add a lightly dressed salad or a green vegetable. Recipes for salads, salad dressings, and rice are in the "Universal Accompaniments" section, below.

BLUE TROUT AND BLACKFISH

Blue Trout? *Truite au Bleu* may be the most famous trout dish of Europe and is surely the most mysterious. The characteristic bluish appearance and body curl result from the trout being kept alive until moments before cooking, when the trout is quickly knocked on the head, gutted, and dropped into boiling water. The mucous layer of the skin, kept intact by skillful handling, gives the trout its vaguely blue tint and the curl, the ultimate proof of freshness, occurs when muscles contract. Any cook who can keep a trout alive until the water is ready can make a Blue Trout, and white-cooking makes a better Blue Trout than the traditional "poaching" in simmering vinegared water. A cook living near a catch-it-yourself trout farm could purchase and transport trout in an aerated live box (see "Caring for the Catch," in the Dining section), and an angler could equip himself with live box, pot, and camp stove. Part of the fame of Blue Trout derived from the sweet, firm wild trout with which it was prepared, and wild-caught trout can no longer be sold in Europe or America. A determined California angler-cook could recreate even this aspect of *Truite au Bleu* by finding and fishing the lakes where Brook Trout, which maintain their populations in the wild, were stocked. Most other trout the angler may catch and keep in California are planted at such large sizes that they scarcely differ from farmed trout.

Chinese Steamed Sacramento Blackfish

The blackfish of this recipe is also cooked by water—water vapor! But while the blue of a Blue Trout has to be developed with considerable care and skill, the black of a Sacramento Blackfish is the natural color of this large, very dark colored minnow. The idea of eating a minnow may seem odd, but in California we have four cyprinid (minnow) species that grow to one-half meter or more in length, and large cyprinids, particularly carp, are eaten in Europe, Africa, and Asia. I've always had a particular interest in cooking cyprinids, and this recipe, given to me by a Chinese-American student's mother, was a feature of ichthyology class "cook-offs" for many years. The original recipe specified a blackfish but any whole fish of two pounds or more can be steamed in this way. You will need a large wok with lid and a bamboo steaming rack.

While an angler who may catch a Sacramento Blackfish is likely to discard it, it is considered one of the better-tasting minnows and is sold live in many Asian fish markets. The Sacramento Blackfish will tolerate very low oxygen levels in water and a single fish will survive for many hours in a five-gallon bucket, so it can be carried home and prepared just before cooking. Kill the fish with a sharp blow to the top of the head and remove scales and internal organs, including the heart-gill complex. If the fish can be arranged on the steamer rack with its head on, leave the head attached.

Steaming Your Blackfish: Place the rack in the wok, add water to an inch below the level of the rack, and bring the water to a rolling boil. Brush oil on the rack, then place the fish on it, cover, and steam the fish until the flesh begins to flake away from the backbone. For a two-pound fish this should take about 15 minutes.

While the fish is steaming, make the sauce. Heat together:

¾ cup vegetable oil
1 Tb chopped ginger root

Also chop:

3 green onions, green only

Season the steamed fish with the ginger oil, onion greens, and:

1 Tb soy sauce

In most Chinese cooking styles, the fish would be rubbed with seasonings and soy, then steamed on a plate so that the seasoned juices are collected as a sauce.

Method 2: Sautéing and Sauté-Poaching

To sauté is to cook quickly in a small amount of very hot fat, and sautéing is one of the best ways to cook good-tasting species. The technique I call "sauté-poaching" is an extension of the process—the partly cooked fish is finished with added liquid or sauce. Strong or muddy flavors can be reduced or masked by sauté-poaching in a highly flavored sauce. Any kind and shape of fish and any cut can be sautéed or sauté-poached, and gravlax is delicious cooked by either method.

Special Advice: The sautéing fat must have a very good flavor. Thin cuts or small fish require close attention but thick whole fish or cross-sections won't overcook easily and may even be finished in the oven.

YOU WILL NEED:

1. A heavy frying pan or shallow sauté pan; a lid which fits it tightly.
2. A spatula for turning the fish. A wire whip is helpful when adding sauce ingredients.
3. Fat that will stand heat. Butter as it comes from the package must be fortified with oil to raise its scorch temperature, but clarified butter, ghee, and oils may be used alone. Olive oil combines well with all forms of butter.

4. For poaching, white wine, vermouth, or citrus juice plus vegetable or chicken stock if more sauce is wanted.
5. Salt, freshly ground pepper, and flavoring herbs as desired: dill, parsley, chives, lemon thyme, etc.
6. A trivet to protect the table from the hot cooking pan.

BASIC PROCEDURE I: SAUTÉING

Preparing the Fish: Cook without coating, pat a *light* coating of flour onto the fish, or pat Awaze Rub (see "Universal Accompaniments"), curry powder, or turmeric onto the raw fish to flavor and color it. Flour helps hold delicate fillets together and will thicken the sauce slightly. I feel that a good fish keeps its sweetness better if not salted before cooking.

Sautéing: Heat the pan and add about a quarter inch of the fats. When very hot (butter will brown, even when mixed with oil), slide in the fish. If the fish exudes juice instead of crackling and browning when it hits the pan, the fat is too cool. Rescue by finishing with liquid. Turn the fish when it shows moisture on the upper surface (most cuts will) or lift to peek at the underside and turn when lightly browned. Pry between the muscle flakes with a fork to test for doneness. Remove pan from heat while the fish is slightly underdone—the muscle flakes should still be slightly translucent and unwilling to separate completely—then let the fish finish in its own heat.

For an "instant" sauced fish: remove the fish to a plate, pour most of the fat out of the pan, add a scant cup of white wine or vermouth, and reduce to about half, stirring to incorporate pan juices as the sauce boils. Adjust salt and return fish to pan to reheat briefly. You may baste it with sauce and finish by sprinkling with minced parsley or dill leaves or a leaf or two of fresh tarragon.

BASIC PROCEDURE II: SAUTÉ-POACHING WITH LIQUID

Briefly sauté the fish as above, just enough to seal the surfaces. As soon as the second side is sealed, hold the fish in the pan with a spatula and tip out almost all of the fat. Return the pan to the stove, turn the heat to high, and add white wine or vermouth to come about halfway up the fish. The liquid should bubble immediately. Slide a spatula under the fish so that liquid flows under it, and spoon boiling liquid over the fish. Cover, reduce heat as low as possible, and simmer for about a minute. Remove from heat before the flakes completely separate from one another. A thick piece may take ten minutes or more to cook. This is the plainest form of sauté-poached fish.

To make more sauce after removing the fish, add more wine, vegetable stock, light chicken stock, or even water. Add a six-ounce can of Thai-style unsweetened coconut milk to make a deliciously rich sauce or to make a dry-fleshed fish moist and succulent. An "extended" sauce can be thickened—stir a teaspoon of cornstarch or arrowroot powder into two to three tablespoons of any cool liquid in a small cup, then add just enough of this suspension to the boiling liquid to thicken it lightly. Add little by little, boiling and stirring between additions. Pour most of the sauce into a pitcher, leaving about a quarter inch in the pan. Return fish to pan and finish as above.

Flavoring vegetables may be added—any or all of chopped onion or scallions (green onions), celery, garlic. Seal the fish surfaces, remove to a plate, and pour off excess fat as above, then add the vegetables to the pan and "wilt" them over moderate heat, turning them until they are translucent but not yet browned. Turn up the heat, add about half a cup of wine, boil hard for a moment, add back the fish, turn down the heat, and finish as above.

Sauce Advice: A bit of orange juice will mellow a sharp sauce. Add turmeric by half-teaspoons for a nice gold color and a touch of flavor.

RECIPES

I have selected a few favorite recipes to demonstrate the flexibility and versatility of the sauté-poaching technique. They are presented in order of their increasing power to reduce or disguise strong or muddy flavors. The longer cooking and strong flavors of the ingredients may overwhelm fish of delicate texture or flavor. Quantities are for about a pound of fish.

Fish in Italian-Style Tomato Sauce

Sauté fish in olive oil, remove from pan, and pour off most of the fat as in Basic Procedure I. The sauce will cover a pound of fish generously, or lightly coat two pounds. Just add more tomatoes to stretch it.

Chop fairly finely—a food processor is quickest, but don't make mush!—and wilt in the remaining fish-cooking fat:

1 small onion
1 large clove of garlic
1 stalk of celery
½ green pepper
[optional: 2–3 Tb fresh mild chile peppers]
[optional: minced parsley to taste]

Add to the vegetables:

about ½ cup of red or white wine or vermouth

Boil hard for a moment to reduce and mellow the wine and add:

about half a large can (28 oz) of tomatoes and juice

Italian plum tomatoes are solid and flavorful. A curl of lemon peel added with the liquids gives a lively accent. Use a potato peeler to take only the yellow peel, and remove before serving.

Bring sauce to a boil, then simmer until vegetables are tender-crisp, add the fish, baste with sauce, cover, and finish by cooking for a few minutes at medium to low heat (don't stir—the fish will break up). Very firm fish can be held on the lowest possible heat for up to half an hour.

Fresh Italian herbs (basil, oregano, rosemary) snipped over the fish with scissors make a pretty, flavorful finish, but remember that fish is easily overpowered with herbs.

Despite the sauce being "Italian," I usually like rice and/or lots of crusty Italian-style bread rather than pasta with this dish.

Ranchero-Style Fish

This is a variation of the preceding recipe. Buy fresh, flavorful cumin seeds where people who appreciate them shop: from the Mexican or Indian food areas of supermarkets, especially the discount supermarkets, from ethnic markets, or from bulk sections of health food stores or supermarkets specializing in natural foods. Develop the flavor of whole cumin by toasting, and avoid flavorless powdered cumin if possible.

Heat a small dry frying pan and add:

1 tsp whole cumin seed

Remove pan from heat as soon as the cumin darkens and begins to release its aroma.

Remove the fish to a plate, pour off most of the fat, and proceed as for Fish in Italian-Style Tomato Sauce, wilting a half cup or more of diced sweet red or green pepper and some mildly hot ones with the flavoring vegetables. Add the toasted cumin seed and tomato, simmer until flavors blend, add back the fish, and finish as for any other sauté-poached fish.

Ross McGinnis's Bachelor Fish Ranchero: After toasting the cumin, add about a cup of ready-made salsa and finish as above. That's it! Ross uses only salsa and says, "It's like cooked ceviche," but others may wish to mellow the Bachelor's style with a pinch of sugar, a spoonful of tomato paste thinned with water, or a spoonful of sweet mild paprika or other mild powdered chile. A mild salsa that isn't strongly vinegary is best.

Garnish either version of Ranchero-Style Fish with fresh cilantro leaves and chopped raw onion.

Accompaniments: Serve corn or flour tortillas to emphasize the Mexican character of Ranchero-Style Fish. A Mexican-style relish plate can be prepared well ahead. Typical ingredients include fresh radishes, carrot and celery sticks, sliced pickled jalapeño peppers, and shredded cabbage moistened with jalapeño liquid. Steamed rice goes very well with Ranchero-Style Fish.

Tunisian Shermula

The next two recipes were inspired by versions contributed by noted cooking and wine expert Lon Hall, many more of whose recipes may be found on the Internet. Shermula, or Charmoula, is a traditional Near Eastern fish dish with many variations—honey and raisins give distinction to this version. Wild as the recipe seems, it's always been well received by guests and the exotic effect is achieved with common ingredients. The instructions may look complex, but Shermula is only a minor expansion of a basic sauté-poach. You may reduce the honey to a less-startling quarter cup. Some cayenne powders are very hot. You may substitute a hot paprika or other unseasoned powdered chile. In either case, adjust to taste by quarter teaspoons.

The quantities given will make sauce for a pound—even two—of fillets, steaks, or small whole fish. If you're in a hurry, mix the rub seasonings with the flour and cook immediately.

Mix together, then rub the fish with:

¼ tsp cayenne pepper
½ tsp ground cumin
½ tsp salt

Let the fish stand in the refrigerator for an hour, then flour lightly. Sear and seal as in Basic Procedure I, using olive oil. After removing the fish and draining most of the oil, wilt until golden:

1 large onion, finely sliced

Add:

½ tsp or more cayenne *(careful!)*

⅓ cup wine vinegar

2 Tb water

a pinch of salt

Cook this mixture for five minutes over medium heat and add:

½ tsp ground cinnamon

½ tsp fresh-ground pepper

½ cup honey

¼ cup raisins

Golden raisins are especially attractive. Simmer until water is nearly evaporated and the sauce is thick, almost pasty. Add the fish and reheat in the sauce before serving; the fish will release some juice and thin the sauce slightly. Serve with rice, which may be garnished with pistachios or pine nuts.

Brazilian Fish with Peanut Sauce

Satay kabobs with peanut sauce are increasingly popular, so the flavors of this distinctive fish dish shouldn't be too unfamiliar. It is a good recipe to make with a fish that you are not sure you are going to like (or one you're sure you won't like) because the sauce is so good and so satisfying that you won't go hungry with a small portion of fish, lots of sauce, and lots of rice.

For added peanut flavor, use an unrefined, cold-pressed peanut oil from a health food store with the butter. I'm unlikely to have coconut flakes, so I substitute half or all of a 16-ounce can of unsweetened Thai-style coconut milk, which I do always have.

Cut into serving pieces and sprinkle with salt, pepper, and paprika:

1 lb fish fillets or steaks

Heat in a heavy frying pan:

¼ inch butter, ghee, or clarified butter

Sear and seal the fish pieces in the butter as in Basic Procedure I, removing while underdone. Wipe the pan with a paper towel.

Dry-cook on medium-low heat, turning frequently to prevent burning:

1 medium onion, chopped

Cook until the edges brown slightly. Add coconut milk now (if using) and reduce to about half, boiling rapidly over medium-high heat. Add:

1 large tomato, chopped
(or ¾ cup canned tomato)
1 tsp minced hot green jalapeño pepper
(or other fresh hot pepper)

Simmer covered for about 10 minutes, stirring occasionally, then stir in:

¼ cup minced scallion, white and green parts
⅓ cup creamy peanut butter
(⅓ cup toasted coconut, if not using coconut milk)
2 tsp fresh ginger, minced
½ tsp turmeric
1 cup water

Cover and simmer about 15 minutes, stirring often, then add back the fish pieces, baste with sauce, and finish slightly underdone as in Basic Procedure II. Note: In the original recipe the sauce was pureed in a processor or blender, then returned to the pan and reheated before adding back the fish. Garnish with:

3 Tb or more chopped cilantro
3 Tb chopped scallions, white and green

Cooking Catfishes and Bullheads

These ictalurids have firm, fine-grained flesh and when caught in cold, pure water have an excellent flavor. Channel Catfish sold in California markets are farmed in southeastern states and almost invariably have an unpleasant musty taste from blue-green algae, which thrive in warm eutrophic water, as do catfish and bullheads that have lived in warm water for a time before being caught. California farms produce excellent catfish, which they sell live from the farm to individuals and to ethnic markets. One of the test fish for this cooking section was a live Channel Catfish purchased in a market specializing in Mexican foods. Sautéed in butter and oil (Method 2), it was delicious: firm, fine-textured, and sweet—and very different from the musty southeastern-farmed packaged fish commonly sold. Catfish and bullheads do have to be skinned—follow the directions under "How to Clean a Fish," in the Dining section.

Good catfish and bullheads are delicious sautéed, pan-fried, or smoked, but here is a recipe in a style we don't often see in the West.

Catfish Couvignon

All over the Deep South, catfish is cooked with the "Cajun trinity" of onion, celery, and green pepper. Couvignon, or "smothered," catfish is one of the simplest dishes you'll ever cook—three vegetables, one catfish, one pan. Slow, covered cooking helps reduce musty flavors in the fish, especially if you use plenty of the vegetables, but the flavors aren't intense enough to rescue a seriously muddy-tasting fish. You could also use sturgeon, but most other freshwater fish are not firm enough for a Couvignon.

To give the fish room to steam, cover your basic nine-inch frying pan with a pie pan (upside down) or a domed lid. Use any fat or oil. Bacon fat is authentic and gives a good flavor. Exact quantities are not important. Chop enough vegetables to nearly fill the pan (they will shrink), or just enough to cover the bottom. The conventional proportion is equal amounts of onion and celery, and about half as much green pepper as there is of onion.

Roughly chop:

1 medium onion
½ bell pepper
3–4 stalks celery
1–2 cloves garlic

Add to pan:

About 1 Tb fat

Add the vegetables. Cut into 2-inch pieces:

1 or 2 small catfish (skinned)

Sprinkle with salt and pepper and, if you like, a pinch of cayenne. Add a tablespoon or two of water, cover the pan, and bring to a boil, then turn the heat down. Simmer over the lowest possible heat for about a half hour, adding water if necessary. Serve with rice (in the South the Couvignon would be served over the rice).

ACCOMPANIMENTS FOR SAUTÉ-POACHED FISH

Salad, cole slaw, marinated cucumbers, or steamed green vegetables go very well with sauté-poached fish. Recipes are in the

"Universal Accompaniments" section below. Rice or a small pasta (tiny bowties or shells, for instance) are good with the sauces, or just serve crusty Italian bread.

Method 3: Pan-Frying

If you pan-fry routinely and are happy with the result, you may prefer to skip this section.

In pan-frying, the fish is cooked by the heat of fat, rather than by the heat of a pan lubricated with a little fat as in sautéing. The fat is used only for cooking the fish and never becomes part of the dish or of a sauce, as the fats used in sautéing and sauté-poaching so often do. Pan-frying is probably the most popular way to prepare smaller species of freshwater fish, and justly so. A perfectly pan-fried trout, perch, or Bluegill is crisp outside, succulent within, and totally delicious. Pan-frying is a rather expensive way to cook the catch, and the cook will have more cleanup than with any other of the methods, but it's all well worth it for a dinner of good pan-fried fish!

Special Advice: Pan-frying is not the best way to cook large whole fish or very thick cuts (the coating will burn before the fish cooks). Fish steaks don't pan-fry very well, and pan-frying does nothing for salmon or sturgeon's distinctive flavor and texture. Any other fillets, chunks, or small whole fish can be pan-fried. Euell Gibbons's 1962 classic *Stalking the Wild Asparagus* gives detailed instructions for quickly cutting tiny, one-bite fillets from Bluegill and similar small panfish, which are then pan-fried.

Pan-frying is the most demanding of the methods and the instructions are correspondingly detailed. Read them carefully if you have not pan-fried fish before or would like to improve your pan-fried fish.

There are three "musts" for good pan-fried fish. The fat must taste good, the coating must taste good, and the fat must be kept very hot throughout the cooking. A mediocre fish can make a very tolerable pan-fry if you attend to these three "musts." The best-eating fish ever caught will be terrible if coated in a premix or other stale coating and slowly boiled in rank grease.

Pan-frying requires close attention and a steady hand. Please read over the whole pan-frying sequence before you start to cook. Everything happens fast when you pan-fry—there's no time to pore over instructions to see what to do next!

YOU WILL NEED:

1. Fat for frying. Make sure that you have enough fat on hand! At least 2 to 2½ cups of frying fats are needed to fill a 12-inch pan to the proper depth for pan-frying. Fats must taste and smell good at all temperatures, and they must withstand high heat without scorching. Oils drain well from the coating, hard fats give a crisper crust, and a combination may be perfect. Cold-pressed peanut, sesame, sunflower, corn, and safflower oils are all good for frying. They taste mildly of the seeds they're pressed from and can add interest to the coating. Buy at health food stores or supermarkets that emphasize natural foods. Use lard, ghee, or clarified butter alone, but mix a mild oil with butter or bacon fat to prevent scorching.
2. A big heavy frying pan, or two. The pans must be large enough so that there is plenty of room around each fish. A 12-inch frying pan can handle about a half dozen small fish at once. It's easier to manage the heat with two pans, and a two-pan assembly line will speed the whole process immensely. Pans with nonstick coatings may release toxic fumes at pan-frying temperature and the thick, stiff, coated spatulas used with them don't slide under the fish well and tend to leave the coating in the pan. Julia Child (in *The Way to Cook*) points out that a pan that is too big for the burner won't keep the fat hot enough. If your pan-fried fish are consistently soggy, you may need to fry fewer fish at a time in a smaller pan.
3. Paper towels—lots of them—on which to drain the fish. Modern newspaper doesn't absorb fat well.
4. A decent metal spatula—flexible and long enough to hold any piece of fish for turning.
5. A box of baking soda (or salt) to smother accidental fires. Water spreads burning grease.
6. Lidded cans for disposing of frying fat are helpful.

BASIC PROCEDURE

Setting Up

Pan-frying is an assembly line process and materials should be laid out in order (from the left): fish, liquid, and coating material; the stove with the pan(s) of fat; one or two plates with draining towels; the towel roll; and the towel-lined serving platter. Few modern kitchens have room for all this in one long row, but do arrange equipment and materials so that you can easily move from each station to the next. Set up the line before starting to heat the fat until you are familiar with the process and can set up quickly.

Coating the Fish

The fish should be ready to put in the pan before coating—cleaned, scaled, and trimmed to fit the pan if necessary.

Coatings should be crisp and tasty. Cornmeal and flour are probably the most commonly used coatings, but a list of successful coatings would be endless. A few are given below. Some cooks salt and pepper the fish, some mix salt and pepper with the coating, and some do neither.

Coatings for Fish to Pan-Fry	
Cornmeal	Crunchy
White cornmeal	Sweeter than yellow
Stone-ground cornmeal	Sweet, "cornier," not so crisp
White flour	Light, crisp
Whole wheat pastry flour	Sweet and wheaty
Rice flour, sweet rice flour	Chewy and crisp
Fine oatmeal	Scottish and Irish standard
Potato flour	Very crisp and brittle
Garbanzo (chickpea) flour	Light, crisp, Indian standard
Rye flour	Firm
Matzo meal	Crisp, fat free
Fine bread crumbs	Delicious
Cracker crumbs	Favorite of many

Step 1. Dampen the fish. Usually, fish to be pan-fried are dipped in liquid before coating. Liquids to use include:

milk
buttermilk
light cream

You can mix an egg with the liquid—half an egg to a cup of liquid. Use a wide bowl or baking pan so you can dampen the fish quickly. Dip each side of the fish, hold over the bowl to drip for a few seconds, and coat immediately. For a very light coating, skip the dip.

Step 2. Coat the fish. Have ready on a plate or in a paper or plastic bag:

the coating material

Put each fish on the flour-filled plate and turn as you pat coating onto it. Or put the coating into a plastic bag and shake a few fish at a time with the coating. If you pan-fry fish frequently, you may freeze leftover coating in the bag for a month or so.

Dip and coat the fish twice, if you like a heavy coating. Some experts direct you to roll the fish in flour, dip, then coat with cornmeal or breadcrumbs or (again) with flour. The double coating is supposed to make the outer coating stick more evenly. I coat a few fish at a time and "rest" them on a bed of coating, turning each fish once as I add to the pan, to dry the upper surface with coating. Most cooks with whom I've discussed this fry the fish immediately after coating, saying that waiting makes the coating soggy. Use whichever sequence works with your stove, pans, and preferred coatings.

Step 3. Fry the first side. Heat:

a wide, heavy pan, or two

When the first pan is quite hot, add:

enough fat to cover the fish halfway—usually about 1 inch

Heating the pan before adding the fat keeps the oil fresher tasting. Add fat to the second pan as the fish in the first begin to brown.

A slight haze over the surface is a sign of the right temperature for pan-frying, but fat that actually smokes is too hot for frying and dangerously near combustion. It's impractical to get a thermometer reading from the shallow fat, but a wooden spoon handle or chopstick end dipped in the fat will send out a fast stream of bubbles when the fat is hot enough. Be prepared to adjust the heat as frying proceeds.

When the fat is hot, slip a fish into the pan. The fish should sizzle vigorously. Don't drop the fish with a splash! A splash on your skin will burn it badly and a splash on a burner may send the whole pan up in flames. If the fish just simmers with a few slow bubbles, discard it, turn up the heat, and wait a bit before trying another fish. Don't crowd the fish, and if the oil shows that it has cooled by producing only slow bubbles rather than a sizzle, stop adding fish and turn up the heat. When all the fish are in and starting to brown, you may need to turn the heat down a bit. Fat that's a little too hot is less ruinous to the fish than slow boiling in cooled fat, but neither overheated fat nor burnt coatings taste good.

You'll need to turn the heat up as you begin turning the fish to brown the other side, because the cooler top side of the fish will cool the oil.

Step 4. Fry the second side. Turn the fish before the upper surface begins to change color. Press the leading edge of the spatula into the pan as you slide it under the fish to pick up the coating with the fish. Some fish and some coatings won't brown much, but the coating should be crisp or at least firm and the fish not soaked with grease, which is achieved by keeping the temperature of the fat high. It is best to turn the fish only once.

Step 5. Drain the fish. When the second side is browned, remove to your towel-lined plates, quickly cover with more toweling, and press down to squeeze the fat from the surface before it soaks in.

Tip: Predrain the fish over the pan, using one of the semicircular racks that hook over the edge of a pan. You'll save a bit of fat and a few towels. Kitchen stores and Asian markets carry the racks.

Serving: Remove the drained fish to a plate lined with several layers of paper towels. It is better not to layer the fish on the plate, even with layers of towel between, but this is less ruinous than trying to keep pan-fried fish warm in an oven.

Frying a Carp

A sumptuous meatless meal, often featuring carp, is a Christmas Eve tradition in many countries, and this luxurious holiday carp dish demonstrates the high regard Europeans have for carp. The carp is essentially pan-fried in clarified butter, but the butter has to be deeper to half-cover a carp than it would be for smaller, flatter fish. Expert cook Heinz Bobek of Bavaria kindly contributed a favorite recipe for this chapter. The title implies, roughly, that the recipe is fit for one of the famous carp from the Aisch River, or that the Aisch carp are so good they're fit for this recipe. Here, in Heinz's own words, is the recipe for

---

Karpfen Aischgruendner Art

My preferred preparation is "Karpfen Aischgruendner Art," called after the river Aisch (tributary of the river Regnitz in Northern Bavaria) where there are particularly lots of carp, those with scales, not the specially bred ones. . . . The recipe is very simple but very good and I'll eat the whole fish; normally it serves two.

1 carp: weight approximately 1.5–2 kg (3–4 lb)
750 g clarified butter [2 lb butter will make about 750 g when clarified—author]
½ head lettuce
½ bunch parsley
1 lemon
salt

Remove scales and clean carp inside and outside under running cold water. Dry it using a paper towel. The fish have to be very dry. Season with salt. Heat up the clarified butter in a big high pot up to 350°F. To check the temperature take a wooden spoon and dip the handle in the fat. If bubbles form, the temperature is nearly that high. Throw the cleaned and dried parsley into the frying fat and fry until crisp, then carefully remove from fat. Then slide the entire carp very very carefully into the fat. Fry 15 minutes while turning carefully around from time to time. Pick apart the lettuce and decorate a prewarmed platter with it. Place the fish on top and drip some frying fat onto it. Slice the lemon into small discs and to decorate the fish sprinkle the fried parsley on top. Serve immediately.

For the rest of the menu, a half grapefruit to start with; the carp with its lettuce and fried parsley garnish, and steamed potatoes sprinkled with parsley; sweet sour pumpkin compote for dessert, Moselle Riesling for wine.

The "carp . . . with scales" would be a Common Carp rather than one of the semiscaleless hybrids. Make sure that the parsley for the fried parsley garnish is absolutely dry or it will spatter horrendously!

A more casual way to fry carp carries the concept of "slashed fish" a step further. Use the fingers to pull chunks of carp meat from the deep flesh on the sides of the fish, then pan-fry without coating. Euell Gibbons has a whole chapter on catching and cooking carp in *Stalking the Wild Asparagus* and reports that guests at his "Wild [Food] Parties" give his deep-fried "pulled" chunks a high rating.

Other recipes to try with carp include Fish Baked in Soy Sauce and Vinegar under Method 4 and Kwablah's Foil-Grilled Carp under Method 5. Any carp dish will have many small forked bones in the meat, though pulling it off the bones before cooking eliminates some. To help deal with the bones, provide a supply of damp napkins for wiping fingers and eat carp by pulling bites of cooked meat from the bones.

ACCOMPANIMENTS FOR PAN-FRIED FISH

You can make a salad or side dish while other fish dishes cook, but not when you pan-fry. All the traditional accompaniments to pan-fried fish are prepared well ahead of cooking time. Serve lemon wedges and a good-sized bowl of Tartar Sauce for the fish, with cole slaw and a potato or macaroni salad as side dishes. Homemade Tartar Sauce and cole slaw (see "Universal Accompaniments," below) will amaze you if you're familiar only with the restaurant versions, and both are very easy to make. Any number of cookbooks have good recipes for potato and macaroni salads—those in *The Way to Cook* are among the best.

Method 4: Oven-Cooking

The oven can cook in many different ways. The hot air itself bakes or roasts, but cover tightly and foods steam or simmer. Broiling, cooking under an oven's flame or grill bars, is yet another way to oven-cook.

Special Advice: Small whole fish are probably better pan-fried. Salmon, striper, and sturgeon are at their best baked, and gravlax baked in milk is superb. See Method 5, "Grilling and Camp Cooking," below, for techniques, bastes, and marinades. They will work just as well for oven-broiling as for grilling over coals.

YOU WILL NEED:

1. Almost any kind of oven. A toaster oven bakes fish very well.
2. A frying pan—check to make sure it fits your oven.
3. A baking pan, if you don't have a suitable frying pan.
4. Pot holders with no holes or thin spots.
5. A trivet to protect the table from the hot pan.

KEEPING OVEN-COOKED FISH MOIST

The commonest complaint about baked fish is that it is dry. Overcooking in dry heat is a common cause—baked fish should be taken from the oven a bit underdone, while muscle flakes are still translucent at their centers and won't quite separate easily. Check frequently—while you are mastering testing techniques, check fillets or steaks every minute or two, thick pieces every 5 minutes. To test, insert a fork or paring knife between the muscle flakes and twist to evaluate, or push a finger at the outside of the fish—as it cooks, the unresilient raw feeling changes to a firm springy one.

Basting and surrounding the fish with liquid will also help keep it moist. Citrus juices, white wine, vermouth, or a mixture of these work well. Mellow acidic additions and extend the pan juices with chicken broth, vegetable stock, or coconut milk. A few "starter recipes" for less-basic bakes with sauces follow the Basic Procedure.

Again, while you are learning to bake fish, *remember to check it often.* Fish will "wait" 10 to 20 minutes in a turned-off oven, but open the door slightly to cool the oven.

BASIC PROCEDURE

A basic fish bake starts with searing on the stovetop or in a very hot oven. This seals surfaces and helps keep the fish moist. After the sealing, the fish may be finished as a roast, or steam-baked with liquid or sauce.

Before You Start: Preheat the oven to 400°F. You may dredge steaks or fillets lightly in flour before searing. Oven thermostats vary greatly—adjust the suggested temperatures as necessary. For sealing surfaces, the oven should be hot enough so that the surface of the fish changes color and seals quickly without leaking liquid.

Starting in a Frying Pan

Heat about ¼ inch of fat in the frying pan and quickly sear and seal the fish on each side. Add ¼ to ½ inch of liquid, which should bubble immediately. Lift the fish to let the liquid flow under it, put it in the preheated oven, and immediately reduce temperature to about 325°F. Test for doneness and finish the fish in its own heat as described above. Cover the fish lightly with foil if it begins to brown too much.

If you will be finishing and serving the fish in a baking dish, slide it from the frying pan into the well-buttered dish, place in the preheated oven, and turn the temperature down to 350°F. Add any liquids to the frying pan now, boiling hard until reduced to about two-thirds. Scrape any browned juices and fish bits into the boiling liquid. Carefully pour the hot liquids over the top of the baking fish. Finish cooking as above.

Roasting

Thickly butter a baking pan (or thinly film with ghee), preheat it for a few minutes at 400°F, then quickly slide the fish into the pan, dot with butter or ghee, shut the oven door, and sear for a few minutes. Cross-sections or tails from large fish such as salmon or Striped Bass can be delicious just roasted with an occasional brushing with butter, ghee, or lemon butter to flavor and brown the skin. Continue to lower the heat. Cover lightly with foil (don't tuck it in!) if the fish seems to be browning too much. A big roast may take 45 minutes or more.

To give a roast fish a nice flavor and help it stay moist, pour a tablespoon or two of wine or mixed lemon and orange juice over the fish as it bakes.

To make a "moist roast" or "oven-steamed" fish, add about a half inch of liquid after the surface seals, then cover the pan with a lid or foil. Be careful! A glass dish can crack violently if cold liquid is poured directly onto it when it is very hot. If the baking dish is glass or ceramic, dribble the added liquid slowly and carefully over the top of the fish.

Oven-Grilling

Oven-grilling is a good way to quickly prepare thin steaks or fillets with a minimum of mess and fuss. Strictly speaking, "grilling" is cooking on a grill, and the oven needs to be very hot to emulate the grill. Preheat your oven to at least 400°F. While the oven heats, butter a wide pan or baking sheet well, season the fish, and, if you wish, dredge it lightly with flour or crumbs. Arrange the prepared fish on the baking sheet or pan with an inch or so between pieces, and bake without added liquid. Not all fish will brown. You may baste once with clarified butter, melted butter, or ghee, or with any of the fats mixed about half and half with wine or citrus juice. Basting adds flavor but tends to prevent browning. Very thin fillets may cook in less than 5 minutes.

RECIPES

You'll find excellent recipes for fish baked with added ingredients in almost any general cookbook. Here are three somewhat unusual treatments.

Gravlax in Milk

Gravlax baked or simmered in milk will be especially appreciated by lovers of Finnan Haddie (a smoked haddock fillet, very expensive if you can get it at all). The bake is started in a cold oven. Place the gravlax in a baking dish that fits it rather closely, add milk almost to the top of the fish, set the thermostat to 250°F to 325°F, and bake. A ½-inch-thick test fillet cooked through in about 25 minutes and was fine after being held for another 20 minutes at 200°F. (Remember, no two ovens are alike.) Gravlax may also be simmered with milk in a frying pan. Another half-inch test fillet was cooked through in about five minutes after the small bubbles signaling "simmer" appeared at the edges of the milk.

Sour Cream Fish Bake

Start this "bake" in a preheated oven. Unlike most fish recipes, this one is expandable: you can bake many pieces of fish in a large pan to feed several guests. Use any good-tasting fresh or frozen fish. Striper, sturgeon, White Bass, catfish, and basa are very suitable. The sauce doesn't have enough neutralizing or masking power for very strong or musty-flavored fish, and we thought salmon too rich. Use a pan big enough to hold the fish in one layer.

Preheat the oven to 400°F and lightly butter the pan. Place fish in pan, grind fresh pepper over it, salt very lightly, and hold at room temperature while preparing the topping.

Tip: Crush garlic with the bottom of a sturdy cup or mug.

Crush together until pasty:

3 or more cloves garlic, peeled
1 tsp salt

Mix with:

1 cup sour cream
(or ½ cup each sour cream and plain yoghurt)
(or ½ cup each yoghurt and mayonnaise)

The topping should cover all the fish to a depth of about a quarter inch.

Spread topping over fish and bake uncovered for about half an hour, checking for doneness after 20 minutes. The fish will hold at 200°F for another half hour after a fork test indicates that it's done. Sprinkle paprika lightly over the topping for added flavor and color.

Soy-Vinegar Fish Bake

The combination of soy, vinegar, and fresh ginger root in this "bake" reduces and disguises strong or muddy flavors and is a good way to cook frozen fish, which will have lost some texture and flavor. Don't attempt this bake with fish with delicate flavor or texture or very thin pieces.

Choose individual, one-serving baking dishes for this dish. The piece (or pieces) of fish should fit rather snugly, so that it doesn't take much of the very concentrated sauce mix to cover the fish. An overly large dish will result in the fish being overly saturated with the strong sauce. An attempt to bake more than two servings in one dish was not very successful for this reason. If you want more sauce, add light chicken or vegetable stock (or even water) rather than just adding more basic mixture.

Preheat oven to 400°F, as in the Basic Procedure.

A processor or blender makes this sauce easy. Add liquids a little at a time, processing each addition with a short burst until the sauce is fairly smooth. The amounts given should nearly cover two steaks or fillets totaling about a pound of fish.

Process in the order given:

1 inch fresh ginger root, peeled, roughly chopped
1 clove garlic
⅓ cup tamari or other soy sauce
⅓ cup rice wine or balsamic vinegar

Part of the vinegar may be replaced with white wine, vermouth, lemon, and orange juice, or straight lemon juice—adjust to taste. Unused sauce will keep for months if refrigerated in a tightly lidded jar.

Place in the baking dish:

sliced onion, 1 layer deep
the fish

Add sauce to cover about two-thirds of each piece of fish. Start fish as in Basic Procedure with about five minutes in a hot oven and finish with low heat, 325°F or less. You can cover the fish for part of the time. Soft or thin fish pieces can take as little as 15 minutes; firm fish can bake long and slowly, for 40 minutes or more at 225°F to 250°F. This dish can be held in a warm oven (200°F or below) for a long time.

Method 5: Grilling and Camp Cooking

Cooking over a strong heat source—grilling—requires less equipment than any other way to cook fish (pl. 105). Camp grilling requires nothing more than the coals of a wood fire and a light, packable wire grill basket or foil. At home, foods can be grilled over wood, charcoal, an electric coil, or gas-heated ceramic

Plate 105. Trout grilling in a folding rack and on a portable grill.

briquettes. Anglers can even grill the catch right on board the boat, using one of the gimbaled hibachi-style barbecue appliances designed to hang over the side of any craft.

In camp, you'll be cooking the day's catch regardless of species or size. Medium or small whole fish, steaks, and fillets are easy with a grill basket, and you can make patties or kabobs of boneless fish meat with the tiniest of fillets or scraps. Chop and season the meat, knead together briefly to consolidate, then pat into shape over a bamboo skewer. Patties can also be grilled in foil.

Special Advice: Marinate or baste fish when grilling to flavor and moisten; add flavor with a "rub."

YOU WILL NEED:

1. A folding grill rack (pl. 105). This two-part wire mesh device folds over the fish to hold it securely for turning, and it is easier to release a grilled fish from its fine wires than from the barbecue's grill. Our favorite folding grill is sold as a hamburger rack, with closely spaced grid wires and a gap between the sides of each of its two compartments, so that it holds but doesn't squeeze the fish. Many folding grill racks are light enough to take backpacking.
2. Bamboo skewers, to push through fillets or steaks so they'll hold together and be easier to handle. Skewers are useful even when grilling in a basket.
3. Heavy-duty or extra heavy aluminum foil for steam-grilling.

4. Nonstick pan spray effortlessly slickens a grill or grill basket.
5. Stainless steel bowls for marinades and bastes won't break or burn.

FISH BASTES AND MARINADES FOR GRILLING

Grilled fish, especially fillets and steaks, will be moister and more flavorful if marinated for a short time and basted while grilling. Baste with the marinade or make a new baste by adding a spoonful of oil to a half cup or so of citrus juice or any of the soy-ginger sauces. Pat Awaze Rub (see "Universal Accompaniments," below), paprika, or a bit of turmeric onto the fish before grilling to add color and flavor.

Basic Lemon Fish Baste and Marinade

One juicy lemon or orange will make marinade and baste for a pound or more of fish. You need only enough marinade to coat the fish, with a little left for basting once the fish is removed to the grill. Marinate in a plastic bag for minimal mess and easy coating. Place fish in bag and add:

juice of 1 lemon, 2 limes, or 1 orange
2 Tb olive oil
1 tsp salt
(or 2 tsp soy sauce)
fresh-ground pepper
[optional: 1 clove minced garlic]

Many variations are possible. You may substitute white wine or dry vermouth for citrus juice, or make a mix. Change the flavor of the marinade with herbs. Start with about a teaspoon of fresh or dried herbs. Herbs to try include dill, fennel, oregano, basil, rosemary, tarragon, saffron, and others—the possibilities are endless.

Unsweetened Thai-style coconut milk is an excellent addition to any marinade, and a remedy for dryness in fish such as salmon or sturgeon. Try adding a 6-ounce can to a marinade for fish such as salmon or sturgeon.

When the marinade is mixed, add the fish to the bag, seal it, and turn it a few times to coat the fish, then place it in a bowl, in case a fin or bone punctures the bag.

BASIC PROCEDURES

Grilling over Coals

Grill over an even bed of mature coals. Oil a folding grill or spray it with a nonstick pan spray, then fold the fish into it, place on the barbecue, and grill on each side. Turn thin pieces once; thick fish or thick cuts may need several turns to avoid scorching. Baste with each turning.

Steam-Grilling in Foil

Flavorings for Steam-Grilled Fish: The selection of chopped flavoring vegetables to wrap with the fish is limited only by your imagination. Onion, celery, green pepper, and sliced tomato are typical. You may also marinate the fish before wrapping.

You can dot the fish with sour cream and butter or add small amounts of liquid ingredients just before closing the wrapping (see directions for wrapping, below). Possibilities include sauce ingredients from any recipe, prepared salsa or tomatillo sauce, citrus juice, or a few spoonfuls of wine.

WRAPPING THE FISH

Use four to six layers of heavy-duty foil or two of extra heavy foil to make a leakproof wrap. The foil piece for each layer should be large enough to wrap around the fish at least one-and-a-half times, with plenty of overlap on the ends. Join foil with a leakproof "French seam" when large sheets are needed. Stack two sheets of foil, matching edges. Make a ½-inch fold along one edge of the double sheet, then press the fold down along its entire length. Repeat several times, pressing the fold down each time, then open out the joined sheets.

To wrap, stack two to four foil sheets, staggering seams. Butter or oil the top sheet, sprinkle part of any flavoring ingredients along the center, position the fish, salt and pepper it if desired, then top with remaining ingredients (a whole fish can be stuffed with, rather than surrounded with, flavorings). Bring the edges of the long seam together and join as above, leaving a small gap if you plan to add liquid, then seal the ends. Close the long seam completely and finish the wrap with one or two more sheets of foil, sealing each layer separately. Keep more foil at hand in case the package is accidentally punctured while grilling.

Kwablah's Foil-Grilled Carp

Mr. Kwablah Attiogbe was a graduate student taking the author's ichthyology course when he cooked the best carp I've eaten in this country. The Common Carp, about a foot long, had just been caught in the lower Stanislaus River, which was very fast, cold, and clear at the time. The onion filling especially complements the flavor of a good carp but is excellent with any large whole fish.

After scaling and cleaning, the fish was sprinkled liberally inside and out with salt and pepper. The body cavity was filled loosely with:

chopped onion (1–2 medium onions)
chopped scallion (3–4 stalks)
minced ginger root (at least 2 inches)
minced garlic (2–3 cloves)

The fish was sealed into foil and the package was placed on a grill above a good bed of coals, where it cooked for about 45 minutes with an occasional turning. The foil-grilled carp was succulent and aromatic, and the steamed flesh was easy to pull from the small forked bones, which are present in even a large cyprinid.

Method 6: Hot-Smoking

Hot-smoking is one of the most delectable ways to prepare fish, and hot-smoking is also a most enjoyable way to extend the life of the catch (pl. 106). Hot-smoked fish that has been well handled and properly refrigerated will remain in good and safe eating condition for up to three weeks.

Special Advice: Almost any fish from small Bluegills to cuts from a hundred-pound sturgeon can be hot-smoked. Smoked gravlax (curing directions in this section) is especially delicious and versatile. Hot-smoking, but not cold-smoking, makes fish safe. Raw fish flesh may have parasites which can infect humans, and they are not killed by light curing and cold-smoking. Freezing the fish for seven days at −4°F (the recommended temperature for home freezing) kills them, with one possible exception. Freezing has not been specifically tested on *Diphyllobothrium,* a tapeworm common in wild salmon. Food and Drug Administration Center for Food Safety and Applied Nutrition (CFSAN) is now

Plate 106. Salmon partway through the smoking process.

testing freezing procedures on this parasite—check CFSAN's "Bad Bug Book" (http://vm.cfsan.fda.gov/~mow/intro.html) for current guidelines.

Most home smokers are designed to hot-smoke foods, essentially a slow cooking process. Test for doneness as you would any cooked fish: there should be no raw, translucent areas between muscle flakes when you pry them apart. Because hot-smoking is slow, a meat thermometer can be used on thick cuts. The recommended temperature for killing parasites is 131°F.

EQUIPMENT FOR SMOKING

You can buy a smoker, make a smoker, or, for a trial or two, improvise by adapting a barbecue grill. Commercially made smokers are packaged with basic directions, but for serious smoking all smoke-cooks should have one or both of two indispensable books: *The Home Book of Smoke Cooking*, by Jack Sleight and Raymond Hull, and Richard Langer's *Where There's Smoke There's Flavor*. Langer's book paid for itself in smoked salmon the first time we used our new Brinkman. *The Home Book of Smoke Cooking*, having paid for itself many times over during our farming and meat-processing years, is now earning dividends as a fish-smoking text. Both books cover ingredients and techniques for smoking very thoroughly. Langer's book focuses on the popular Brinkman smokers and presents a large and adventurous collection of recipes for smoked foods and dishes to accompany them, while Sleight and Hull concentrate on brines, cures, and different smoking techniques. *The Home Book of Smoke Cooking*'s formula for "Basic Fish Brine" is an excellent basis from which to develop your own favorite brine, as I have done.

Tip for Successful Smoking: Keep a log! Include date, species of fish, size or weight and shape of pieces, brine recipe used, brining time, smoking time, and temperature range.

BRINING AND SMOKING

Brining fish for a short time before hot-smoking firms and flavors it. "Brines" for curing meat and fish include sugar and flavorings as well as salt and water. It is best to use a pure salt such as the white crystalline "sea salt" sold in bulk in many markets and health food stores or the (more expensive) packaged "pickling salt." Though table salt could be used for a first experiment, the many additives in it may cloud solutions, give a "chemical" taste, soften the fish, or all the above. Results will be better with a pure salt.

The formula given below has an ideal salt concentration of 10 percent and makes enough solution to brine a trial quantity of fish—a pound or a little more. The "fill the jar" mixing method makes it easy to vary the flavorings and the kinds of liquid used without altering the brine concentration. Liquor is a popular addition—start by adding a tablespoon or two of gin, rum, whiskey, or wine to the dry ingredients, increasing to taste in subsequent trials.

Starter Brine

Measure into a quart Mason jar:

½ level cup salt
2 Tb dark or light brown sugar
juice of half a lemon (1½ Tb)
(or 1½ Tb vinegar)
1–2 cloves garlic, peeled and smashed
(or 1 tsp garlic powder)

Fill the jar to the shoulder with water, tighten the cap down well, and shake to dissolve and mix the dry ingredients. Then (this is important) add water to the very top of the jar and shake again. You will now have added 4 cups of water for an ideal brine concentration of 10 percent. Always use enough brine to cover and surround the fish. Brining in a heavy plastic bag makes for much easier handling—the kind with a sliding "zipper" is the most secure. Trim sharp fins, add fish and brine to the bag, zip closed, then place the bag in a bowl in case of a puncture. Shift the bag from time to time so that all pieces come in contact with the brine. Don't hesitate to adjust the brining times to taste—you are brining mainly to flavor the fish and only secondarily for the slight preservative effect.

TIMING BRINING

All brining times assume that you are brining at ambient temperature and using a 10 percent brine like the one above. Brining times reference the weight of *each* piece of fish. That is, if you had eight skinless fillets of about a quarter pound each, you would brine them according to the "¼–½ lb" chart time (45 minutes), not the time for the total weight of all the fillets. Just be sure that there is enough brine to generously surround each piece of fish. Brining times are always approximate and vary according to the shape of the pieces (for instance, a thin fillet has more surface area than a thick chunk of the same weight and will absorb brine more quickly), the species and composition of the fish, and even the temperature. Change the brining time, not the brine recipe, to adjust the flavor of the hot-smoked fish to your taste. Be sure to record times in the smoking log!

Guidelines for Brining Skinless Fish

Weight	Time
Under ¼ lb	30 min
¼-½ lb	45 min
½-1 lb	1 hr
1-2 lb	2 hr
2-5 lb	1 hr/lb
Above 5 lb	(Untried!)

Fish with Skin On: Increase brining time by about one-fourth. For example, if your ¼-pound fillets have the skin on, you would divide the chart time of 45 minutes by 4 and add about 11 minutes, for a total time of—call it 55 minutes plus.

Oily Fish: Salmon and fat trout may need an increase in time. Add up to another 25 percent of the minimum time. For instance, if the skinless fillets in the example above were from very fat trout, you would brine for 55 minutes. If they were skin-on, you would add another 11 minutes, brining for a total of 65 to 70 minutes.

Don't try to reuse brine! Fish juices will dilute it, bacteria will contaminate it, and it will be unsafe as well as ineffective.

Most smoking instructions recommend air-drying brined fish before smoking, but a fish patted dry with paper towels will taste just as good, and it will be free of dust and fly leavings.

Smoke the brined fish according to your smoker's instructions. It will keep at meat compartment temperatures (32°F to 34F°) for up to three weeks if it has been cleanly handled. To avoid contaminating a whole batch of hot-smoked fish with "germs" when removing portions, pack several small bags with two to three servings each.

GRAVLAX: A SCANDINAVIAN TRADITION

This delectable Scandinavian preparation of lightly cured salmon is normally served in thin, uncooked slices. As explained in the introduction to the hot-smoking section, raw fish may carry parasites which can infect humans. However, you don't need to deny yourself this marvelous dish for fear of parasites. Hot-smoke your gravlax! It is, we found, delicious when cooked by any method, and the author and I have found that the soft gravlax made from flabby farmed salmon is better cooked than when served sliced raw in the usual way. You will find suggestions for using cooked gravlax following the curing instructions and a special recipe for Gravlax in Milk following Method 5. Our experiments with other species suggest that many of the larger freshwater fishes, such as trout, catfish, carp, and even Largemouth Bass, could also be delicious when prepared as cooked gravlax.

Curing Procedures

The traditional cut of fish is a skin-on fillet of salmon. Before applying the cure, give the skin side a lemon juice scrub to remove mucus and improve flavor.

Mixing the Cure: Choose the kinds of ingredients you want to use, then choose your proportions. The amounts given will cure about two-and-a-half pounds of salmon, and all you have to do to make up a stock of a favorite cure is use a larger measuring device. Use a half-measure each of pepper and dried dill or dill seed for each measure of salt used. (Don't add fresh dill until curing time.)

Salt: Use a pure salt, as you would for a brine cure.

Sugar: White or light brown sugars are most commonly used. Americans sometimes use maple syrup.

Pepper: The traditional white peppercorns have a distinctive flavor, but black is an adequate substitute. Grind or crush peppercorns (use a stout piece of wood or the bottom of a strong cup) before adding. A tablespoon of peppercorns makes about a tablespoon of crushed pepper, so you can measure before crushing.

Dill: Fresh heads and leaves, dried leaves, or seeds may be used. Fresh dill has the best flavor and is most traditional. Use a handful of fresh dill or dried leaves with the cure amounts given, and about a tablespoonful of the stronger-flavored dried seeds.

The times and cure amounts given below are appropriate for raw-sliced gravlax. If you plan to cook the gravlax in dry heat, as in roasting or smoking, you may use half as much cure or cure the fish for half as long.

For all formulas:

1 Tb pepper
dill (see above)

Equal amounts of salt and sugar are most traditional:

2 Tb sugar and 2 Tb salt

Overfill (round) the spoonfuls to make about 5 Tb total.

We prefer a less sweet gravlax:

2 Tb sugar and 3 Tb salt

But a sweeter mix is liked by some:

3 Tb sugar and 2 Tb salt

Traditionally, cure-coated fish pieces are layered with fresh whole dill sprigs on a plate and weighted. The weighting presses the seasoning ingredients into the fish and firms it. The fish pieces are turned periodically so that seasonings and cure ingredients penetrate thoroughly. This makes an attractive raw-sliced gravlax, but gravlax for cooking will be rinsed and wiped off and can be cured much more easily using plastic bags and any form of dill. Mix salt, sugar, pepper, and dried dill in a small bag or jar, then measure 2 Tb of mix for each pound of fish into a large, heavyweight zip-seal bag. If using fresh dill, add it to the large bag with the fish, shake to coat with cure, then seal the bag, pressing air out. Work from the outside of the bag to keep sticky cure off your hands as you stack fillets so that the thick end of one fillet is over the thin end of the other. Place in a pan (a large bread pan is ideal), place a weight on the bagged fish, and refrigerate. The exact amount of weight is not important, and weights may be "found objects" from the refrigerator shelves. My favorite is an unopened ½-gallon of milk laid on its side, and jars of mayonnaise also work well. You could also weight with a brick or a collection of marbles or large pebbles bagged in plastic. The sides of the pan help stabilize the weight and contain any juices that escape. Weights tend to tip—place the pan close to a side wall.

Turn the fish once a day to redistribute the cure. Remove weights, turn the bag over a time or two, and replace weights. Gravlax steaks or filets for cooking may be ready in two days. A tail needs another day or two and another turn or two. Gravlax may be stored in the meat compartment for at least five days after the minimum curing time.

Leftover cooked gravlax can be used in many ways. Smoked gravlax makes an elegant, easy spread or dip. Just flake leftovers and mix with cream cheese, sour cream, or yoghurt, adding chives, dill, or other herbs as desired. For main dishes, use flaked cooked gravlax (hot-smoked gravlax is especially good) in any recipe that calls for canned tuna or salmon. Gravlax patties are good for breakfast, lunch, and dinner. Mix flaked cooked gravlax with cooked potato and a bit of grated onion, then pat into thick rounds and brown in butter.

Universal Accompaniments

The condiments, side dishes, and other accompaniments included in this section are good with fish cooked by any of the methods.

Tartar Sauce: A French Classic

Tartar Sauce is an ancient and popular French "emulsion" sauce, probably older than the mayonnaise (the emulsion) with which it's now made. It is simply mayonnaise enhanced with a selection of tart or spicy additions, and is not at all like the sweet sticky glop served in most California restaurants. Pan-fried fish and Tartar Sauce are traditional partners, but any fish not cooked in sauce is good with Tartar Sauce. Make the best of a "catch of the day emergency," when a caught or bought fish is already cooked and turns out to be terrible, by heaping it with a Tartar Sauce loaded with additions.

Cheap brands of mayonnaise usually won't hold up well with the juicy additions. Use a good quality mayonnaise, unsweetened if you can find it. Choose the pickle component to suit your taste. You may use sour, sweet, bread and butter, or dill pickles, or even pickle relish. Don't be afraid to increase the proportion of onion and pickle in the starter recipe!

Chop fine:

half a medium onion
pickles to make 2 Tb

Press excess juice from the chopped pickles, add to serving bowl, and mix in:

about ¾ cup mayonnaise

That's it! Raw onion is omitted from many recipes, though we consider it essential. Vary and personalize your Tartar Sauce by adding one, several, or all ingredients from the list below. For a "rescue" sauce, add lots of celery with the other ingredients, until the sauce is almost a salad, and heap it high on the fish pieces as you serve them.

Optional additions—chop fine:

hard-boiled egg
horseradish (no need to chop; grate fresh, or buy minced in bottle)
scallions
lemon flesh (insides of sections)
capers
½–1 anchovy
fresh herbs: parsley, chives, tarragon
celery (chop or process to a fine mince)

Tarragon is a strong herb—try 1 or 2 leaves.

To further explore the classic *Sauce Tartare, Sauce Remoulade,* and others of this fascinating family of emulsion sauces, see the discussions and recipes in *Mastering the Art of French Cooking* (Volume I) and *The Way to Cook.* Most of the sauces are as simple as any above, once you've made or acquired the mayonnaise.

RUBS

Add flavor and color to any fish by patting a "rub" into the surface before cooking. Curry powder, paprika, or turmeric make good rubs by themselves, and the following easily made mixture not only performs outstandingly as a fish (and meat) rub but is a fine cooking seasoning or table condiment for any food.

Awaze Rub

Awaze (Ah-Wah-Zee) is an Ethiopian mixture dominated by powdered chile or paprika and ajowan, a spice much like celery seed. Adding sugar transforms Awaze from a stew seasoning to a useful rub and all-around condiment. Though I've substituted celery seed, which can be bought in any market, for ajowan (also spelled "ajwain"), this tiny seed is sold in any Near Eastern or Indian market and adds an interesting celery-thyme flavor to many dishes. I use whole spices and fresh garlic because that's what I have on hand, but most rub recipes make use of semi-prepared ingredients such as garlic powder, celery salt, and dry mustard, and there's no reason not to try such substitutions. Paprika and other ground chiles vary greatly in hotness. Taste powdered chile before adding, and don't add the cayenne until everything else has been mixed. Increase or decrease the suggested amounts of cayenne or omit to adjust the hotness to your taste. The easiest way to mix is to shake ingredients in a sturdy plastic bag or tightly closed jar. Store in a cool dark place, as you would any other spice.

Grind together:

1 tsp peppercorns
1 tsp celery seed
2 tsp mustard seed

Mix with:

¼ cup powdered chile or paprika
3 Tb brown or white sugar
2 cloves garlic, very finely minced
(or about 1 tsp garlic powder)
[optional: 1–2 tsp salt]
[optional: cayenne to taste]

Sprinkle the rub on the fish and pat in gently. If possible, let the "rubbed" fish stand for a half hour before cooking.

SIDE DISHES I: SALADS

Slaws and Dressings for Slaw

Cole slaw, a salad of finely sliced or grated cabbage, goes well with most fish dinners and is a tradition with pan-fried fish. Slaw is very fast and easy to make using the slicing and grating discs of a food processor and not much more trouble using a knife and hand grater. The cabbage is often enhanced with carrots, raisins, or other fruits and vegetables and you can even buy "slaw mix" with green and red cabbage and carrots in about the same proportions as in the recipes below. Any supermarket carries it, but slaw made from fresh ingredients tastes much better! Choose small, tight heads of cabbage for slaw. A proportion of red cabbage is often added, but red cabbage is generally too tough to use alone. Most slaws are even better the day after they're made. Many can be kept (refrigerated) for several days, and some slaws may even improve through part of that time.

A typical slaw dressing is creamy and slightly sweet. I have given directions for mixing a slaw with our favorite creamy dressing of mayonnaise and yoghurt, and suggestions for a variety of even lighter dressings follow the recipe.

Carroty Slaw

The carrots and raisins in this slaw make the sweetening usually used in slaw dressings unnecessary. Grate cabbage and carrots with a coarse grater, or slice the cabbage fine and grate the carrots. The amount of grated cabbage should be about four cups. Cut out the core (the stem inside the folded leaves) with a sharp paring knife before grating.

Shred into processor bowl:

½ head green cabbage
¼ head red cabbage
3 carrots

All slaws are better if they are mixed in sections. It is easiest to mix with your hands. Put a big handful of shredded vegetables in the serving bowl, add a spoonful or two of dressing ingredients, toss, and repeat until all the vegetables are used up. Mix in sections as described above. With each "section" of vegetables, add a big spoonful each of:

mayonnaise and yoghurt

The slaw should be well moistened but not swimming in dressing. Toward the end of the mixing toss in:

1 cup golden raisins

Golden raisins are a bit tangier than black, and they make an especially pretty slaw. You can find them in any supermarket.

Other Slaws: Shred or grate about four cups of cabbage and toss (in sections, as described above) with one of the following dressings.

Mix a tablespoon of sugar and a tablespoon of vinegar into a half cup of plain yoghurt. Use alone or add alternately with equal amounts of mayonnaise.

Slaw made with this salt-free dressing is one of our favorites. Mix a scant teaspoonful of wasabi powder into a half cup of plain yoghurt, then grate about half a small onion into the mix. Adjust the heat level by tasting as you toss, alternating a spoonful of plain yoghurt with a spoonful of the wasabi mixture. Wasabi is a horseradish relative, and many creamy slaw dressings include horseradish, which could be substituted here.

Shredded cabbage is also very good with either of the dressings given for Marinated Cucumbers (below) with Zingy Lemon Dressing (below), or with any of the variations in the recipe for Soy-Ginger Sauces (following Method 1). It's debatable whether shredded cabbage dressed with citrus and soy should be called slaw, and the delicious Mexican cabbage salad or relish made by mixing chopped or shredded cabbage with juice from pickled jalapeño peppers seems to have no name, but all are very good with pan-fried fish.

Green Salads

Why use bottled dressings when it's so easy to dress a salad yourself, and so much better? Salads made with either of these quickly made dressings are particularly good with fish dishes.

Zingy Lemon Dressing

Anyone who has enjoyed a salad in a Thai, Indian, or Chinese restaurant will recognize the style of this oil-free dressing with its perfect balance of sweet and tart. Use it to dress any combination of raw or cooked vegetables.

Measure into a small jar:

juice of 1 lemon or 2–3 limes
½ tsp salt
1 Tb honey
1 Tb sugar
a pinch of cayenne

Put the lid on the jar and shake to moisten and mix. Let the mixture rest for a few minutes and shake until dry ingredients dissolve. Some cayenne is essential to enliven the vegetables, and some cayennes are much hotter than others. Start with a pinch and increase as you learn to use the dressing, if you wish, but don't omit cayenne entirely! The dressing is very concentrated. Use just enough to lightly coat the greens or vegetables.

Use Zingy Lemon Dressing to make a crunchy, substantial salad that is particularly good with fish. Toss mixed greens and chunks of raw vegetables such as cauliflower, radish, raw peas, and carrots with Zingy Lemon Dressing. An exhausted angler can just toss a bag of prewashed greens with a small spoonful of the dressing. Use mixed baby lettuce, baby spinach, or any combination of prebagged greens.

Head lettuce isn't regarded with much respect any more, possibly just because it is reliable, inexpensive, always available, and quick and easy to prepare. Thin wedges of head lettuce sprinkled with Zingy Lemon Dressing make a nearly instant salad and one of the very best to accompany a fish dinner.

Salad Dressed with Olive Oil and Lemon

Salads dressed with lemon juice go especially well with fish dinners, since unlike vinegar dressings they don't clash with wine and lemon used in or served with fish dishes. The amount of oil and lemon juice needed to coat a large bowl of greens is amazingly small. A teaspoon recipe will dress up to 5 cups of greens. Any kind of oil will make a good salad as long as it tastes good. This salad is also "nearly instant" when made with prewashed greens.

Peel, crush, and finely chop:

1 clove garlic

Add to serving bowl with:

about 1 tsp very good, fruity olive oil
the salad greens

Toss the greens gently to coat them. Just before serving, toss with:

a pinch of salt
fresh lemon juice to taste
(or mixed citrus juices)
(or rice, balsamic, or wine vinegar)

Taste as you add the juice or vinegar—a teaspoon to a tablespoon may be enough. Dry off a wet salad by tossing with a half slice of dry or fresh bread or a paper towel (remove before serving). Five cups of greens will yield about three cups of salad.

Cucumber Accompaniments

Cucumber dishes complement fish especially well and are quick and easy to prepare. The recipes below were tested on half a large "English" cucumber and serve two to four people. English cucumbers are tender, seedless, and never bitter, and never need peeling. They compare favorably in price with American slicing cucumbers, because there is so little waste—no tough peels or bitter ends to cut off, and no hard seeds to remove. A cut cucumber will keep for many days if you allow the cut end to dry and seal itself before refrigerating.

Marinated Cucumbers

Cucumbers in sweetened vinegar are a favorite dish in American and Japanese cultures alike. Take a tip from the Japanese and use their mild rice wine vinegar straight from the bottle to dress both styles of marinated cucumbers. Distilled or cider vinegar from a supermarket should be diluted with about an equal amount of water. Cucumbers marinated in vinegar solutions, without oil, are best made several hours or a day before serving, and can be stored in the refrigerator for several days.

Mix in serving bowl:

2 Tb each vinegar, water, and sugar
(or 3 Tb each rice wine vinegar and sugar)

Toss gently with:

½ English cucumber, thinly sliced

Add salt to taste if you wish. You may double the amount of vinegar for a more tangy pickle, and home gardeners can add sliced tomatoes and a thinly sliced onion for a classic summer salad.

To transform your marinated cucumbers into a Japanese "fresh pickle," add:

about 1 tsp fresh ginger root, grated
a few drops toasted sesame oil
(or a sprinkle of toasted sesame seed)

Black sesame seed, found where Asian foods are sold, would be most authentic. Japanese-style cucumbers tend to be noticeably sweet. Increase the sugar if you wish. There is no absolute rule with marinated cucumbers!

Cucumbers in Cream

These richly dressed cucumbers complement white-cooked or plainly baked fish well. Mix:

2 Tb each sour cream, vinegar, and sugar

Toss with the sliced cucumbers. Salt is optional. You may lighten the dressing by using half yoghurt. Cucumbers in Cream should be eaten the day they are made.

SIDE DISHES II: RICE

Rice goes well with all fish dishes with the possible exception of pan-fried fish. Three types of rice are commonly prepared for the table: dry, fluffy long grained; delicate, sticky medium grained; and chewy, nutty-flavored wild rice. Rice has become so popular that the not-very-sophisticated supermarket in my small town now regularly carries at least 20 different types and brands!

Types of Rice: Basmati and jasmine rice are aromatic long-grain rices, with a distinct aroma of popcorn, especially when they're cooking. Basmati is firmer than jasmine rice and the grain is (or should be) longer. Buy in discount groceries or Asian or Near Eastern markets for the best price and quality, but freeze the rice for a week or two to kill eggs of pantry pests. Other long-grained rices have a delicately neutral flavor.

The subtly flavored fatter, stickier grains of medium- and short-grain rices are Japanese, Korean, and Cambodian staples. Look for any rice with "rose" in the name, such as Calrose or Kokuho Rose.

Cooking Rice

In general, one 1 of raw rice serves two to four people, but I cook all rice in a rice cooker, which has its own one-serving measuring cup (its ⅔ cup measure makes a large serving—rice cookers are made in rice-eating countries). I should have said "we." Even a new cook can make good rice in the rice cooker. Mine cost much less than most of my saucepans and I've used it at least four times a week for more than ten years. It takes only moments to throw two of its cups of rice into the cooker, add water to line 2, and push the switches (though washing the rice takes a minute or two).

Washing: We're often advised not to wash rice and some brands don't need it, but check the label: much rice is coated with glucose or starch. Imported jasmine and basmati rice always seem to need washing.

To wash, measure the rice into the cooking pot. Nearly fill the pot with cold water and rub the rice lightly between the hands while lifting and dropping it under the water. Pour off the water and repeat until the water is nearly clear, usually three to five rinses. Turn the rice into a sieve to drain for a few moments and return to the cooking pot. (This step is not necessary with a rice cooker—you just add water to the appropriate measuring line after the last wash.)

Cooking: Each variety and batch of rice cooks differently, and the size of the pan relative to the amount of rice affects the cooking as well. I consulted a dozen cookbooks while researching this section and found at least 20 different proportions and procedures. If these instructions don't work for you after slight adjustments of quantity and temperature, look in any cookbook and try one of the other methods. Or—buy a rice cooker!

Proportions: I use 1⅓ cups water to each cup of raw rice (again, a cup serves two to four people) when making rice on a stove top. Short- and medium-grained varieties (the ones with "rose" names) require less water than long-grained types, and the more recently the rice was harvested, the less water it takes to cook it. You may find that your rice cooks up best with 1¼ cups water, or even less. If you have forgotten to measure before washing, equal amounts of washed rice and water are very close to 1 unit of rice plus 1⅓ units water. I have always "steamed" rice (so does the rice cooker). In a heavy, covered pan, bring rice and water to a boil over medium heat (steam puffs from under the lid when the water boils), then immediately turn the heat to low or very low and simmer until all water is absorbed. Don't lift the lid until done or nearly done—in about 20 minutes for most types of rice. A timer may help you remember to check before the rice burns.

Lemon Rice

Enliven a bland fish meal with Lemon Rice. It can be custom-made hot cooked rice made especially for making Lemon Rice, or it can be made with yesterday's cold cooked rice. Heat a tablespoon (or more) of ghee or butter over medium heat in a heavy frying pan and add hot or cold cooked rice, as much as you need for the meal. Turn with a fork to distribute the fat and fluff the grains, squeezing a half lemon over the rice as it dries and separates. This Indian dish is often further enhanced by toasting a half teaspoon of cumin and/or a teaspoon of mustard seed in the fat before adding the rice. For a further Indian touch, add about a teaspoon of powdered turmeric to color and mildly flavor the rice. A handful or two of frozen peas will make the dish pretty and add another taste and texture.

Wild Rice

Wild rice is more flavorful than white rice, and chewier. It is especially good with salmon or trout, but its nutty flavor will complement any fish dish. Wild rice is especially good with oven-cooked or sauté-poached fish. Measure for measure, wild rice makes more cooked rice than white rice, and at three dollars a pound for cultivated "wild" rice, the price per serving is comparable to that of basmati.

Cooking Wild Rice: Package directions often work. The very large-grained hybrid types are cooked like white rice, using 2 cups of water to 1 cup of rice (or whatever the package says). Smaller-grained types tend to be firmer (and tastier) and usually make more cooked rice per unit of raw rice than the hybrids. Soak for a few hours to speed cooking, and to make even more cooked rice. A cup of many smaller-grained wild rices will absorb up to 4 cups of water, and you can add hot water as the rice cooks if it absorbs all the water before the grains open up. Wild rice won't go gluey as white rice would. To keep the water from foaming over, hang a bacon strip over the inside side of the pan and lock it in place with the lid. The rice cooker makes exceptionally good wild rice (use the proportions above). Cook in diluted chicken broth for even more flavor. Wild rice seems to need a bit of salt, perhaps half a teaspoon per cup of dry rice.

Special Ingredients

BUTTER, GHEE, AND CLARIFIED BUTTER

Only 80 percent of the butter in a cube of butter is fat. The other 20 percent is "overrun," water and milk residues, and the milk residues burn when heated. Ghee and clarified butter are made by removing the residues and withstand heat better, keep better, and lubricate a pan better than butter. Well-made ghee will keep for months in a dark cupboard and its flavor may even improve and intensify. Clarified butter keeps for a long time if refrigerated. Ghee is used for every purpose throughout India and the Near East, and clarified butter is a cooking essential over much of Europe. Ghee and clarified butter withstand high heat and can be used for any kind of sautéing or frying. Use them alone, or lighten the flavor with oil. Ghee has an especially "nutty" flavor and a small amount will give a good buttery flavor to a mixture.

Ghee and clarified butter are easy to make, but if you decide to try ghee by buying some, make sure the label says "pure butter." Some ghees contain other fats.

Making Ghee

Ghee is easiest to make with two or more pounds of butter. I use 4 to 6 pounds of butter to make a whole year's supply.

You Will Need:

1. A deep, heavy pan large enough so that the melted butter fills it no more than halfway. The butter may foam to twice its height or more as it boils.
2. Heatproof jars with tight lids for your ghee or clarified butter. The straight-sided 8- and 12-ounce Mason jars sold for jelly work well—it's easy to get all the ghee out of a straight-sided jar.
3. A wide-mouthed metal canning funnel is useful but not essential (plastic might melt).

Melt the butter over medium heat and boil gently, stirring down foam as it threatens to overflow the pan. Skim the top if necessary to prevent running over. As the liquid in the bottom of the pan condenses into thick-looking strands, lower the heat and continue heating, scraping the residue off the bottom of the pan occasionally, until what's left is dry and light brown. Stop cooking before the butter gets so hot it loses all color.

Let the ghee cool somewhat (it will be hot enough to crack glass jars at first), and carefully pour into the jars. The jars must be perfectly dry when you pour the ghee into them, or they will crack. A bit of thoroughly dried and toasted residue won't ruin ghee, but a thick layer might make it spoil faster. Scrape the last bit of fat and solids into a small jar to solidify into a salty paste. The paste keeps forever if refrigerated, and can be used to season cooked vegetables and sauces in place of butter and salt. Nothing need be wasted. You can use skimmings for seasoning, and you can make deliciously buttery-flavored potatoes or rice by cooking them in the ghee pot before you wash it.

Store ghee in a dark cupboard, where if properly made it will intensify in flavor over the months. I do sometimes store the last of an opened jar in the refrigerator if I am using it slowly, for fear that air on the increasing surface area will degrade the flavor.

Clarified butter is just underdone ghee. Start as for ghee and cook until the milky fluids thicken and color somewhat. Cool slightly and pour the hot fat off the residue, leaving every scrap of the liquid behind—it will make the clarified butter go rancid even if refrigerated. Ghee is really easier and handier in the long run.

You're now ready to start exploring the possibilities of your stock of ghee or clarified butter. Use them for any kind of sautéing or frying and to baste baked and grilled fish (remember, these butters are semi-liquid at room temperature). A thin film of ghee or clarified butter will keep tortillas, pancakes, or reheating rice from sticking. Substitute for melted butter in doughs and batters and use for making any kind of flour-thickened sauce. Finally, soothe dry skin with a film of ghee.

CITRUS

Lemons and limes are commonly served with fish and used in cooking it, but orange juice is also useful. Orange juice concentrate or bottled juice may be used, but the acrid, chemical taste of bottled lemon or lime juice will ruin any dish. If there is no fresh lemon or lime available, rice wine vinegar is a better substitute than bottled juice. Use orange juice to mellow the acidity of lemon or lime juice or vinegar, or just for a change of flavor. Squeeze an orange half into the sauce or dressing, or add frozen concentrate by teaspoons. Lemons can be expensive—look for big bags of lemons in "big box" stores like Costco, or buy lemons and limes in Asian groceries.

Sour Citrus: Sour wild citrus and especially developed sour varieties are used for cooking in all countries where citrus grows. If you are a gardener and are lucky enough to grow calamondins, Rangpur limes, kumquats, or limequats, by all means use the fruit and juice in your fish cookery! If you have oranges or tangerines but no sour citrus, pick them while unripe and tart for juice and sliced fruit to use in fish and salad recipes.

HERBS AND SPICES

Dried herbs and spices are not imperishable! Most deteriorate rather quickly if exposed to light and air. Spices and herbs will keep flavor and aroma for months if stored in airtight containers and kept in the dark, and they'll maintain quality for years if refrigerated or frozen. The tiny jars of spices supermarkets sell are expensive and seldom fresh. Look for cumin, coriander, and other spices in Mexican, Indian, or Near Eastern markets, or in the bulk sections of health food stores and markets specializing in natural foods.

Fresh herbs for snipping over fish dishes can be bought in most supermarkets. Leftover sprigs dry easily and keep their flavor for several weeks if stored out of the light.

SOY SAUCES

Dark soy sauces in the Japanese style are now much more familiar to most Americans than the lighter Chinese soys. Japanese-style soys made in America can be bought in any California supermarket, and are quite adequate, though Japanese shoyu from Japan has more flavor. "Imported" is usually prominent on the label. Tamari, a rich-flavored seasoning soy, is also sold by many supermarkets, and may always be found in a health food store or a supermarket specializing in natural foods. Light Chinese soys seem to be carried only by Asian markets these days. Any type of soy sauce may be used in any recipe here.

WINES AND VINEGARS

Fish is usually cooked in white wine, as red wine turns fish blue and may be overpowering. Dry vermouth is a good choice. Because it is "fortified" with added alcohol, vermouth will keep indefinitely even after the bottle has been opened, and it won't overwhelm a dish with acid or "off" flavors. Dry California vermouths are widely available, and they are very inexpensive and reliable. Red wines and balsamic and red wine vinegars can be used in dark tomato sauces or soy-based sauces. When red wine is used for cooking salmon, wine-blue and salmon-orange combine to produce a quite acceptable brown surface, particularly on bright orange farmed salmon.

Specialty vinegars may often be substituted for wine or citrus juice (supermarket distilled and cider vinegars don't work very well). You'll find red and white wine vinegars, red (and maybe white) balsamic vinegars, and rice wine vinegar in most supermarkets. Rice wine vinegar's pleasant mild flavor makes it a good choice for all cooking, but be sure to check the fine print on the label to make sure the vinegar you buy isn't salted, sugared, or otherwise flavored!

In an emergency, when you have nothing but distilled or supermarket cider vinegar, use half or two-thirds the amount suggested in recipes, adding water or broth to make up the quantity.

YOGHURT

Whenever yoghurt is called for in a recipe, plain, unflavored yoghurt is meant. Yoghurts that are made only from milk (some of the milk may be powdered milk, but that's not a problem) and cultures work best for use as an ingredient, and they taste better, too. Supermarket brands tend to include gels, starches, and strange additions like citric acid, and these modified yoghurts won't make nearly as good a dressing or sauce as yoghurts made only from milk, though they're not unusable in an emergency. Not all supermarkets carry milk-only yoghurt, but they can generally be found in health food stores, specialty markets such as Trader Joe's, and supermarkets specializing in natural foods. Plain yoghurt keeps for many weeks if the container is not opened, and it will keep for a few weeks after opening if it is never touched with anything but a very clean spoon. Yoghurt is useful in many dishes. Recipes in this chapter show the way: yoghurt mixed with or substituted for sour cream or mayonnaise is an ingredient or suggested variation in many dishes. Yoghurt can often, though not always, be used in place of milk in sauces, salad dressings, and batters. You can serve yoghurt with a fish dinner first and last and anywhere in between: as a salad dressing, as a fish dish ingredient, and topped with fruit or a drizzle of honey for a refreshing dessert!

Further Reading and Exploration

WHERE TO GO FOR FREE INFORMATION

Your public library is a wonderful resource for information on cooking and preparing fish. Libraries often have useful government pamphlets and local publications not available anywhere else.

The Internet and Worldwide Web are becoming more and more important as sources of up-to-date information. The Web is a great place to look for books and to comparison-shop for kitchen equipment. Sites and search engines change too fast to suggest many addresses, but Google (http://www.google.com) is currently one of the most useful search utilities. Amazon.com (http://www.amazon.com) and many other sites have reviews and detailed descriptions of books and even, for some books, pictures of pages and searchable content. It is less well known that Amazon .com gives you access to one of the most comprehensive collections of kitchenware and small appliances to be found anywhere.

INTERNET REFERENCES

Web sites are in a constant state of flux, and so is the material they offer. The addresses below, mostly of government agency sites, have proved to be fairly stable.

It is often easier to use a search engine to search on titles of agencies or documents than to try to keep up with the rapidly changing URLs. For instance, the document location http://oehha.ca.gov/fish.html might be changed, but typing "oehha California" into a search engine should bring up a link. A good way to narrow down a search that produces a lot of irrelevant links is to repeat the address, but put it in quotes (" "). That tells the search engine to only retrieve sites that have the two words just as you typed them. Californians can't often just type "ca" into a search expression. Many California sites will be missed with "ca" instead of "California," and "ca" is the location code for Canada, so you'll have a lot of Canadian links among the sites you do pull up.

Most sites require the full address as given. Most browsers automatically run "autofill" and "URL guesser" utilities that fill in the "http://" and "www."—but they don't always work. If an address won't work and you haven't typed in all the boring stuff, try it again and put the boring stuff in.

For Californians

California Department of Fish and Game (CDFG)
http://www.dfg.ca.gov/
Click on "Fish and Game" on DFG's home page, and then follow the links for up-to-date information on fishing.

Office of Environmental Health Hazard Assessment
http://oehha.ca.gov/fish.html/
This is where site-specific information for contaminants and recommendations for eating California-caught fish are currently collected.

Other Sites

CFSAN: U.S. Food and Drug Administration Center for Food Safety and Applied Nutrition
http://www.cfsan.fda.gov/
CFSAN's massive database includes up-to-date links to hundreds (maybe thousands) of government agencies, and it's quite easy to use. Of special interest to anglers are links to the "Bad Bug Book" or "Foodborne Illness" on the index list. "Fish Images" takes you to the Regulatory Fish Encyclopedia—an identification resource for shoppers as well as inspectors, though it's mostly ocean fish at this point. The excellent overview, What You Need to Know About Mercury in Fish and Shellfish, contains sensible and detailed recommendations for eating adequate quantities of fish safely. Many other sites link to this document.

Morbidity and Mortality Weekly Report
http://www.cdc.gov/mmwr/
MMWR reports on outbreaks of all reportable diseases, including foodborne illnesses, and on many other foodborne illnesses. Don't depend on scare headlines—search MMWR.

The American Heart Association
http://www.americanheart.org/
The AHA site isn't exactly for fishermen or cooks, but it's a great place to get links to the latest information on foods that may affect your heart—and theAHA is very enthusiastic about eating fish. The site is very easy to search.

BOOKS

The books mentioned below are "classics," first published some decades ago and never out of print since. At the time of writing, Amazon.com listed "availability" for all the books as "24-hour shipping," which means that any bookstore should be able to get them quickly. And since these books have been in print for decades, there's an excellent chance of finding a copy through one of the Internet sites for used and discounted books.

Three All-Purpose Classics

Child, Julia. 1989. *The Way to Cook.* Knopf.

Child, Julia, Simone Beck, and Louisette Bertholle. 1964–2001. *Mastering the Art of French Cooking* (two volumes). Knopf.

Cookbook enthusiasts prefer the 1964 and earlier editions of *The Joy of Cooking* to the 1975 one, because the recipes are better and better tested, and the pithy Rombauer commentary and anecdotes are intact. Look for them on used-book web sites:

Rombauer, Irma S. *Joy of Cooking.* 1931. *Facsimile Edition: A Facsimile of the First Edition 1931.* Scribner, 1998.

Rombauer, Irma. 1964. *The Joy of Cooking.* Scribner.

Rombauer, Irma S., and Marion Rombauer Becker. 1984. *The Joy of Cooking.* Macmillan.

Rombauer, Irma, and others. 1997. *The Joy of Cooking: The American Household Classic.* Plume (Penguin Putnam).

Specialty Cookbooks

Gibbons, Euell. 1987. *Stalking the Wild Asparagus.* Reprint Edition. Chambersburg, Penn.: Alan C. Hood & Co.

Langer, Richard W. 2001. *Where There's Smoke There's Flavor.* Little, Brown & Company.

Sleight, Jack, and Raymond Hull. 1981. *The Home Book of Smoke Cooking.* Mechanicsburg, Penn.: Stackpole Books.

CHECKLIST OF CALIFORNIA FRESHWATER FISH SPECIES AND SUBSPECIES

LAMPREYS (PETROMYZONTIDAE)

- ☐ Pacific Lamprey *(Lampetra tridentata)*, N
- ☐ River Lamprey *(Lampetra ayresi)*, N, CSSC
- ☐ Pit-Klamath Brook Lamprey *(Lampetra lethophaga)*, N
- ☐ Klamath River Lamprey *(Lampetra similis)*, N, CSSC
- ☐ Kern Brook Lamprey *(Lampetra hubbsi)*, N, CSSC
- ☐ Western Brook Lamprey *(Lampetra richardsoni)*, N

STURGEONS (ACIPENSERIDAE)

- ☐ Green Sturgeon *(Acipenser medirostris)*, N, CSSC
- ☐ White Sturgeon *(Acipenser transmontanus)*, N

SHAD (CLUPEIDAE)

- ☐ Threadfin Shad *(Dorosoma petenense)*, I
- ☐ American Shad *(Alosa sapidissima)*, I

TROUT AND SALMON (SALMONIDAE)

- ☐ Pink Salmon *(Oncorhynchus gorbuscha)*, N
- ☐ Chum Salmon *(Oncorhynchus keta)*, N
- ☐ Chinook Salmon or King Salmon *(Oncorhynchus tshawytscha)*, N, SE, FE, ST, FT (area dependent)
- ☐ Coho Salmon or Silver Salmon *(Oncorhynchus kisutch)*, N, SE, FT (area dependent)
- ☐ Kokanee Salmon or Sockeye Salmon *(Oncorhynchus nerka)*, N (coastal streams), I (lakes)
- ☐ Coastal Rainbow Trout *(Oncorhynchus mykiss irideus)*, N
- ☐ Steelhead *(Oncorhynchus mykiss irideus)*, N, FT, FE (area dependent)
- ☐ California Golden Trout *(Oncorhynchus mykiss aguabonita)*, N

- ☐ Little Kern Golden Trout *(Oncorhynchus mykiss whitei),* N, FT
- ☐ Kern River Rainbow Trout *(Oncorhynchus mykiss gilberti),* N, CSSC
- ☐ Sacramento Redband Trout *(Oncorhynchus mykiss stonei),* N, CSSC (area dependent)
- ☐ Eagle Lake Rainbow Trout *(Oncorhynchus mykiss aquilarum),* N, CSSC
- ☐ Coastal Cutthroat Trout *(Oncorhynchus clarki clarki),* N, CSSC
- ☐ Lahontan Cutthroat Trout *(Oncorhynchus clarki henshawi),* N, FT
- ☐ Piute Cutthroat Trout *(Oncorhynchus clarki seleniris),* N, FT
- ☐ Brown Trout *(Salmo trutta),* I
- ☐ Bull Trout *(Salvelinus confluentus),* N, SE, FT, extinct in California
- ☐ Brook Trout *(Salvelinus fontinalis),* I
- ☐ Lake Trout or Mackinaw Trout *(Salvelinus namaycush),* I
- ☐ Mountain Whitefish *(Prosopium williamsoni),* N

SMELTS (OSMERIDAE)

- ☐ Delta Smelt *(Hypomesus transpacificus),* N, ST, FT
- ☐ Longfin Smelt *(Spirinchus thaleichthys),* N, CSSC
- ☐ Eulachon *(Thaleichthys pacificus),* N, CSSC
- ☐ Wakasagi *(Hypomesus nipponensis),* I

MINNOWS (CYPRINIDAE)

- ☐ California Roach *(Lavinia symmetricus),* N, CSSC
- ☐ Hitch *(Lavinia exilicauda),* N
- ☐ Lahontan Redside *(Richardsonius egregius),* N
- ☐ Speckled Dace *(Rhinichthys osculus),* N
- ☐ Tui Chub *(Siphateles bicolor),* N
- ☐ Owens Tui Chub *(Siphateles bicolor snyderi),* N, SE, FE, FPS
- ☐ Mojave Tui Chub *(Siphateles bicolor mohavensis),* N, SE, FE, FPS
- ☐ Blue Chub *(Gila coerulea),* N
- ☐ Arroyo Chub *(Gila orcutti),* N, CSSC
- ☐ Thicktail Chub *(Gila crassicauda),* N, extinct

- ☐ Bonytail *(Gila elegans)*, N, SE, FE, possibly extinct in California
- ☐ Sacramento Splittail *(Pogonichthys macrolepidotus)*, N, FT, CSSC
- ☐ Clear Lake Splittail *(Pogonichthys ciscoides)*, N, extinct
- ☐ Hardhead *(Mylopharodon conocephalus)*, N, CSSC
- ☐ Sacramento Blackfish *(Orthodon microlepidotus)*, N
- ☐ Colorado Pikeminnow *(Ptychocheilus lucius)*, N, SE, FE, FPS
- ☐ Sacramento Pikeminnow *(Ptychocheilus grandis)*, N
- ☐ Common Carp *(Cyprinus carpio)*, I
- ☐ Goldfish *(Carassius auratus)*, I
- ☐ Golden Shiner *(Notemigonus crysoleucas)*, I
- ☐ Red Shiner *(Cyprinella lutrensis)*, I
- ☐ Fathead Minnow *(Pimephales promelas)*, I
- ☐ Grass Carp *(Ctenopharyngodon idella)*, I
- ☐ Tench *(Tinca tinca)*, I

SUCKERS (CATOSTOMIDAE)

- ☐ Sacramento Sucker *(Catostomus occidentalis)*, N
- ☐ Tahoe Sucker *(Catostomus tahoensis)*, N
- ☐ Owens Sucker *(Catostomus fumeiventris)*, N, CSSC
- ☐ Lost River Sucker *(Catostomus luxatus)*, N, SE, FE, FPS
- ☐ Klamath Largescale Sucker *(Catostomus snyderi)*, N, CSSC
- ☐ Klamath Smallscale Sucker *(Catostomus rimiculus)*, N
- ☐ Modoc Sucker *(Catostomus microps)*, N, SE, FE, FPS
- ☐ Mountain Sucker *(Catostomus platyrhynchus)*, N, CSSC
- ☐ Santa Ana Sucker *(Catostomus santaanae)*, N, FT
- ☐ Razorback Sucker *(Xyrauchen texanus)* N, SE, FE, FPS
- ☐ Shortnose Sucker *(Chasmistes brevirostris)*, N, SE, FE, FPS

CATFISHES AND BULLHEADS (ICTALURIDAE)

- ☐ Brown Bullhead *(Ameiurus nebulosus)*, I
- ☐ Black Bullhead *(Ameiurus melas)*, I
- ☐ Yellow Bullhead *(Ameiurus natalis)*, I
- ☐ Channel Catfish *(Ictalurus punctatus)*, I
- ☐ Blue Catfish *(Ictalurus furcatus)*, I
- ☐ White Catfish *(Ameirurus catus)*, I
- ☐ Flathead Catfish *(Pylodictis olivaris)*, I

PUPFISHES (CYPRINODONTIDAE)

- ☐ Amargosa Pupfish *(Cyprinodon nevadensis)*, N, CSSC
- ☐ Desert Pupfish *(Cyprinodon macularius)*, N, SE, FE
- ☐ Salt Creek Pupfish *(Cyprinodon salinus)*, N
- ☐ Cottonball Marsh Pupfish *(Cyprinodon salinus milleri)*, N, ST
- ☐ Owens Pupfish *(Cyprinodon radiosus)*, N, SE, FE, FPS

KILLIFISHES (FUNDULIDAE)

- ☐ California Killifish *(Fundulus parvipinnis)*, N
- ☐ Rainwater Killifish *(Lucania parva)*, I

LIVEBEARERS (POECILIIDAE)

- ☐ Western Mosquitofish *(Gambusia affinis)*, I
- ☐ Sailfin Molly *(Poecilia latipinna)*, I
- ☐ Shortfin Molly *(Poecilia mexicana)*, I
- ☐ Porthole Livebearer *(Poeciliopis gracilis)*, I

SILVERSIDES (ATHERINIDAE)

- ☐ Inland Silverside *(Menidia beryllina)*, I

STICKLEBACKS (GASTEROSTEIDAE)

- ☐ Threespine Stickleback *(Gasterosteus aculeatus)*, N
- ☐ Unarmored Threespine Stickleback *(Gasterosteus aculeatus williamsoni)*, N, SE, FE, FPS
- ☐ Brook Stickleback *(Culaea inconstans)*, I

TEMPERATE BASS (MORONIDAE)

- ☐ Striped Bass *(Morone saxatilis)*, I
- ☐ White Bass *(Morone chrysops)*, I

SUNFISHES, CRAPPIES, AND "BLACK" BASSES (CENTRARCHIDAE)

- ☐ Sacramento Perch *(Archoplites interruptus)*, N, CSSC
- ☐ Bluegill *(Lepomis macrochirus)*, I
- ☐ Pumpkinseed *(Lepomis gibbosus)*, I
- ☐ Green Sunfish *(Lepomis cyanellus)*, I
- ☐ Redear Sunfish *(Lepomis microlophus)*, I
- ☐ Warmouth *(Lepomis gulosus)*, I
- ☐ Black Crappie *(Pomoxis nigromaculatus)*, I

☐ White Crappie *(Pomoxis annularis)*, I
☐ Largemouth Bass *(Micropterus salmoides)*, I
☐ Smallmouth Bass *(Micropterus dolomieu)*, I
☐ Spotted Bass *(Micropterus punctulatus)*, I
☐ Redeye Bass *(Micropterus coosae)*, I

PERCHES (PERCIDAE)

☐ Yellow Perch *(Perca flavescens)*, I
☐ Bigscale Logperch *(Percina macrolepida)*, I

CICHLIDS (CICHLIDAE)

☐ Mozambique Tilapia *(Oreochromis mossambicus)*, I
☐ Redbelly Tilapia *(Tilapia zillii)*, I
☐ Blue Tilapia *(Oreochromis aureus)*, I
☐ Nile Tilapia *(Oreochromis niloticus)*, I

VIVIPAROUS PERCH (EMBIOTOCIDAE)

☐ Shiner Perch *(Cymatogaster aggregata)*, N
☐ Tuleperch *(Hysterocarpus traski)*, N
☐ Russian River Tuleperch *(Hysterocarpus traski pomo)*, N, CSSC

MULLET (MUGILIDAE)

☐ Striped Mullet *(Mugil cephalus)*, N

GOBIES (GOBIIDAE)

☐ Yellowfin Goby *(Acanthogobius flavimanus)*, I
☐ Tidewater Goby *(Eucyclogobius newberryi)*, N, FE (area dependent)
☐ Shimofuri Goby *(Tridentiger bifasciatus)*, I

SCULPINS (COTTIDAE)

☐ Prickly Sculpin *(Cottus asper)*, N
☐ Riffle Sculpin *(Cottus gulosus)*, N
☐ Pit Sculpin *(Cottus pitensis)*, N
☐ Coastrange Sculpin *(Cottus aleuticus)*, N
☐ Reticulate Sculpin *(Cottus perplexus)*, N, CSSC
☐ Marbled Sculpin *(Cottus klamathensis)*, N
☐ Rough Sculpin *(Cottus asperrimus)*, N, ST

☐ Paiute Sculpin *(Cottus beldingi)*, N
☐ Pacific Staghorn Sculpin *(Leptocottus armatus)*, N

RIGHTEYED FLOUNDER (PLEURONECTIDAE)

☐ Starry Flounder *(Platichthys stellatus)*, N

PIKE (ESOCIDAE)

☐ Northern Pike *(Esox lucius)*, I

Summary

Total number of nonextinct, native California freshwater species: 63*
Total number of established, introduced freshwater species: 50
Total number of anadromous or catadromous native California species: 15
Total number of introduced anadromous or catadromous species: 4

Introductions that May Lead to Permanently Established Populations in the Los Angeles Basin

FAMILY COBITIDAE

☐ Oriental Weatherfish *(Misgurnus anguillicaudatus)*

FAMILY POECILIIDAE

☐ Guppy *(Lebistes reticulatus)*
☐ Green Swordtail *(Xiphophorus helleri)*

* Subspecies are not counted in any of these totals with one exception: The Kokanee Salmon has been tallied as an introduced "species," and its true species affiliation, the Sockeye Salmon, is counted as a native. Also, the four species that now appear to be extinct in California (Thicktailed Chub, Clear Lake Splittail, Colorado Pikeminnow, and Bull Trout) are not included in this total.

ABBREVIATIONS

TL	Total length
FL	Fork length
SL	Standard length
LL	Lateral line
AF	Anal fin
DF	Dorsal fin
PelF	Pelvic fins
PecF	Pectoral fins
C	Centigrade
F	Fahrenheit
mm	Millimeter
cm	Centimeter
m	Meter
gm	Gram
Kg	Kilogram
L	Liter
in.	Inch
ft	Foot
ppb	Parts per billion
ppm	Parts per million
ppt	Parts per thousand
CDFG	California Department of Fish and Game
USFWS	United States Fish and Wildlife Service
ESA	Endangered Species Act
FE	Federal Endangered species
FT	Federal Threatened species
SE	State Endangered species
ST	State Threatened species
FPS	Fully Protected Species
CSSC	California Species of Special Concern
N	Native California Species
I	Introduced California Species

GLOSSARY

Acclimation Physiological adaptation to changes in ambient conditions such as temperature or salinity.

Adipose fin A small, fleshy fin lacking rays and located between the dorsal and caudal fins.

Alevin The yolk-sac fry stage of anadromous salmonids.

Alkaline Having a pH level greater than 7.

Ammocete A lamprey larva.

Anadromous Referring to a life history pattern in which fishes spawn in freshwater, then the fry migrate to saltwater to mature.

Axillary process A fleshy or scaly appendage at the base of the pelvic fin.

Barbel A long, fleshy projection or "whisker" attached to the upper or lower jaw of some fishes.

Brackish Having a low to medium salt content, in contrast to seawater.

Carnivore An organism that feeds primarily on flesh.

Carrying capacity The number of individuals of a species that a habitat can support on a permanent basis without injury to either the habitat or the species.

Catadromous Referring to a life history pattern in which fishes spawn in saltwater, then the fry migrate to freshwater to mature.

Caudal fin The tail fin.

Caudal peduncle The narrow posterior end of the body to which the caudal fin is attached.

Centrarchid A member of the black bass, crappie, and sunfish family (Centrarchidae).

Cloaca The common chamber into which the reproductive, excretory, and digestive systems empty.

Cultigen A species or subspecies cultivated by humans and not occurring naturally.

Cyprinid A member of the true minnow family (Cyprinidae).

Delta The meandering channels of the Sacramento and San Joaquin Rivers east of Suisun Bay.

Desmid A member of the algae family Desmidiaceae.

Detritus Disintegrated bottom material in ponds, lakes, and streams.

Diatom A type of algae with a siliceous coat or shell.

Diurnal Active during the day.

Ecosystem All plants and animals present and interacting within a given habitat.

Endangered Being or relating to any species in danger of extinction throughout all or a significant portion of its range. It is granted full protection by state law, federal law, or both.

Endemic Originating in, and occurring in, a specific habitat or geographic location.

Epilimnion The upper, warm-water zone in a thermally stratified lake.

Euryhaline Able to live in waters with a wide range of salinities.

Eutrophic Describes a highly productive habitat.

Exotic Imported, not native.

Family Taxonomic grouping containing closely related genera.

Fauna The animal life found in a region.

Filter feeding Employment of structures such as gill rakers to collect small food items from the water.

Fingerling A large fry, approximately the length of a human finger, usually denoting a minimum size for hatchery stocking.

Food chain A concept describing the interrelated feeding habits of animals within an ecosystem.

Fry A general term for any young fish.

Gill arch Bony support structure for the gill filament and arch.

Gill raker Comprises the bony projections on the gill arch that prevent food from exiting through the opercular slit.

Gonopodium An anal fin modified for copulation.

Hellgrammite The Dobsonfly's *(Corydalis cornutus)* aquatic larva, used as bait.

Heterocercal Having the vertebral column extend into the upper half of the caudal fin.

Homocercal Having the vertebral column end at the base of the caudal fin.

Hypertonic Containing a higher concentration of solutes than the surrounding medium.

Hypolimnion The lower, cool-water zone in a thermally stratified lake.

Hypotonic Containing a lesser concentration of solutes than the surrounding medium.

Ichthyology The study of fishes.

Ictalurid A member of the catfish and bullhead family (Ictaluridae).

Indigenous Being native to a given region.

Inferior Describes a fish mouth that opens ventrally.

Instinct Genetically endowed behavior requiring no learning process.

Introduced Placed into a habitat directly or indirectly by human beings; nonnative.

Larva A posthatching developmental animal form, usually not resembling the adult. The plural is *larvae.*

Lateral line The pressure-sensing organ on the lateral body wall and dorsal head area of fishes.

Littoral zone The shallow inshore area of an aquatic habitat.

Livebearer A fish species that gives birth to developed fry.

Maxilla One of the bones of the upper jaw of fishes. The plural is *maxillae.*

Metabolism The process by which organisms utilize energy.

Minnow The common name for members of the family Cyprinidae.

Native Occurring naturally in a geographic area.

Nocturnal Active during the night.

Olfactory Referring to the sense of smell.

Oligotrophic Minimally productive; usually referring to aquatic habitats.

Operculum The bony covering of the gill.

Oviduct Tubular portion of the female reproductive system, which in fishes leads from the ovary to the cloaca.

Oviparous Egg-laying.

Ovoviviparous Retaining and incubating eggs with large yolks within the ovary or oviduct. In ovoviviparous species,the young receive little or no nourishment from the mother's blood.

Palatine Refers to teeth positioned on the palate or roof of the mouth.

Panfish Any small game fish, such as the Bluegill *(Lepomis macrochirus)*, that is normally cooked by frying in an open pan.

Papilla Any small fleshy bump or projection, as on the lips of suckers. The plural is *papillae.*

Parr The stage in anadromous salmonids between alevin and smolt, when parr marks are present.

Parr marks Dark round or oval markings on the lateral body surface of anadromous salmonid fry.

Pectoral fin A paired lateral fin behind the operculum.

Pelvic fin A paired ventral fin anterior to the anal fin.

Pharyngeal teeth Teeth borne on the last gill arch and located in the throat area of some fishes.

Photosynthesis The process by which plants utilize solar energy to produce glucose from carbon dioxide and water.

Photosynthetic zone The upper aquatic zone where maximal light penetration permits good photosynthetic activity.

Physiological Referring to life functions such as respiration and circulation.

Phytoplankton Suspended or floating microscopic algae.

Piscivore An animal that feeds primarily on fishes.

Planktivore An animal that feeds primarily on zooplankton.

Predaceous Feeding, or preying, upon other animals.

Primary consumer A food-chain category encompassing organisms that eat plants.

Primary freshwater fish A member of a group of fishes that can tolerate only freshwater and lives its life entirely in freshwater.

Protected Being or relating to any species the collection or killing of which is forbidden by law.

Ray The flexible supportive element of fish fins.

Rotenone A selective poison for fishes that reduces the blood supply to the gills, causing suffocation.

Salmonid A member of the trout, salmon, char, whitefish and grayling family (Salmonidae).

San Francisco Estuary The Delta, Suisun Bay, and San Francisco Bay.

Secondary consumer A food-chain category describing animals that feed on primary consumers.

Sexual dimorphism A condition in which the sexes of the same species differ in morphology, size, color, and so forth.

Smolt A stage in anadromous salmonid development when parr lose their markings and begin their seaward migration.

Spawning Egg laying and external fertilization in an aquatic habitat.

Species A genetically isolated population; the smallest major taxon.

Spine The bony, rigid, sharp supportive element in fish fins.

Spiracle An opening on the top or side of the head leading to the mouth cavity.

Spoon An angler's term for a metallic, spoon-shaped lure.

Subspecies Two or more populations within a species that (1) differ from each other in morphology, physiology, or behavior; (2) occupy different localities within the species range; and (3) may interbreed.

Substrate The material making up the floor or bottom of an aquatic habitat.

Superior Describes a fish mouth that opens dorsally.

Swim bladder The air-filled buoyancy organ in fishes.

Sympatric Occurring in the same habitat.

Taxonomy The study of the classification of organisms.

Terminal Describes a fish mouth that opens horizontally.

Thermocline A zone of steep temperature gradation in a thermally stratified lake.

Threatened Being not presently threatened with extinction but likely to become endangered if environmental conditions worsen.

Trolling An angling method in which a bait or lure is pulled behind a moving boat.

Tubercle A small bump or projection, usually hard, on the skin.

Viviparity Condition in which fertilized eggs are retained within the uterus or oviduct and derive nourishment via a placenta, a placenta-like structure, or maternal secretions.

Yolk sac A structure that nourishes the embryo of an egg-laying animal during development and shortly after hatching.

Zooplankton Collective term for all small, pelagic animals, usually of microscopic size.

STATE FISH HATCHERIES

The California Department of Fish and Game operates 12 trout hatcheries, eight salmon and steelhead hatcheries, and two fish-planting bases. Visit their hatchery list Web site (www.dfg.ca.gov/lands/fish1.html) for updated information about California fish hatcheries. Besides allowing you to view large numbers of fishes at close hand, a visit to a California hatchery gives you an overview of the massive task of producing the trout and salmon to maintain California's thriving sport and commercial salmonid fisheries. Most hatcheries have a number of rearing ponds available for public viewing during normal operation hours, and some give short tours of the facilities. It is usually best to call ahead when planning a visit.

Iron Gate Hatchery (Salmon/Steelhead)

8638 Lakeview Road
Hornbrook, CA 96044
E-mail: irongate@dfg.ca.gov
Directions: Exit Interstate 5 at the Henley Hornbrook turnoff onto Copco Road. Follow hatchery signs east for 8 miles. Hatchery is on the right, across the river. Siskiyou County.

Mad River Hatchery (Salmon/Steelhead)

1660 Hatchery Road
Arcata, CA 95521
E-mail: wcartwri@dfg.ca.gov
Directions: From Hwy. 101, drive east on Hwy. 299 and take Blue Lake exit south. Hatchery is 2 miles south of Blue Lake on Hatchery Road. Humboldt County.

Trinity River Hatchery (Salmon/Steelhead)

P.O. Box 162
Lewiston, CA 96052
E-mail: trinityriver@dfg.ca.gov
Directions: Take Hwy. 299. At 24 miles west of Redding turn north on the highway leading to Lewiston and Trinity Dam and follow it for for 6 miles. Hatchery is 1 mile north of Lewiston on the east side of the Trinity River. Trinity County.

Feather River Hatchery (Salmon/Steelhead)

5 Table Mountain Boulevard
Oroville, CA 95965
E-mail: featherriver@dfg.ca.gov
Directions: Take the Grand Nelson exit from Hwy. 70, and go east on Nelson Ave. 2 miles. Go south on Table Mountain Blvd. about half a mile to the hatchery entrance. Butte County.

Warm Springs Hatchery (Salmon/Steelhead)

3246 Skaggs Springs Road
Geyserville, CA 95441
E-mail: bawilson@dfg.ca.gov
Directions: Take Hwy. 101 at Healdsburg. Go west on Dry Creek Road 11 miles to the Visitor Center below Warm Springs Dam. Sonoma County.

Nimbus Hatchery (Salmon/Steelhead)

(No address or contact information on Web site as of January 2003)
Directions: Located off Hwy. 50, 18 miles east of Sacramento. Travel north on Hazel Ave. 0.7 mile, turn left at Gold Country Blvd. Travel 200 yards to the parking lot entrance on the right. Sacramento County.

Mokelumne River Hatchery (Salmon/Steelhead)

25800 N. McIntire Road
P.O. Box 158
Clements, CA 95227
E-mail: mokelumneriver@dfg.ca.gov
Directions: On Hwy. 12 about 2 miles east of Clements, go north on McIntire Road 1 mile to hatchery entrance. At the base of Comanche Dam. San Joaquin County.

Merced River Hatchery (Salmon/Steelhead)

P.O. Box 94
Snelling, CA 95369
E-mail: mercedriver@dfg.ca.gov
Directions: On Hwy. 59, 19 miles northeast of Merced, turn south on Snelling Road, cross the river, turn east on Robinson Road, pass Allen Road to canal. Turn left, follow canal, at sign follow channel to hatchery. Merced County.

Mount Shasta Hatchery (Trout)

3 North Old Stage Road
Mt. Shasta, CA 96067
E-mail: mtshasta@dfg.ca.gov
Directions: On Interstate 5 take Central Mt. Shasta exit, turn west one-half mile to stop sign. Cross North Old Stage Road to hatchery entrance. Siskiyou County.

Darrah Springs Hatchery (Trout)

29661 Wildcat Road
Paynes Creek, CA 96075
E-mail: darrahsprings@dfg.ca.gov
Directions: On Hwy. 36, 10 miles east of Red Bluff to Dales Station. Follow hatchery signs north on County Road A6 for 8 miles. Cross Battle Creek, turn left onto Wildcat Road, proceed 3 miles to hatchery entrance. Near Manton, Shasta County.

American River Hatchery (Trout)

2101 Nimbus Road
Rancho Cordova, CA 95670
Directions: On Hwy. 50, 18 miles east of Sacramento north on Hazel Ave. 0.7 mile, west at signal on Gold Country Blvd., about 200 yards, turn right into parking lot. Sacramento County.

Silverado Fisheries Base (Egg Quarantine and Trout Planting)

7329 Silverado Trail
P.O. Box 47
Yountville, CA 94599
E-mail: spoe@cwo.com
Directions: On west side of Silverado Trail, 9 miles north of Napa. Napa County.

Moccasin Creek Hatchery (Trout)

P.O. Box 159
Moccasin, CA 95347
E-mail: moccasincreek@dfg.ca.gov
Directions: Off Hwy. 49 at junction of Hwy. 49 and Hwy. 120, 20 miles south of Sonora. Tuolumne County.

Hot Creek Hatchery (Trout)

Star Route 1
P.O. Box 208
Mammoth Lakes, CA 93546
E-mail: hotcreek@dfg.ca.gov
Directions: About 37 mi north of Bishop, or 3.5 mi south of junction of Hwy. 395 and Hwy. 203. One mile north of Hwy. 395. Near airport, watch for sign. Mono County.

Fish Springs Hatchery (Trout)

P.O. Box 910
Big Pine, CA 93513
E-mail: fishsprings@dfg.ca.gov
Directions: West off Hwy. 395, 6 mi south of Big Pine on Fish Springs Road. Inyo County.

Black Rock Rearing Ponds (Trout)

Star Route 1
P.O. Box 100
Independence, CA 93526
E-mail: blackrock@dfg.ca.gov
Directions: On Hwy. 395, about 9 mi north of Independence. Turn east at sign, go about 1 mile to hatchery. Inyo County.

Mount Whitney Hatchery (Trout)

HCR 67, P.O. Box 26
Independence, CA 93526
E-mail: mtwhitney@dfg.ca.gov
Directions: On Hwy. 395 about 2 mi north of Independence. Turn west at sign, go about 1 mile to hatchery. Inyo County.

San Joaquin Hatchery (Trout)

17372 Brook Trout Drive
P.O. Box 247
Friant, CA 93626
E-mail: sanjoaquin@dfg.ca.gov
Directions: East of Hwy. 41 at Friant, from Friant Road, turn west onto Flemming Road, follow into hatchery. One mile below Lake Millerton dam about 12 miles north of Fresno. Fresno County.

Kern River Planting Base (Trout)

P.O. Box 1908
Kernville, CA 93238
E-mail: kernriver@dfg.ca.gov
Directions: Off west side of Sierra Way (Johnsondale Road), 1 mile north of Kernville. Kern County.

Fillmore Hatchery (Trout)

P.O. Box 666
Fillmore, CA 93016
E-mail: fillmore@dfg.ca.gov
Directions: South side of Hwy. 126, about 1 mile east of Fillmore. Ventura County.

Mojave River Hatchery (Trout)

12550 Jacaranda Avenue
Victorville, CA 92392
E-mail: mojave@dfg.ca.gov
Directions: From Hwy. 395, go east on Bear Valley Road about 6 miles to hatchery road, turn north, go half a mile. Next to college. San Bernardino County.

Crystal Lake Hatchery (Trout)

Route 2, Box 113
Burney, CA 96013
Phone: 530-335-4111 (no E-mail listed)
Directions: Take Hwy. 299E east out of Burney for 8 miles, turn right on Cassel Rd. and follow the signs for another 4 miles to the hatchery, north of the town of Cassel. Shasta County.

WORLD AND CALIFORNIA SPORTFISHING WEIGHT RECORDS

Unlike most other vertebrates, fishes are indeterminate growers and therefore, when habitat conditions are favorable, continue to grow throughout their lives. Perhaps the most common desire anglers share is to catch "the biggest one," and the biological trait of continuous growth adds a realistic dimension, ensuring that indeed, an "even bigger one" is out there. Record data are also of ecological interest because when coupled with geographic data, they may provide a general idea of the optimal climate and habitat that promote maximum growth of a species.

The following sportfishing weight records were compiled from data from the International Game Fish Association and the California Department of Fish and Game for 2005. Like all records, these are continuously subject to change, and in future years one should do a computer search on the key words "fishing records," "freshwater," and "California/world" for the most recent records. Space is given for you to update the records or to list and compare your own personal records of best catches against those of California and the world. Note that the common basis for fish size records is weight, not length. Because records are traditionally filed in pounds and ounces, a practice not likely to change soon, metric equivalents of the weights are not given here.

Anglers interested in pursuing personal or official sportfishing weight records should keep in mind that fishes lose body water rapidly after death when kept out of water. To help avoid this, pack your trophy fish in crushed ice until weighed. The certified scales found in supermarkets are a convenient choice for weighing your fish, and checkout clerks usually welcome such requests as an interesting departure from their normal duties. If you believe that your trophy is a new record, obtain the names, addresses, and phone numbers of two witnesses to the weigh-in and include this information with your application. For information on filing a new record, contact the California Department of Fish and Game, 1416 Ninth Street, Sacramento, California 95814. Freezing your trophy fish is also recommended because another weighing may be necessary. Note that an unwrapped fish placed on a freezer shelf will lose substantial body water to evaporation; to avoid this, seal your trophy in a plastic bag before placing it in the freezer.

World and California Sportfishing Weight Records

Species	World record			California record			Your record
	LB	OZ	LOCATION	LB	OZ	LOCATION	
Rainbow Trout (incl. Steelhead)	42	2	Alaska	27	4	Del Norte Co.	
Golden Trout	11	0	Wyoming	9	8	Fresno Co.	
Lahontan Cutthroat Trout	41	0	Nevada	31	8	Lake Tahoe	
Brown Trout	40	4	Arkansas	26	8	Mono Co.	
Brook Trout	14	8	Ontario, Can.	9	12	Mono Co.	
Bull Trout	32	0	Idaho	9	11	Siskiyou Co.	
Lake Trout	72	0	Northwest Territories, Can.	37	6	Lake Tahoe	
Mountain Whitefish	5	8	Alberta, Can.	2	7	Truckee River	
Arctic Grayling	5	15	Northwest Territories, Can.	1	12	Mono Co.	
Kokanee Salmon	9	6	British Columbia, Can.	4	13	Lake Tahoe	
Coho Salmon	33	4	New York	22	0	Marin Co.	
Chinook Salmon	97	4	Alaska	88	0	Tehama Co.	
Striped Bass	78	8	New Jersey	67	8	Merced Co.	
White Bass	6	13	Virginia	5	5	Colorado River	
White Sturgeon	468	0	California	468*	0	San Pablo Bay	
Largemouth Bass	22	4	Georgia	21	12	Los Angeles Co.	
Smallmouth Bass	11	15	Kentucky	9	1	Trinity Co.	

* Caught in brackish water and therefore not recognized as a “freshwater” record.

continued ➤

World and California Sportfishing Weight Records *(continued)*

Species	World record			California record			Your record
	LB	*OZ*	*LOCATION*	*LB*	*OZ*	*LOCATION*	
Spotted Bass	9	9	California	10	4	Fresno Co.	
Redeye Bass	8	12	Florida		No record		
Bluegill	4	12	Alabama	3	8	Merced Co.	
Warmouth	2	7	Florida	0	12	Sacramento Co.	
Sacramento Perch	3	10	California	3	10	Mono Co.	
White Crappie	5	3	Mississippi	4	8	Lake Co.	
Black Crappie	4	8	Virginia	4	1	Calaveras Co.	
Redear Sunfish	5	3	Virginia	5	3	Sacramento Co.	
Green Sunfish	2	2	Missouri	1	12	Shasta Co.	
Yellow Perch	4	3.5	New Jersey		No record		
Channel Catfish	58	0	South Carolina	52	10	Orange Co.	
Blue Catfish	111	0	Tennessee	101	0	San Diego Co.	
Flathead Catfish	123	9	Kansas	60	0	Colorado River	
White Catfish	22	0	Florida	22	0	Sacramento Co.	
Black Bullhead	8	0	New York		No record		
Brown Bullhead	5	11	Florida	4	8	Trinity Co.	
American Shad	11	4	Maine	7	5	Feather River	
Common Carp	75	11	France	52	0	San Luis Obispo Co.	

SELECTED REFERENCES

Ichthyological literature, though extensive, is for the most part widely scattered. Much pertinent information is contained in short notes and papers, which often appear in hard-to-find journals and reports. Only a relatively few authors, many of whom are listed in the following references, have been ambitious enough to pull together regional information into a single volume. First and foremost with respect to his knowledge of California freshwater fishes is Dr. Peter B. Moyle, whose landmark revision of *Inland Fishes of California* was published in 2002. I was able to review a copy of this 500-page work while completing the manuscript for the revision of my own book, and it was invaluable for checking ichthyological fine points concerning the current taxonomy and distribution of certain species, several of which I had missed in the journal literature. Two books by Dr. Robert J. Behnke were also indispensable. His most recent, *Trout and Salmon of North America,* is the most complete book ever written on this subject, and his American Fisheries Society Monograph, *Native Trout of Western North America,* is also a landmark work. William A. Dill and Almo J. Cordone's *History and Status of Introduced Fishes in California, 1871–1996,* provided a wealth of invaluable information for the "native versus introduced species" theme of this book.

Although I consulted over 300 additional references during the writing of this book, only a small selection is listed below. Most are readily obtainable from college, university, and larger public libraries, and from the Sacramento offices of the California Resources Agency and the California Department of Fish and Game.

Akihito, and K. Sakamoto. 1989. Reexamination of the status of the striped goby. *J. Ichthyol.* 36:100–112.

American Fisheries Society. 1987. *Carp in North America.* Bethesda, Md.: American Fisheries Society.

Axelrod, H.R. 1970. *Photography for aquarists.* Jersey City, N.J.: TFH Publications.

Baxter, J.L. 1966. *Inshore fishes of California.* Sacramento: California Resources Agency, California Department of Fish and Game.

Behnke, R.J. 1992. *Native trout of western North America.* Monograph 6. Bethesda, Md.: American Fish Society.

———. 2002. *Trout and salmon of North America.* New York: Free Press.

Bond, C.E. 1979. *Biology of fishes.* Philadelphia: W.B. Saunders.

California Department of Fish and Game. 1965. *Warmwater game fishes.* Sacramento: California Resources Agency.

———. 1969. *Trout of California.* Sacramento: California Resources Agency.

———. 2005. *Freshwater sport fishing: California regulations.* Sacramento: California Resources Agency.

Carlander, K.D. 1969. *Handbook of freshwater fishery biology.* Vol. 1. Ames: Iowa State University Press.

———. 1977. *Handbook of freshwater fishery biology.* Vol. 2. Ames: Iowa State University Press.

Courtois, L.A. 1979. *Status of the Owens Pupfish,* Cyprinodon radiosus *(Miller), in California.* Inland Fisheries Endangered Species Program Special Publication 76-3. Sacramento: California Resources Agency, Department of Fish and Game.

Dewees, C.M. 1984. *The printer's catch: An artist's guide to Pacific Coast edible marine life.* Monterey, Calif.: Sea Challengers.

Dill, W.A., and A.J. Cordone. 1997. *History and status of introduced fishes of California, 1871–1996.* Fish Bulletin 178. Sacramento: California Resources Agency, Department of Fish and Game.

Eddy, S., and J.C. Underhill. 1969. *How to know the freshwater fishes.* Dubuque, Iowa: Wm. C. Brown.

———. 1974. *Northern fishes, with special reference to the upper Mississippi Valley.* Minneapolis: University of Minnesota Press.

Eschmeyer, W.N., and E.S. Herald. 1983. *A field guide to Pacific Coast fishes of North America.* Boston: Houghton Mifflin.

Fitch, J.E., and R.J. Lavenberg. 1975. *Tidepool and nearshore fishes of California.* Berkeley and Los Angeles: University of California Press.

Fry, D.H. 1973. *Anadromous fishes of California.* Sacramento: California Resources Agency, California Department of Fish and Game.

Hart, J.L. 1973. *Pacific fishes of Canada.* Bulletin 180. Ottawa: Fisheries Research Board of Canada.

Hasler, A.D., E.S. Gardella, H.F. Henderson, and R.M. Horrall. 1969. Open-water orientation of white bass, *Roccus chrysops,* as determined by ultrasonic tracking methods. *J. Fish. Res. Board Can.* 26: 2173–92.

Hopkirk, J.D. 1973. *Endemism in fishes of the Clear Lake region of central California.* University of California Publication, Zoology 96. Berkeley and Los Angeles: University of California Press.

Lagler, K.F. 1964. *Freshwater fishery biology.* Dubuque, Iowa: Wm. C. Brown.

Lagler, K.F., J.E. Bardach, R.R. Miller, and D.R.M. Passino. 1977. *Ichthyology.* New York: John Wiley and Sons.

Leidy, R.A. 1984. Distribution and ecology of stream fishes in the San Francisco Bay drainage. *Hilgardia* 52:1–175.

Love, M.S., and G.M. Cailliet. 1979. *Readings in ichthyology.* Santa Monica, Calif.: Goodyear Publishing.

Matern, S.A. 2001. Using temperature and salinity tolerance to predict the success of the Shimofuri goby, a recent invader into California. *Trans. Am. Fish. Soc.* 130:592–99.

McPhee, J. 2002. *The founding fish.* New York: Farrar, Straus, and Giroux.

Miller, D.J., and R.N. Lea. 1972. *Guide to the coastal marine fishes of California.* Fish Bulletin 157. Sacramento: California Resources Agency, Department of Fish and Game.

Moyle, P.B. 2002. *Inland fishes of California.* Berkeley and Los Angeles: University of California Press.

Moyle, P.B., and J.J. Cech Jr. 2002. *Fishes: An introduction to ichthyology.* Englewood Cliffs, N.J.: Prentice Hall.

Needham, J.G., and P.R. Needham. 1973. *A guide to the study of freshwater biology.* San Francisco: Holden-Day.

Page, L.M., and B.M. Burr. 1991. A *field guide to freshwater fishes of North America north of Mexico.* Boston: Houghton Mifflin.

Pister, E.P. 1993. Species in a bucket. *Nat. Hist.* 102(1):14–19.

Richey, D. 1979. *How to catch trophy freshwater gamefish.* New York: Crown Publishing.

Robins, C.R., R.M. Bailey, C.E. Bond, J.R. Brooker, E.A. Lachner, R.N. Lea, and W.B. Scott. 1991. *A list of common and scientific names of fishes from the United States and Canada* (5th ed.). Special publication 20. Bethesda, Md.: American Fisheries Society.

Soltz, D.L., and R.J. Naiman. 1978. *The natural history of native fishes in the Death Valley system.* Los Angeles: Los Angeles Natural History Museum.

Steinhart, P. 1990. *California's wild heritage: Threatened and endangered species of the Golden State.* San Francisco: Sierra Club Books.

Tinbergen, N. 1951. *The study of instinct.* Oxford, U.K.: Clarendon Press.

Turner, J.L., and D.W. Kelly. 1966. *Ecological studies of the Sacramento–San Joaquin Delta, Part II.* Fish Bulletin 136. Sacramento: California Resources Agency, Department of Fish and Game.

Usinger, R.L. 1963. *Aquatic insects of California.* Berkeley and Los Angeles: University of California Press.

Vladykov, V.D., and W.I. Follett. 1965. *Lampetra richardsoni,* a new nonparasitic species of lamprey (Petromyzontidae) from western North America. *J. Fish. Res. Board Can.* 22(1):139–58.

———. 1967. The teeth of lamprey (Petromyzontidae): Their terminology and use in a key to the Holarctic genera. *J. Fish. Res. Board Can.* 24(5):1067–75.

Ward, H.B., and G.C. Whipple. 1963. *Freshwater biology.* New York: John Wiley and Sons.

Wydoski, R.S., and R.R. Whitney. 1979. *Inland fishes of Washington.* Seattle: University of Washington Press.

ILLUSTRATION REFERENCES

All illustrations are by Doris Alcorn with a few exceptions: the maps (figs. 1 and 8); the lake stratification chart (fig. 2); and the Gyotaku print (fig. 9), which was created by the author, Sam McGinnis. Illustrations were largely based on live or preserved specimens, and on photos by the author and illustrator. Written color descriptions were relied upon for a few species when photos and actual specimens could not be obtained.

During preparation of black-and-white illustrations for the original edition, the Department of Ichthyology at the California Academy of Sciences loaned specimens and reference materials. Dr. John Hopkirk assisted with suggestions and materials during preparation of some illustrations for the families Cottidae, Ictaluridae, Salmonidae, and Centrarchidae. The lamprey mouth drawings were based partly on Vladykov and Follett (1965 and 1967). Adult salmon were partly based on photographs in Hart (1973).

During preparation of the color illustrations for the second edition, descriptions in Moyle (2002) and in Page and Burr (1991) were especially helpful. Male cichlid colors and patterns were largely based on photographs of *Tilapia zillii, Oreochromis niloticus,* and *Oreochromis mossambicus* from the USGS Web site on Nonindigenous Aquatic Species (nas.er.usgs.gov).

INDEX

Page numbers in **bold** indicate the main discussion of the taxon.

ABOUT THE AUTHOR

Photo by Molly McGinnis

Sam McGinnis is Professor Emeritus of biology at California State University East Bay. He grew up fishing the lakes and streams of Wisconsin, and while attaining his B.S. degree in zoology at the University of Wisconsin–Madison, he worked as a fisheries research assistant on the most studied lake in the world, Lake Mendota. After receiving his Ph.D. in zoology from the University of California, Berkeley, he taught an annual field course in ichthyology for nearly four decades. The first edition of this book, published in 1984, was prompted by his need for an adequate field guide for that course. His research has included a study of competition for food between juvenile Striped Bass and other introduced species, fish use of marsh habitats in the Lower Sacramento–San Joaquin Delta, and surveys of Steelhead populations in coastal creeks. With Dr. Robert C. Stebbins, he is coauthoring the forthcoming *Field Guide to Amphibians and Reptiles of California,* also from UC Press.

Series Design:	Barbara Jellow
Design Enhancements:	Beth Hansen
Design Development:	Jane Tenenbaum
Illustrator:	Doris Alcorn
Composition:	Jane Tenenbaum
Text:	9/10.5 Minion
Display:	Franklin Gothic Book and Demi
Printer and binder:	Golden Cup Printing Company Limited

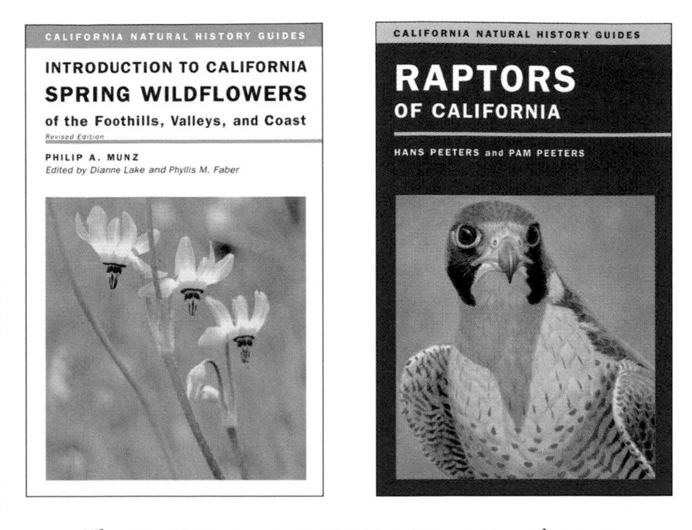

The **CALIFORNIA NATURAL HISTORY GUIDES** are the most authoritative resource on the state's flora and fauna. These short, inexpensive, and easy-to-use books help outdoor enthusiasts make the most of California's abundant natural resources. The series is divided into two groups: **INTRODUCTIONS** for beginners and **FIELD GUIDES** for more experienced naturalists. Please visit our web site for announcements, a regular natural history column, and the most up-to-date list of books. To hear about new guides through UC Press E-News, fill out and return this card, or sign up online at www.californianaturalhistory.com.*

Name ______________________________

Address ______________________________

City/State/Zip ______________________________

Email ______________________________

Which book did this card come from? ______________________________

____

Where did you buy this book? ______________________________

____

What is your profession? ______________________________

Comments ______________________________

WE'D LOVE TO HEAR FROM YOU!

* UC Press will not share your information with any other organization.

POST OFFICE
WILL NOT
DELIVER
WITHOUT
POSTAGE

Return to:

University of California Press
Attn: Natural History Editor
2120 Berkeley Way
Berkeley, California 94704-1012